Conscription, Conscientious Objection, and Draft Resistance in American History

Studies in Peace History

VOLUME 1

The titles published in this series are listed at *brill.com/sph*

Conscription, Conscientious Objection, and Draft Resistance in American History

By

Jerry Elmer

BRILL

LEIDEN | BOSTON

Cover illustration: Anti-draft demonstration in Boston on October 16, 1967. On that date, more than 1,200 men in 30 cities committed felonies by publicly burning their draft cards. This was the first nationally coordinated event sponsored by the Resistance, which deliberately sought to invoke the French Resistance to the Nazis during World War II (see Chapter 7). Copyright Peter Simon; used by gracious permission from Ronni Simon and the Peter Simon Collection, Robert S. Cox Special Collections and University Archives Research Center, UMass Amherst Libraries.

The Library of Congress Cataloging-in-Publication Data is available online at https://catalog.loc.gov

Typeface for the Latin, Greek, and Cyrillic scripts: "Brill". See and download: brill.com/brill-typeface.

ISSN 2772-9826
ISBN 978-90-04-51528-4 (hardback)
ISBN 978-90-04-54668-4 (e-book)

This book is printed on acid-free paper and produced in a sustainable manner.

For Anita

Contents

Acknowledgements

The seed that later grew into this book was planted in 1969, when I read Jim Forest's essay, *In Time of War*.[1] I was active in the Vietnam-era anti-draft movement, and I was very intrigued by one of Jim's main points: that the then-current opposition to conscription had venerable antecedents in American history. Over the decades, in fits and starts, I did some research on the subject when time permitted. When I retired from the practice of law in June 2020, I turned full time to researching the topic and writing this book. The further I got into the research materials, the more astonished I became that no one had previously written a volume that dealt with conscription—and opposition to it—over the entire sweep of US history. Toward the end of his life, Jim knew I was working on this book; unfortunately, he died on January 13, 2022. I am grateful to Jim for having planted the seed.

I am deeply grateful to Dr. Scott Bennett, one of the co-editors of Brill's new series, *Studies in Peace History*, for correcting out-and-out errors in the manuscript; for editing advice and suggestions; and, above all, for extensive mentorship. I have no formal training in history, and this volume owes much to Scott's mentorship. His advice was always excellent (even when I was unhappy to receive it).

I am grateful to Professor Laura Weinrib of Harvard Law School (for providing a copy of Walter Nelles's *amicus curiæ* brief in the World War I *Arver* case); Steve Suffett (for providing FBI documents he had previously obtained through a Freedom of Information Act request); Duane Stolzfus of Goshen College (for sharing with me some of his previously published work); Olivia Olafsson (for excellent research assistance); Polly Kummel (for wise and sensitive copyediting and for her labors conforming my anarchic footnotes to the required format); Frank Peachey and Fern Habecker of the Mennonite Central Committee (for generous assistance in navigating the MCC archive in Akron, Pennsylvania); to the staff of the Swarthmore College Peace Collection, especially former Interim Director Anne Yoder, Director Rachel Mattson, and Archivist Vicki Russo (for guidance in navigating the SCPC); Professor Robert Griswold and librarian Laurie Scrivener of University of Oklahoma (for finding and providing a 1932 master's thesis on the Green Corn Rebellion discussed in Chapter 4); and my son-in-law, historian David Mislin (for general advice throughout).

I am especially grateful to my wife, Anita Kestin, who read the manuscript, made helpful suggestions, and was always supportive. This volume is for her.

1 James M. Forest, "In Time of War," in Daniel Berrigan, et al., *Delivered into Resistance* (New Haven, CT: Advocate Press, 1969), 1–10.

Illustrations

Figures

Tables

Abbreviations

AFSC	American Friends Service Committee
APL	American Protective League
AUAM	American Union Against Militarism
BLA	Bureau of Legal Advice
BSC	Brethren Service Committee/Commission
CCC	Civilian Conservation Corps
CCCO	Central Committee for Conscientious Objectors
CNVA	Committee for Nonviolent Action
CO	Conscientious objector
CORE	Congress of Racial Equality
CPS	Civilian Public Service
FOR	Fellowship of Reconciliation
LSCO	Legal Services for Conscientious Objectors
MCC	Mennonite Central Committee
NCAC	National Council Against Conscription
NCLB	National Civil Liberties Bureau
NSBRO	National Service Board for Religious Objectors
NCCO	National Committee on Conscientious Objectors
UMT	Universal military training
UMTSA	Universal Military Training and Service Act
WRL	War Resisters League

Commonly Used Selective Service Classifications

I-A	Immediately eligible to be drafted
I-A-O	Conscientious objector available for noncombatant military duty
I-O	Conscientious objector available for alternative civilian service (after 1951)
IV-E	Conscientious objector available for alternative civilian service (under the 1940 statute)
IV-F	Unavailable for military service due to physical or moral problem

Introduction

- In June 1863, when conscription came to the Union during the Civil War, a band of a thousand armed draft opponents drove conscription officers out of Rutland, Vermont. In Indiana antidraft vigilantes killed the conscription officers John Stevens and Fletcher Freeman. In Hancock County, Illinois, conscription officers who had nearly completed registering young men for the draft had their homes raided at night and their draft records destroyed; the loss was nearly total.[1]
- When conscription came to the Confederacy during the Civil War, draft opponents formed well-organized and heavily armed bands to resist it. Many bands numbered in the hundreds, and some had a thousand men or more. One Confederate officer reported that the resisters were "organized [and] drilling regularly" and said that they were "determined to kill in avoiding apprehension." One reason that these armed bands were successful was that they enjoyed widespread support of relatives and neighbors.[2]
- In August 1917, when the World War I draft commenced, an interracial group of about five hundred poor and largely illiterate sharecroppers and farmers in rural Oklahoma organized and armed themselves to resist the draft. They cut telephone and telegraph lines, burned down railroad bridges, and dynamited a bridge over a river. Subduing these draft opponents required posses of a thousand men; one man was killed and another wounded in the ensuing gunfights.[3]

Readers may find these descriptions surprising, but these events were by no means outliers. Stevens and Freeman were only two of the thirty-eight Union conscription officers murdered while performing their duties during the Civil War. Sixty more were shot and wounded but survived.[4] In many states of the

1 Eugene C. Murdock, *One Million Men: The Civil War Draft in the North* (Madison: State Historical Society of Wisconsin, 1971), 43; US War Department, *The War of the Rebellion: A Compilation of the Official Records of the Union and Confederate Armies* (Washington, DC: Government Printing Office, 1880–1901), ser. 3, 3:245; Robert E. Sterling, "Civil War Draft Resistance in the Middle West" (PhD diss., Northern Illinois University, 1974), 244, 225–26.

2 US War Department, *War of the Rebellion*, ser. 4, 2:783–85.

3 Charles C. Bush, "The Green Corn Rebellion" (master's thesis, University of Oklahoma, 1932).

4 US Office of the Provost Marshal General, *Final Report Made to the Secretary of War, by the Provost Marshal General, of the operations of the Bureau of the Provost Marshal General of*

Confederacy, draft refusal was the norm, not the exception. On the eve of draft registration for World War I, spontaneous antidraft protests erupted in large cities and small towns across America, hundreds of thousands—perhaps millions—of men illegally evaded the draft, and some who openly confronted and resisted the draft were sentenced to terms as long as fifty years or even death. During World War II, thousands of men went to prison for resisting the draft, and tens of thousands were recognized as conscientious objectors. During the relatively quiet conformist period of the 1950s, thousands of men became conscientious objectors (more than during World War II), hundreds went to prison for publicly resisting the draft, and tens of thousands quietly (and illegally) declined to register. In addition, during this period, scores of local anticonscription groups sprang up nationwide.

Perhaps the reason these facts are not more widely known is that there has never been a complete, single-volume history of conscription in the United States across the centuries. To be sure, there have been many studies of conscription in a particular conflict. Eugene C. Murdock's *One Million Men: The Civil War Draft in the North* (1971) addressed the Civil War draft but only in the Union. Albert Burton Moore's *Conscription and Conflict in the Confederacy* (1924) is a classic but addressed the Civil War draft only in the Confederacy. Ella Lonn's *Desertion during the Civil War* (1928) discussed conscription in both the Union and the Confederacy but was limited to the Civil War. Gilbert C. Fite and H. C. Peterson's *Opponents of War: 1917–1918* (1957) looked only at World War I, as did many more recent books, including Christopher Capozzola's *Uncle Sam Wants YOU: World War I and the Making of the Modern American Citizen* (2008) and David M. Kennedy's *Over Here: The First World War and American Society* (1980). Needless to say, there have been enough books published about the Vietnam era draft to fill a small library. John Whiteclay Chambers's *To Raise an Army: The Draft Comes to Modern America* (1987) was mainly about the World War I era, although it contained brief discussions of what preceded and what followed it. Even Jack Franklin Leach's *Conscription in the United States: Historical Background* (1952) missed much of the early history and, of course, most of the post–World War II Cold War period and all of the Vietnam era.

This is the first book that describes conscription across all of US history. Each chapter is divided into three broad parts (although the chapter subheads do not precisely track these broader categories). First is a discussion of the provisions of the draft law of that era—based on the actual language of the statute. Scholars have not always done this. For example, Charles DeBenedetti

---

the US: [...] *March 17, 1863 to March 17, 1866,* [...] (Washington, DC: Government Printing Office, 1866), pt. 1, 143, and pt. 2, 352.

used secondary sources in *The Peace Reform in American History* (1980) to describe the provisions of the World War I draft law, and Capozzola did not quote or cite the actual language of the statute even once.[5] Charles Chatfield used secondary sources in *For Peace and Justice: Pacifism in America, 1914–1941* (1971) to describe the provisions of the World War II draft law. Even Stephen M. Kohn, an attorney and part-time law professor, relied in part on secondary sources, rather than the actual statutes, to describe the World War I, World War II, and Vietnam era draft laws in *Jailed for Peace: The History of American Draft Law Violators, 1658–1985* (1986). But statutory language is important, and to understand any law, one must refer to the actual language of the statute and the regulations implementing the statute.

Second, each chapter includes detailed discussion of the opposition to conscription during the relevant era. This includes opposition that was public and clandestine, legal and illegal, religious and secular, violent and nonviolent. Throughout US history, draft laws have engendered opposition from many quarters and in widely disparate forms; to fully apprehend the depth and breadth of this opposition, it is necessary to understand the various component parts (and how those parts relate to one another).

Third, every chapter includes a discussion of the era's major court cases that addressed some of the questions and issues posed by both conscription and opposition to it. The law of conscription in the United States has evolved, in significant part, through court rulings; thus, this is a necessary component of the broader discussion.

A word must also be said about what this book is *not*. This is not a military history, and it does not address or discuss wars, battles, or military strategy. Nor is this a history of the peace movement in the United States, many of which have been written.[6] Indeed, in researching this book in archives and elsewhere, I came across much interesting (and exciting) information about various peace movements in US history that I deliberately omitted because—while peace movements and antidraft activity are related—they are not the same thing. This book discusses peace movements only insofar as that

---

5 Charles DeBenedetti, *The Peace Reform in American History* (Bloomington: Indiana University Press, 1980), 102n45.

6 See, for example, Charles DeBenedetti, *Origins of the Modern American Peace Movement, 1915–1929* (Millwood, NY: KTO Press, 1978); Charles Chatfield, *For Peace and Justice: Pacifism in America, 1914–1941* (Knoxville: University of Tennessee Press, 1971); DeBenedetti, *Peace Reform in American History*; Lawrence S. Wittner, *Rebels against War: The American Peace Movement, 1933–1983* (Philadelphia: Temple University Press, 1984); Scott H. Bennett, *Radical Pacifism: The War Resisters League and Gandhian Nonviolence in America, 1915–1963* (Syracuse, NY: Syracuse University Press, 2003).

discussion is directly relevant to conscription. Indeed, this book is not even about antimilitarism in the United States, another related but not identical topic; a book on that topic, Arthur A. Ekirch, Jr.'s *The Civilian and the Military: A History of the American Antimilitarist Tradition* (1972) nicely illustrates the distinction I am drawing. While Ekirch's book was about antimilitarism, it included only half a paragraph about the provisions of the World War I draft and only two and a half paragraphs describing the provisions of the World War II draft.

My principal thesis is that opposition to conscription in the United States has been far more widespread and active than is generally recognized, even by historians. The New York City antidraft riots in July 1863 and the burning of draft cards during the Vietnam era may be widely known today, but the extent to which these examples are emblematic of a deep and consistent vein of antidraft sentiment in the American psyche is less generally understood.

One of the benefits of looking at conscription across the centuries is the ability to discern similarities (and differences) in events that occurred during different eras. For example, during the Vietnam War, antidraft activists were stunned when they learned of a memorandum, known as "Channeling," that was written by General Lewis B. Hershey, the head of the Selective Service (1941–70). In it, Hershey said that only one relatively small, function of the draft system was to conscript a few thousand men into the military each month; a much larger function was—through the elaborate system of deferments—to channel the entire male population into areas that the government believed aided the war effort or were otherwise in the national interest.[7] When Hershey's memo was revealed in 1967, the outcry was huge; the public thought that Hershey's views were novel and shocking. But General Enoch Crowder, the head of the Selective Service during World War I, had said much the same thing, as had other individuals, when Congress was considering the draft in 1917 and in 1918 when the draft was a reality. In 1940, when Congress was considering the World War II draft, General William Shedd had said the same thing when testifying before Congress in support of conscription.[8] Throughout the long years of the Cold War, Hershey repeated the channeling

7 US Selective Service System, "Channeling," a document dated 1965 that was included in the service's orientation kit for draft boards nationwide, in CDGA Resistance box, folder labeled "The Resistance," Swarthmore College Peace Collection, Swarthmore College, Swarthmore, Pennsylvania.

8 Michael S. Foley, *Confronting the War Machine: Draft Resistance during the Vietnam War* (Chapel Hill: University of North Carolina Press, 2003), 62; US Office of the Provost Marshal General, *Report of the Provost Marshal General to the Secretary of War on the First Draft Under the Selective-Service Act, 1917, 1918* (Washington DC: Government Printing Office, 1918), 33;

mantra many times. By the time the antidraft activists of the Vietnam era learned of the Hershey memorandum, the concept had guided US policy on conscription literally for generations.

Another example: in October 1967, the Reverend Philip Berrigan mutilated draft records at a Baltimore draft board by pouring blood on them, and in May 1968 he destroyed draft files at a draft board in a Baltimore suburb by burning them with homemade napalm. Berrigan, and the Vietnam era activists who followed his example (and there were many), believed that they had come up with a new and novel way to undercut the draft. They were mistaken. A widely used tactic of draft opponents in the North during the Civil War was destruction of draft files.[9]

Readers will also see a consistent thread running through the discussions of court cases that appear in every chapter. Whenever the United States has had conscription, courts have had to address complex issues of jurisdiction. Congress, in response to what it believed to be overly solicitous views of draft objectors by courts, has repeatedly attempted to divest federal courts of jurisdiction to hear draft-related cases. Intermittently, courts have pushed back against these attempts. In some cases, federal courts have rendered sweeping rulings about conscription despite congressional prohibitions that courts even be allowed to *hear* draft cases.

I must make one more point. I am not (and do not pretend to be) a neutral observer of the matters covered in this book. I was active in the Vietnam era peace movement, and my previous book was *Felon for Peace: The Memoir of a Vietnam-Era Draft Resister* (2005). When I later decided to attend law school, I was—as a result of my antidraft activity—a convicted felon, the only felon in my graduating class at Harvard Law School. This book is a product of the unusual perspective that I bring to the subject.

Hearings before the Committee on Military Affairs of the United States House of Representatives on H.R. 10132, 76th Cong., 3d sess. (1940), 536.

9 "Blood Poured on Baltimore Draft Files; Priest, Minister, 2 Others Seized," *New York Times*, October 28, 1967.

CHAPTER 1

Background and Context

On June 2, 1784, the Continental Congress disbanded the Continental Army. The War of Independence had been won (the Treaty of Paris had been signed on September 3, 1783, and ratified by the Continental Congress on January 14, 1784). What use could the newly independent states possibly have for a standing army? Notably, the Continental Congress had twice refused to approve General George Washington's proposal for the creation of a small peacetime standing army: first, on May 12, 1783, before the signing of the Treaty of Paris and again on October 18, 1783, after the treaty was signed. The colonies had just finished fighting the War of Independence against Britain's standing army; Americans had good cause to look skeptically upon standing armies.[1]

The Continental Congress's disbanding of the Continental Army is not surprising for another reason. State constitutions adopted after the signing of the Declaration of Independence—those of Pennsylvania, Vermont, and North Carolina—all contained this identical language: "As standing armies in the time of peace are dangerous to liberty, they ought not to be kept up." The language in the New Hampshire Constitution was a bit more qualified, but the sentiment was similar: "Standing armies are dangerous to liberty, and ought not to be raised, or kept up, without the consent of the Legislature." The Ohio Constitution of 1802 was a variation on the theme: "As standing armies in time of peace are dangerous to liberty, they shall not be kept up." By the middle of the nineteenth century, the language used in the Kansas Constitution had become a bit more strident on the point: "Standing armies, in time of peace, are dangerous to liberty, and shall not be tolerated."[2]

When the states were asked to ratify the proposed new constitution that emerged from the Philadelphia Convention in 1787, they expressed similar skepticism to standing armies. It is well known that several states were reluctant to ratify the constitution because it lacked a bill of rights. Less well known is that several states were also reluctant to ratify the constitution because it did not include an express ban on standing armies and an explicit provision

1 Robert K. Wright, Jr., *The Continental Army* (Washington, DC: Center of Military History, US Army, 2006), 180–82; John Whiteclay Chambers II, *To Raise an Army: The Draft Comes to Modern America* (New York: Free Press, 1987), 13.

2 Pa. Const. § 13; Vt. Const. chap. 1, § 15; N.C. Const. § 17; N.H. Const. pt. 1, art. 25; Ohio Const. art. 8, § 20; Kans. Bill of Rights § 4.

to account for the beliefs of men who were conscientious objectors (COs) to militia service. The Rhode Island ratification convention suggested that both provisions be added to the newly proposed federal constitution: "That standing armies, in time of peace, are dangerous to liberty, and ought not to be kept up, except in cases of necessity" and "that any Person religiously scrupulous of bearing arms ought to be exempted upon payment of an equivalent to employ another to bear arms in his stead."[3]

The New York ratifying convention copied verbatim the Rhode Island language pertaining to standing armies.[4] The Virginia ratifying convention made the same two recommendations as Rhode Island, except that the wording of the text opposing standing armies was slightly altered: "That standing armies, in time of peace, are dangerous to liberty, and therefore ought to be avoided, as far as the circumstances and protection of the community will admit." The North Carolina ratifying convention adopted the Virginia language verbatim.[5]

The foregoing provisions were not the voices of individual dissenters who held minority views. These provisions represented the people of several states, speaking through their elected representatives. Skepticism of standing armies was a majoritarian viewpoint.

The maximum size the Continental Army achieved at any one time was 16,800 men in 1778.[6] More than eighty years later, at the beginning of the Civil War in April 1861, the standing army of the United States remained minuscule: the actual strength of the entire US Army was only 14,657 men, smaller than the Continental Army had been at its highest point, while the population of the United States had increased eightfold, from 3.2 million in the 1790 US Census to more than 26 million in the 1860 US Census.[7]

From the colonial era through the Civil War, first the colonies, then the independent states, then the states under the Constitution of 1789 all had

---

3 Jonathan Elliot, *The Debates in the Several State Conventions, on the Adoption of the Federal Constitution, on the Adoption of the Federal Constitution, as Recommended by the General Convention at Philadelphia, in 1787* (Washington: Published Under the Sanction of Congress, 1836), 335 (hereafter, *Elliot's Debates*). See also *United States v. Emerson*, 270 F.3d 203, 244 (5th Cir. 2001), which cites *Elliot's Debates*.

4 District of Columbia v. Heller, 554 U.S. 570, 656 (2008) (Stevens, J., dissenting).

5 *Elliot's Debates*, 659, 244.

6 Chambers, *To Raise an Army*, 23.

7 US Office of the Provost Marshal General, *Final report made to the Secretary of War, by the Provost Marshal General, of the operations of the Bureau of the Provost Marshal General of the US:* [...] *March 17, 1863 to March 17, 1866,* [...] (Washington, DC: Government Printing Office, 1866), pt. 1, 6; US Census Bureau, *Population of the United States in 1860: Recapitulation of the Tables of Population Nativity and Occupation* (Washington, DC: Government Printing Office, 2021), 593, 600.

state militias. These were comprised of so-called citizen-soldiers. Typically, every free man aged eighteen to forty-five was a member of the state militia, although he might be called into service for only a few weeks or months every year or every several years.[8]

Many states made provisions for conscientious objectors (cos) with regard to their state militias. These provisions fell into three categories. In the first (and smallest) category, one state, Maine, had an absolute constitutional exemption for certain cos: "Persons of the denominations of Quakers and Shakers ... may be exempted from military duty."[9]

In the second category, the co provision appeared in the state constitutions but required the exempted man to pay a fee. The Indiana Constitution of 1851 was an exemplar; it provided: "No person conscientiously opposed to bearing arms shall be compelled to do militia duty; but such person shall pay an equivalent for exemption; the amount to be prescribed by law." Other states in this category were Indiana, Iowa, Illinois, Kansas, Kentucky, and Pennsylvania.[10] (Chapter 2 discusses some of the problems associated with this fee.)

In the third category are state constitutions that recognized the issue of cos but left it to the legislature to enact statutes after determining what accommodation would be made. The Michigan Constitution of 1851 was typical of those in this category; it provided that all "citizens, of any religious denomination whatever, who, from scruples of conscience, may be averse to bearing arms, shall be excused therefrom, upon such conditions as shall be prescribed by law." Other states in this third category were Minnesota, New York, Ohio, and Vermont.[11]

Two conclusions flow from these facts. First, although the United States had no *federal* conscription until 1863, there certainly was near-universal state conscription of men for state militias. That is, opponents of conscription cannot accurately say that the United States had no conscription at all until the Civil War, only that there was no federal conscription by the federal government. Second, that some men were religiously or morally opposed to military service (and to the bearing of arms) was widely recognized, although there was

8 Alan Hirsch, "The Militia Clauses of the Constitution and the National Guard," *University of Cincinnati Law Review* 56, no. 3 (1988): 919, 921, 923; Chambers, *To Raise an Army*, 14.

9 Maine Const. of 1819 art. 7, § 5.

10 Ind. Const. of 1851 art. 12, § 6; Edward Needles Wright, *Conscientious Objectors in the Civil War* (New York: A. S. Barnes, 1931), 39.

11 Mich. Const. of 1850 art. 17, § 1; E. N. Wright, *Conscientious Objectors in the Civil War*, 39.

considerable variation among the states about how such people were to be accommodated.[12]

1 The United States Constitution

The Constitutional Convention in Philadelphia concluded on September 17, 1787, and the new constitution came into effect on March 4, 1789, after nine states ratified it.[13]

The US Constitution gave Congress two specific powers directly addressing military matters, and one additional clause is relevant. Article 1, section 8, lists the powers of the US Congress in a series of eighteen clauses. The first direct reference to military manpower is clause 12, which gives Congress the power "to raise and support Armies, but no Appropriation of Money to that Use shall be for a longer term than two Years." The second direct reference appears in clause 15, which states that another of the powers of Congress is "to provide for calling forth the Militia to execute the Laws of the Union, suppress insurrections and repel Invasions."

That is, the Constitution contains a fundamental bifurcation: It bestows on Congress two distinct ways to procure a military force. Clause 12 allows Congress to create its own professional standing army of the type that Washington believed was necessary but that the Continental Congress refused to approve. Clause 15 allows Congress to second the state militias into federal service but only for three specifically stated reasons: to enforce federal laws ("execute the Laws of the Union"), suppress insurrections, and repel invasions. The Continental Army was created and fought the War of Independence made up entirely of seconded state militias from the thirteen colonies.

One historian explains the ideological underpinning of this basic constitutional bifurcation this way: "In their emphasis upon individual liberty, republicanism, and civic virtue, the colonists idealized the citizen-soldier of

12 For a useful collection of original documents pertaining to conscientious objectors from colonial times through the Civil War, see Peter Brock, ed., *Liberty and Conscience: A Documentary History of the Experiences of Conscientious Objectors in America through the Civil War* (New York: Oxford University Press, 2002).

13 Max Farrand, ed., *The Records of the Federal Convention of 1787* (New Haven, CT: Yale University Press, 1911), 650; U.S. Const. art. 7; Joseph Gales, ed., *Debates and Proceedings in the Congress of the United States* (Washington, DC: Gales and Seaton, 1834), 15, 99 (hereafter, Annals of Cong.).

the militia; hostility toward standing armies became an integral part of the American ideology."[14]

The Constitution also confers on Congress a more generalized power. Article 1, section 8, clause 18, states that Congress has the power "to make all laws which shall be necessary and proper for carrying into Execution the foregoing Powers."

When someone asks whether the Constitution specifically permits *conscription* in order to raise a professional standing army (as opposed to seeking only *volunteers*), the standard response is to refer to article 1, section 8, clause 12, which allows Congress "to raise and support Armies." That was certainly Abraham Lincoln's view. When his resort to conscription during the Civil War brought a constitutional challenge, Lincoln wrote: "The constitution declares that 'The congress shall have the power ... To raise and support armies'; the whole scope of the conscription act is 'to raise and support armies.' There is nothing else in it.... It is a law in litteral [*sic*] pursuance of this part of the United States Constitution." Lincoln concluded, "The power is given fully, completely, unconditionally."[15]

This was the view of the federal courts, including the US Supreme Court, throughout the twentieth century whenever federal conscription laws were challenged on constitutional grounds. The leading US Supreme Court case for the proposition that conscription is constitutional, Selective Draft Law Cases (*Arver*), decided in 1918, arose under the World War I conscription statute. In that case, the court recited that Congress has the power "to raise and support armies" and the authority "to make all laws necessary and proper for carrying into execution the foregoing powers." Then the court continued, "As the mind cannot conceive an army without the men to compose it, on the face of the Constitution the objection that it does not give power to provide for such men would seem to be too frivolous to further notice."[16]

The same result obtained in a challenge to the constitutionality of the 1940 draft law, both before and after US entry into World War II, including the postwar peacetime draft enacted in 1948 and the successor statutes in effect during the Vietnam War.[17] In every case, during every war (and in peacetime), federal

14 Chambers, *To Raise an Army*, 19.

15 Abraham Lincoln, "Opinion on the Draft," in *Collected Works of Abraham Lincoln*, ed. Roy Basler (New Brunswick, NJ: Rutgers University Press, 1953), 6:446.

16 Selective Draft Law Cases, 245 U.S. 366, 377 (1918) (hereafter, Arver).

17 United States v. Cornell, 36 F. Supp. 81 (D. Idaho 1940); United States v. Newman, 44 F. Supp. 817 (E.D. Ill. 1942); United States *ex rel.* Zucker v. Osborne, 54 F. Supp. 984 (W.D.N.Y. 1944); United States v. Henderson, 180 F.2d 711 (7th Cir. 1950); United States v. O'Brien, 391 U.S. 367 (1968).

courts have held that the various conscription statutes are constitutional. In one of these cases, the reviewing court stated simply, "Congress was expressly given the power, 'To raise and support Armies.' ... This is an unqualified power given to Congress in order that it may protect the very existence of government. There is neither express nor implied limitation in the Constitution to this power."[18]

Nevertheless, abundant evidence shows that when the framers wrote the Constitution, they did not mean that the power to raise and support armies meant the power to *conscript* those armies. Article 1, section 8, gives Congress many powers over many matters but never mentions—nor even hints at—carrying out any of those powers by *conscription*.

Article 1, section 8, gives Congress the power to establish post offices, but it does not give Congress the power to conscript letter carriers or post office clerks. Article 1, section 8, gives Congress the power to coin money, but no one believes that Congress has the power to conscript the staff of the US Mint; Congress can raise money through taxes and hire people who voluntarily choose to work for the mint. Congress has the power to protect intellectual property by enacting patent and copyright laws, but this does not confer on Congress the power to impress citizens into involuntary employment in the US Patent Office.

In the *Federalist*, no. 23, Alexander Hamilton wrote that "the Union ought to be invested with full power to levy troops; to build and equip fleets, and to raise the revenues, which will be required for the formation and support of an army and navy in the customary and ordinary modes practiced by other governments." Today, the word *levy* carries the connotation of "the compulsory enlistment of personnel, as for military service," but at the time of the *Federalist*, dictionaries defined the word as simply meaning "to raise."[19] Moreover, when these words were written in 1787, conscription did not exist in England (except as punishment, after trial, for "idle and disorderly persons") or in the rest of Europe. Conscription was unknown in Europe until 1798 and did not become

---

18 Henderson, 180 F.2d at 715–16.

19 *Federalist*, no. 23 (Alexander Hamilton), ed. Clinton Rossiter (New York: Signet, 1961), 154–55. Compare the definition of *levy* given by Noah Webster, *Webster's New World Dictionary of the American Language* (New York: World, 1964), 842, and that by Samuel Johnson, *Dictionary of the English Language,* 2d ed. (London: Knapton, Hitch, Hawes, Millar, Dodsley and Longman, 1755), n.p. Charles A. Lofgren, "Compulsory Military Service under the Constitution: The Original Understanding," *William and Mary Quarterly* 33, no. 1 (January 1976): 69–70, relies on Johnson's definition.

common until after the Napoleonic era in the early nineteenth century.[20] When Hamilton spoke in 1789 of "the customary and ordinary modes" of procuring a professional standing army, he could not have meant conscription, because conscription—far from being "customary and ordinary"—was virtually unknown.

This also helps to explain the meaning (and importance) of the limitation in article 1, section 8, that "no Appropriation of Money to that Use [an army] shall be for a longer Term than two Years." The framers were not proposing to give the newly designed federal government unprecedented powers (conscription) that other countries did not have; they were proposing something quite customary and ordinary at the time (taxation): the ability to raise money to pay soldiers in order to induce voluntary enlistments.

In the *Federalist*, no. 31, Hamilton addressed specifically the way in which the proposed new federal government would be able to discharge its "duties of superintending the national defense and of securing the public peace against foreign or domestic violence." Hamilton explained that the national defense was to be secured by the constitutional power "to lay and collect Taxes" because "revenue is the essential engine by which the means of answering the national exigencies must be procured."[21] With the power to raise money, Congress could raise a professional, albeit volunteer, army. The two-year constitutional limit on this "essential engine" was an accommodation to the view, widespread at the time, that standing armies were dangerous and unwise.

James Madison, another framer, had the same understanding as Hamilton: that the constitutional power "to raise and support armies" did not include the right to conscript those armies. In the first session of the first Congress, on June 8, 1789, then-representative Madison introduced into the House of Representatives a set of proposals for what would eventually become the Bill of Rights. As originally introduced, one of Madison's proposals (which would be amended and eventually become what we know today as the Second Amendment) said, in its entirety, "The right of the people to keep and bear arms shall not be infringed; a well-armed and well-regulated militia being the best security of a free country; but no person religiously scrupulous of bearing arms shall be compelled to render military service in person."[22] The only conceivable reason

20 For details see Harrop Freeman, "The Constitutionality of Peacetime Conscription," *Virginia Law Review* 31, no. 1 (1944): 68; Lofgren, "Compulsory Military Service," 80–81; Leon Friedman, "Conscription and the Constitution: The Original Understanding," *Michigan Law Review* 67, no. 8 (June 1969): 1493, 1498–99.

21 U.S. Const. art. 1, § 8, cl. 1; *Federalist*, no. 31 (Alexander Hamilton), 194–95.

22 1 Annals of Cong. 452 (1789).

why Madison included an exemption for COs for service in the militia—but not for the professional standing army permitted in article 1, section 8—is that part-time citizen service in the state militias was conscripted, but service in the professional army was voluntary. No CO provision was necessary for a force that was strictly voluntary; COs could simply choose not to enlist. This is the only reading that makes sense of Madison's original language.

Madison's understanding continued beyond 1789. Almost a decade later, in May 1798, Congress considered a proposal to create a provisional (professional) army. During the floor debate on the bill, Senator Samuel W. Dana of Connecticut made a statement that reflected the prevailing understanding of the time: "If there be any characteristic marks by which to distinguish a militia from regulars, it is that the former are compelled to serve, and the latter enter voluntarily into the service." To Dana, it was clear that the power to raise armies did not include the power to conscript those armies but to raise an army of men who "enter voluntarily into the service." Senator Robert G. Harper of South Carolina expressed the same view: "The distinct property of the militia is this, that they are obliged to perform military duty," while the service by men in a standing army was strictly voluntary.[23] Senator Harrison Otis of Massachusetts also was at pains to explain the difference between a militia and a professional standing army: "The striking characteristic difference is this: that militia are *bound* to serve by the laws of the country, without any respect for personal consent, and the volunteers [in the proposed provisional professional army] will be bound only by contract; they will consist only of those who come forward to defend their country."[24]

In 1803, Congress enacted a statute pertaining to the construction of some federal arsenals. The statute made the same basic bifurcation clear by referring to "detachments of militia and the volunteer corps."[25] The militia were the part-time, state-organized citizen-soldiers referred to in clause 15; the volunteers were the professional standing army referred to in clause 12. The latter were volunteers because the Constitution provided no authority to conscript them.

In the years immediately following the ratification of the Constitution, while several authors of the Constitution were sitting members of Congress, Congress passed a number of statutes pertaining to the military. All these

23 8 Annals of Cong. 1705 (1798).
24 8 Annals of Cong. 1733 (1798), emphasis added.
25 An Act directing a detachment from the Militia of the United States, and for erecting certain Arsenals, 2 Stat. 241 (1803).

statutes were consistent with the understanding that the Constitution had given Congress two specific military manpower authorities.

In 1790, Congress exercised its clause 12 powers to create a professional standing army by authorizing the recruitment by voluntary service of 1,216 soldiers for three-year terms of enlistment. The statute made no mention of Congress's clause 15 power to second state militias into federal service, and the statute certainly made no reference to any conscripted soldiers for the standing army.[26]

On May 8, 1792, Congress passed the Militia Act of 1792. Here, Congress was acting under its clause 15 powers pertaining to federal seconding of state militias. The law called for state enrollment (registration) of eligible men for the various part-time, citizen-soldier state militias and provided that when any state militiaman was wounded while seconded to federal service "he shall be taken care of and provided for at the public expense."[27]

In the wake of the Whiskey Rebellion in western Pennsylvania in 1794, on February 28, 1795, Congress passed the Militia Act of 1795. This statute, too, related only to the clause 15 militia power, which expressly permits the federal government to call forth the state militias for three (and only three) purposes. The act tracked the constitutional requirement with admirable clarity. It authorized the president to call forth the state militias "whenever the United States shall be invaded or in imminent danger of invasion," or "in case of an insurrection in any state," or "whenever the laws of the United States shall be opposed, or ... obstructed ... by combinations too powerful to be suppressed by the ordinary course of judicial proceedings."[28] The law made no reference to the professional standing army.

Especially pertinent to the present discussion is a statute enacted in 1803 pertaining to both the clause 15 part-time citizen-soldier state militias and the clause 12 professional standing army of volunteers. Section 1 of the 1803 statute pertained to the federal secondment of state militias by the president in time of emergencies. Section 2 of the statute referred to the clause 12 professional standing army that may be recruited from time to time; significantly, the law expressly referred to the standing army as a "corps of volunteers." And section 3 referred to both "the detachments of militia and volunteer corps as aforesaid" and specified how officers for the combined force were to be selected. The language of the statute made clear that the professional standing

26 An Act for regulating the Military Establishment of the United States, 1 Stat. 119 (1790).

27 An Act more effectually to provide for the National Defence by establishing an Uniform Militia throughout the United States, 1 Stat. 271 (1792).

28 An Act to provide for calling forth the Militia to execute the laws of the Union, suppress insurrections, and repel invasions; and to repeal the Act now in force for those purposes, 1 Stat. 424 (1795). The authority granted the president is in § 1.

army contemplated by the constitutional provision that allows the federal government "to raise and support armies" meant the raising of a *volunteer* force, not a *conscripted* one.[29]

∴

The United States fought the War of 1812 with the small federal volunteer army and state militias seconded under the Militia Act of 1795. There was no federal draft. In 1814, a federal draft was proposed, but it never made it through Congress.

The 1814 proposal had some support in Congress but also significant opposition. On the House floor, Representative Cyrus King of Massachusetts asked how Congress could "exercise the Constitutional power of raising armies." His answer: "By voluntary enlistment of persons capable of contracting; for enlistment is only a contract." During floor debate, Morris S. Miller of New York agreed: "I object, sir, to the whole system of force and coercion; and contend that under this Constitution you have no right to raise armies except by voluntary enlistment." Daniel Sheffey of Virginia echoed the theme that the federal government could raise and support armies by contract or voluntary enlistment but not by coercion.[30]

On the Senate side, Jeremiah Mason of New York clarified that "the military power ... given by the Constitution to this Government is of two sorts.... There have always been in this country important distinctions between the militia and the regular Army." With regard to the militia, the Constitution provides that state militias can be seconded into federal service "in three specified cases only, and for a service within the limits of the United States." With regard to a standing army, the Constitution allows Congress to raise and support the regular army, but "voluntary enlistment [is] the only method by which a regular Army could be raised."[31]

Even the use of seconded state militias in the War of 1812 was controversial because the Madison administration meant to take the war across the northern US border and fight British troops in Canada.[32] Governors Caleb Strong of Massachusetts and Roger Griswold of Connecticut refused to allow their states' militias to be used in the conflict because the plan to send the militias

29 An Act directing a detachment from the Militia, 2 Stat. 241 (1803).

30 28 Annals of Cong. 723, 851 (1814).

31 Ibid., 77–84.

32 Donald R. Hickey, *The War of 1812: A Forgotten Conflict* (Urbana: University of Illinois Press, 1989). He writes, "The War of 1812 was America's most unpopular war. It generated more intense opposition than any other war in the nation's history, including the war in Vietnam" (255).

into Canada violated the constitutional mandate that seconded state militias could be used only to enforce federal laws, suppress insurrections, and repel invasions. Strong even asked the Supreme Court of Massachusetts for an advisory opinion on the question of whether he, as governor, had authority under the US Constitution to determine whether one or more of the three required conditions existed. In response to Strong's request, the court ruled in the affirmative—that, yes, the governor does have that power and implied that he also had the power to not send his state militia if he determined that the prerequisites had not been met.[33]

In addition, there were repeated instances during the War of 1812 when federally seconded state militias refused to obey illegal orders from their commanders to cross the border to fight in Canada. In July 1812, two hundred Ohio state militia refused orders from General William Hull to cross the Detroit River into Canada in order to attack Fort Malden on the Canadian side. In October 1812, New York State militia refused orders from Major Stephen Van Rensselaer to cross the Niagara River into Canada. In November 1812, in perhaps the best known case of a seconded state militia's refusing an unlawful order to cross into Canada, militiamen under General Henry Dearborn refused to cross the border from Plattsburgh, New York, to attack Montreal. The following year, in October 1813, one thousand militiamen commanded by General Wade Hampton refused to enter Canada at a similar spot on the Chateauguay River near Montreal.[34]

The United States fought the Mexican-American War of 1848 exclusively with volunteers, not seconded state militias. It could not have been otherwise, because a foreign military adventure to seize another country's territory was not a constitutionally permissible purpose for the federal government to second state militias.[35]

2 The Issue of Enumerated Powers

During the Vietnam War, a quotation from the early US statesman Daniel Webster appeared on posters and in magazines: "Where is it written in the Constitution, in what article or section is it contained, that you may take children

33 Hickey, *War of 1812*, 259–60; Opinion of the Justices, 8 Mass. (8 Tyng) 548 (1812).
34 Hickey, *War of 1812*, 81, 86–87, 88, 144–50.
35 Chambers, *To Raise an Army*, 3.

from their parents, & parents from their children, & compel them to fight the battles of any war, in which the folly or the wickedness of Government may engage it?" The words came from an 1814 antidraft speech that Webster had given in the House of Representatives in response to the proposal to impose a draft for the War of 1812. To Vietnam era peace activists, the quotation read as a pæan to the cruelties of war, but the backstory to Webster's quotation implicates a serious constitutional debate, one far wider than the narrow issue of conscription.[36] This debate continues today both in the courts and throughout the wider society.

On October 17, 1814, toward the end of the War of 1812, Secretary of War James Monroe, unhappy with the results of using state militias seconded to federal service, sent a proposal to Congress for conscription. Monroe understood that it would be controversial and would be denounced as unconstitutional, and he was right. As a result, Monroe felt constrained to preface his unprecedented proposal for a draft with a constitutional argument as to why Congress had the power to conscript: "The idea that the United States cannot raise a regular army in any other mode than by accepting voluntary service of individuals, is believed to be repugnant to the uniform construction of all grants of power.… An unqualified grant of power gives the means necessary to carry it into effect."[37]

The issue Monroe and Webster debated in 1814 addressed what today is known as the enumerated powers doctrine. The doctrine holds that the federal government is one of enumerated powers only, and it cannot exercise any powers that the Constitution does not expressly give it. This is a concept of venerable provenance that goes back to a US Supreme Court case decided in 1819 in which Chief Justice John Marshal wrote, "This government is acknowledged by all to be one of enumerated powers." To put the same point another way, as the court did two centuries later, "Every law enacted by Congress must be based on one or more of its powers enumerated in the Constitution."[38] The gist of Webster's argument on the House floor in 1814 was that the Constitution contains no specifically enumerated power to conscript. The reason Monroe felt forced to address that question at such length is that he himself was unsure that was the case.

---

36 "Speech of Daniel Webster in the House of Representatives December 9, 1814," in Martin Anderson, ed., *The Military Draft* (Stanford, CA: Hoover Institution Press, 1982), 639. See also *WIN Magazine* 6, no. 21 (December 15, 1970): 27.

37 Monroe's proposal included four alternatives for effecting federal conscription, but he clearly favored the first of the four. See the text of his communication in Anderson, *Military Draft*, 503–13.

38 McCulloch v. Maryland, 17 U.S. 316, 405 (1819); United States v. Morrison, 529 U.S. 598, 607 (2000).

Controversies pertaining to the enumerated powers doctrine are still very much with us today. For example, in 1964, after decades of delay, evasion, opposition, and filibuster, Congress passed, and President Lyndon Johnson signed, the landmark Civil Rights Act of 1964. The moral issue presented by the legislation was whether the United States wanted to provide equal rights to all its people. The political issue presented was whether, a century after the Civil War, the United States had sufficiently overcome centuries of entrenched racism to muster the collective political will to pass this legislation. But the Civil Rights Act also presented a constitutional question emanating from the enumerated powers doctrine: Did the US Constitution grant Congress the specifically enumerated power to dictate to individuals to whom they may legally sell their home and whom they were legally obliged to serve in their restaurants?

That someone would file a federal lawsuit challenging the constitutionality of the Civil Rights Act was utterly predictable, as was the constitutional basis of the challenge: that no enumerated power in the Constitution gives Congress the authority to mandate civil rights in public accommodations and other putatively private relations between private individuals. In the event, the actual lawsuit was brought in the US District Court for the Northern District of Alabama by one Ollie McClung, Jr., owner of Ollie's Barbecue, a restaurant in Birmingham. Stripped of its lawyerly mumbo jumbo, McClung's argument was so simple that it can be stated in a single sentence: there is no power enumerated in the Constitution that allows the federal government to tell me whom I must serve in my own restaurant.

The federal court agreed.[39] And it didn't take long, either. Johnson had signed the Civil Rights Act on July 2, 1964, and the US District Court declared the statute unconstitutional on September 17, 1964, based on the enumerated powers doctrine: "The federal government is a union of delegated powers. It has only such as are reasonably implied from those granted." The federal court in Alabama decided that there was no reason to conclude that "the national government has the power to control the conduct of people at the local level."[40]

The case went immediately to the US Supreme Court, which promptly and unanimously upheld the constitutionality of the Civil Rights Act. Interestingly, the court held that Congress got its power to enact the Civil Rights Act from—of all things—the commerce clause, which gives Congress the power "to regulate Commerce ... among the several States."[41] The Supreme Court based its

39 McClung v. Katzenbach, 233 F. Supp. 815 (1964). The decision was rendered by a three-judge panel.

40 McClung, 233 F. Supp. at 824–25.

41 Katzenbach v. McClung, 379 U.S. 294 (1964); U.S. Const. art. 1, § 8, cl. 3.

decision on two facts: Ollie's Barbecue was eleven blocks from an interstate highway, some of its customers were truckers who drove across state lines, and Ollie bought 46 percent of the meat he served from a supplier who procured the meat from out of state. Thus, Ollie's little local barbecue joint was really involved in interstate commerce, said the Supreme Court, and Congress had the power to tell Ollie whom he must serve because the Constitution gave Congress that power in the commerce clause.

It fell to Justices Arthur Goldberg and William O. Douglas to file concurring opinions in which each said that the decision should instead have been grounded in the equal protection clause of the Fourteenth Amendment, which says that no state will "deny to any person within its jurisdiction the equal protection of the law."[42]

Note, however, that *all* these judges and justices—the three-judge district court panel that found the law unconstitutional, the Supreme Court justices who found the law constitutional based on the commerce clause, and the concurring justices who would have preferred that the decision be based on the equal protection clause of the Fourteenth Amendment—grounded their very different rulings on the enumerated powers doctrine. All agreed that the legal questions that needed to be answered were these: Does the Constitution give this power to the federal government? If yes, exactly where does it do that?

In 2012, when the Supreme Court addressed the constitutionality of the Affordable Care Act (ACA, colloquially known as Obamacare), the outcome of the case again hinged on the enumerated powers doctrine; once again the decisive questions were whether the Constitution gives this power to the federal government and, if so, exactly where?[43]

Four liberal justices said yes, Congress has the power to enact the ACA, and it derived that power from the commerce clause. Four conservative justices disagreed and said that the commerce clause does not go that far and that nothing else gives Congress the power to enact the ACA. Chief Justice John Roberts broke the tie and cast the deciding vote. Roberts disagreed with both the liberals and the conservatives. Roberts agreed that Congress has the power, but it comes from its power to "lay and collect Taxes [and] Duties," because, in certain circumstances, the ACA provides a tax-like penalty for noncompliance.[44] Thus, although the justices could not agree on *where* the Constitution confers

42 U.S. Const. Amend. 13, § 1. Justices Arthur Goldberg and William O. Douglas also cited section 5 of the amendment, which says that "Congress shall have power to enforce, by appropriate legislation, the provisions of this article."

43 National Federation of Independent Business v. Sebelius, 567 U.S. 519 (2010).

44 Ibid., 589, 646, 529.

the power to pass the ACA, a five-justice majority agreed that Congress does have the power to enact the ACA. I also will address the enumerated powers doctrine in discussions of constitutional challenges to later conscription statutes.

3 Civil War Conscription Antecedents

On April 15, 1861, three days after the attack on Fort Sumter, Lincoln signed a proclamation seconding into federal service seventy-five thousand state militia for ninety days.[45] Lincoln's legal authority for doing so was the Militia Act of 1795, which was still in effect. The ninety-day period of service specified by the proclamation was fully consistent with the widespread belief in the North that the war would be short and glorious—ninety days was plenty of time to bring it to a successful conclusion.[46]

Lincoln's proclamation was legally sound: it tracked the underlying statutory authority closely. The 1795 Militia Act permitted the president to second the state militias into federal service whenever "the laws of the United States shall be opposed or the execution thereof obstructed in any state by combinations too powerful to be suppressed by the ordinary course of judicial proceedings." Accordingly, Lincoln's proclamation recited that "the laws of the United States have been for some time past, and now are opposed, and the execution thereof obstructed, in South Carolina, Georgia, Alabama, Florida, Mississippi, Louisiana, and Texas, by combinations too powerful to be suppressed by the ordinary course of judicial proceedings."

Second, the Militia Act of 1795 was itself constitutional under article 1, section 8, clause 15, which permitted federal secondment of state militias to "execute the Laws of the Union." In short, Lincoln's proclamation of April 15, 1861, carefully conformed to the requirements of the underlying statute, which itself was constitutional.

On May 3, 1861, Lincoln issued his Proclamation 83 in which he called for 42,034 new volunteers for the Union army.[47] Note that under article 1, section 8, clause 12, *Congress*, not the president, has the power "to raise and support

45 Abraham Lincoln, "Proclamation calling 75,000 Militia, and convening Congress in extra Session, April 15, 1861," in *The Complete Works of Abraham Lincoln*, ed. John George Nicolay and John Hay (New York: Francis D. Tandy, 1894), 6:246. See also James M. McPherson, *Battle Cry of Freedom* (New York: Oxford University Press, 1988), 339–68.

46 McPherson, *Battle Cry of Freedom*, 322.

47 Abraham Lincoln, "A Proclamation," in *Collected Works of Abraham Lincoln*, 4:353–54.

Armies" by means of seeking volunteers. Lincoln purported to do this under his constitutional authority as commander in chief, though historians now recognize that Lincoln had no authority to do this.[48] However, when Congress came back into session—far from objecting to this bald presidential usurpation of congressional authority—it ratified Lincoln's Proclamation 83 retroactively and went considerably beyond it. On July 22, 1861, one day after the Confederacy routed Union forces at the First Battle of Bull Run, Congress approved the recruitment of 500,000 volunteers for the Union army.[49] Three days later, on July 25, 1861, Congress approved the recruitment of an additional half-million volunteers.[50] There can be no doubt that both laws were constitutional. Article 1, section 8, clause 12, gives Congress the power to "raise and support armies." While there is an argument as to whether that power includes the power to *conscript* those armies, the power to recruit volunteers is clear. Because both statutes called for only volunteers, they were constitutional.

On July 17, 1862, Lincoln signed into law the Militia Act of 1862. Section 1 allowed the president to second into federal service an unlimited number of state militiamen for nine months of service. This was constitutional. Article 1, section 8, clause 15, gives the federal government the right to call the state militias into federal service, but the Militia Act of 1795—which had formed the statutory basis for Lincoln's April 15 state militia secondment—limited the term of state militiamen in federal service to three months. Now that Congress deemed that longer service was needed, it passed a new law allowing the president to second state militias into federal service for nine-month terms.[51]

Section 3 provided for the president to recruit as many as 100,000 volunteers for nine-month terms and also provided that these nine-month men would receive bounties (bonuses) of twenty-five dollars. Section 4 allowed the president to recruit an unlimited number of volunteers for one-year terms, and it provided that these one-year men would receive a bounty of fifty dollars (half payable immediately, half at the expiration of their enlistment). Both provisions seeking volunteers were constitutional under article 1, section 8, clause 12, which allowed Congress to raise and support an army.

Sections 12, 13, and 15 of the Militia Act of 1862 addressed the matter of Blacks in the military. Section 12 condescendingly allowed Blacks to serve in menial

48 For details see the U.S. Const. art. 2, § 2, cl. 1, and McPherson, *Battle Cry of Freedom*, 3.

49 An Act to authorize the Employment of Volunteers to aid in enforcing the Laws and protecting Public Property, 12 Stat. 268 (1861); McPherson, *Battle Cry of Freedom*, 322, 348.

50 An Act in addition to the "Act to authorize the Employment of Volunteers," 12 Stat. 274 (1861); McPherson, *Battle Cry of Freedom*, 322, 348.

51 The Militia Act, 12 Stat. 597 (1862); McPherson, *Battle Cry of Freedom*, 492–93.

roles—digging trenches, performing camp services, "or any other labor … for which they may be found competent." Section 15 provided that, for their labor, Black troops would receive 61 percent of the pay that white troops received.

Section 13 provided that enslaved men who fought for the Union—but *only* those from the Confederate states—would be freed upon completion of service, as would the former slave's mother, wife, and children. (The statute made no mention of the former slave's father.) Tellingly, this manumission provision did not apply to enslaved men from the slave states in the Union, the four so-called border states: Maryland, Missouri, Delaware, and Kentucky.[52] In carefully and deliberately not offering freedom to enslaved men who fought in the Union army from these Northern slave states, Congress was no doubt thinking of the critical importance of maintaining support from the border states both for the Union side in general and for the Militia Act in particular. Nevertheless, the unequivocal message sent by this hypocrisy was that slavery, manumission, and abolition are by no means moral issues; these matters are merely cudgels with which to punish the recalcitrant successionists in a way we deem will hurt them financially—by taking what they regard as their property.

The failure of the Militia Act of 1862 to meet the perceived military manpower needs of the Union army led to the enactment of the federal government's first true conscription statute early the following year (see chapter 2).

For those who wanted to see the auguries, signs of opposition to the Militia Act of 1862 presaged the far more widespread opposition to the conscription statute that was enacted the following year. For example, the Militia Act required seven companies of soldiers from Wisconsin's Thirtieth Regiment to keep conscription officers safe in Dodge County, Wisconsin. And even seven companies were not enough. (These conscription officers were conscripting local men for the state militia as contemplated by section 1 of the Militia Act.) One conscription officer was shot while signing men up. The shooter was quite popular with his neighbors; when he made bail, he went home accompanied by a supportive parade with a marching band. Another conscription officer in Dodge County was shot as well. In Washington County, Wisconsin, the county conscription officer was attacked by a mob and had to flee town.[53]

Two cases of interest arising under the Militia Act of 1862 were decided by the Wisconsin Supreme Court. The first case, *Druecker v. Salomon*, arose from the Wisconsin state militia draft that was conducted in November 1862 in Port Washington in Ozaukee County. When the state draft (in response to

52 McPherson, *Battle Cry of Freedom*, 284.

53 Lynn Ira Schoonover, "The History of the Civil War Draft in Wisconsin" (MA thesis, University of Wisconsin, 1915), chap. 5.

the federal Militia Act) was conducted in Port Washington, an enormous riot broke out. The plaintiff in this case, someone named Druecker, was arrested for his part in the riot. The defendant was Edward Salomon, the governor of Wisconsin. This was a wrongful imprisonment lawsuit. Plaintiff Druecker, who admitted his role and participation in the antidraft riot, was suing the governor because he (Druecker) had been held in jail for too long after his arrest.[54]

The Wisconsin Supreme Court report of the case is remarkable for the lengthy and detailed account it provides of the riot in Port Washington on November 10, 1862:

> On that day, before the draft commissioner had commenced drafting or attempted to, a large body of men marched through the street, carrying a flag, on which, in large letters, were the words "No Draft;" [and] when the commissioner was about to commence drafting, he was assaulted, stoned, badly bruised and beaten, and compelled to run for his life; and he, being hunted by the mob, secretly left the town. Those engaged in the riot or insurrection then went to his house and destroyed his furniture, and the doors and windows of the house, and otherwise injured it; and did the same to the houses and furniture of several other loyal citizens. The mob continued ... carrying a large part of the time the flag with "No Draft" on it; and some of them were armed with guns.[55]

In response to the rioting in Port Washington on November 10 and 11, Salomon sent the Twenty-Eighth Regiment "to enforce the draft and arrest those resisting it." Perhaps the most interesting part of the Wisconsin Supreme Court case report is the discussion of exactly why sending troops to Port Washington was necessary—because the antidraft rioters enjoyed support from the local civil authorities. According to the Wisconsin Supreme Court, this is what happened after the Twenty-Eighth Regiment arrived in Port Washington:

> Those following the "No Draft" flag at times amounted to several hundred; and ... a detachment of soldiers sent out to make arrests after the arrival of the troops, came within sight of a body of the rioters, from 100 to 200 ... most of them armed with guns, and some of them fired on the soldiers, who captured about fifty of them with their guns. The evidence also tends to prove that the civil authorities sympathized with the movement to resist the draft, and were utterly powerless to suppress

54 Druecker v. Salomon, 21 Wis. 621 (1867).

55 Ibid., 623–24.

> the riot, if they had desired to do it; and ... none of the rioters have ever been punished by the civil authorities.[56]

Needless to say, Druecker lost his false imprisonment lawsuit against Salomon.

The second Wisconsin Supreme Court case deals with the situation of a man named Nicholas Kemp, who was one of the antidraft rioters arrested in Port Washington in November 1862. On December 4, 1862, three weeks after his arrest, Kemp was still in jail and sought his release by bringing a writ of habeas corpus action against Brigadier General W. L. Elliott, who was holding Kemp.[57] The writ of habeas corpus dates to the Magna Carta in 1215; for centuries English and American courts have used the writ to determine whether a government action to hold a prisoner is proper and lawful. Elliott told the court that Lincoln had suspended the writ of habeas corpus on September 24, 1862, for all those held in any "place of confinement by any military authority."[58]

Elliott was correct. Indeed, Lincoln had issued his proclamation of September 25 pretty much for cases exactly like Nicholas Kemp's: violent resistance to Union efforts to recruit an army under the Militia Act of 1862.

On January 13, 1863, the Wisconsin Supreme Court ruled in favor of Kemp and set him free. Chief Justice Luther S. Dixon wrote the main opinion. The basis of the decision was that only Congress—not the president—has the power to suspend the writ of habeas corpus. In coming to this conclusion, the court relied on the fact that the provision pertaining to when the writ can be suspended appears in article 1 of the Constitution, which sets forth the powers of Congress, not in article 2, which sets forth the powers of the president. Article 1, section 9, clause 2, says, in full: "The Privilege of the Writ of Habeas Corpus shall not be suspended, unless when in Cases of Rebellion or Invasion the public Safety may require it." The Civil War was certainly a case of rebellion, but Congress had not acted to suspend the writ, and the president did not have the authority to do so on his own, the Wisconsin Supreme Court said.

The ruling that under the US Constitution only Congress has the power to suspend the writ of habeas corpus did not go unnoticed in Washington. On March 3, 1863, less than two months after the ruling in *In re Kemp* and on the same day Congress enacted the nation's first conscription statute, it passed

56 Ibid., 624–25.

57 For details see *In re Kemp*, 16 Wisc. 359 (1863).

58 Abraham Lincoln, "Proclamation 94 of September 24, 1862, by President Abraham Lincoln suspending the writ of Habeas Corpus," National Archives Catalog, RG 11, Series: Presidential Proclamations, 1791–2016, https://catalog.archives.gov/id/299959, last accessed December 5, 2022.

the Habeas Corpus Act. The first sentence of the act stated: "During the present Rebellion, the President of the United States, whenever, in his judgment, the public safety may require it, is authorized to suspend the privilege of the writ of habeas corpus in any case throughout the United States, or any part thereof."[59] As chapter 2 relates, Lincoln took advantage of that power on September 15, 1863.

59 An Act relating to Habeas Corpus, and regulating Judicial Proceedings in Certain Cases, 12 Stat. 755 (1863).

CHAPTER 2

The Civil War: the Union

In early 1863, Congress passed the first federal conscription law in the history of the United States. President Abraham Lincoln signed the bill into law on March 3, 1863.[1]

The immediate reasons why the unprecedented law was needed are fairly plain. The Civil War was not proving to be as brief as many had assumed back in April 1861 when Fort Sumter was attacked. The initial war fervor that so frequently accompanies the start of hostilities had waned substantially, so voluntary enlistments were declining. The soldiers who had volunteered for two-year stints in response to Lincoln's Proclamation 83 in May 1861, as well as those seconded into federal service for nine months under the Militia Act of 1862, were all coming to the end of their tours. The Union needed more troops, and not all were going to be volunteers.[2]

Many (but by no means all) observers—including, surprisingly, the Selective Service System itself—have, over the decades, viewed the Civil War draft in the North as an abject failure. Significantly, however, even the harshest critics of the Civil War draft have consistently understated the full extent of the disaster. Scholars state that the number of draft refusers was 20 percent, an amazingly high figure in itself (especially considering that these men risked being shot), but this chapter will show that the correct figure for draft refusers in the North was more than 50 percent. The carnage of the New York City antidraft riots is widely known, but the near-ubiquitous violence against conscription officers throughout the Union is not as well known. Injustice stood at the very core of the conscription apparatus—rich men legally bought their way out of the draft by paying a fee, while the poor died at the front—but even the degree of injustice is often understated. For example, the separate bounty (bonus) system made a farce of the entire conscription apparatus by creating an incentive for men to enlist and desert—then do so repeatedly—in order to collect multiple bounties.

1 An Act for enrolling and calling out the national Forces, and for other Purposes, 12 Stat. 731 (1863) (hereafter, Conscription statute).

2 Bruce Catton, *The Civil War* (Boston: Houghton Mifflin, 2005), 111–25.

 | DOI:10.1163/9789004546684_004

1 The Statute

The 1863 conscription statute created a two-stage process. The first stage would enroll all men aged twenty to forty-five. (*Enrollment* was the term used during the Civil War for what, in later drafts, would be called *registration*—the collection of names so the government would know who was eligible to be drafted. As War Department Circular Number 17, dated June 2, 1863, put it, "This enrollment is ... simply a census of all male citizens.")[3] An enrollment officer was assigned to each congressional district, charged with enrolling all the eligible men in that district.[4]

In the second stage, the president could issue draft calls for certain numbers of men from the lists of all those enrolled. In the event, Lincoln issued four separate draft calls (see Table 2.1). The first draft, in 1863, was for a flat 20 percent of the number of men enrolled.[5] In the remaining three drafts, the number of men required in each call was allocated to congressional districts proportionately, based on the relative numbers of men enrolled in each district. In all four drafts, lotteries determined who would be called.[6]

The 1863 conscription statute contained some additional details.

The law included a list of categories of men to be exempted from military duty, including those (rather vaguely described as) "physically or mentally unfit" for service, only sons of widowed mothers dependent on those sons,

3 US War Department, *The War of the Rebellion: A Compilation of the Official Records of the Union and Confederate Armies*, 128 vol. (Washington, DC: Government Printing Office, 1880–1901), ser. 3, 3:245 (hereafter, *OR*).

4 Conscription statute, section 1, specifies the ages of enrollees; see section 4 for the division by congressional districts, and section 9 regarding enrollment officers. See, generally, Eugene C. Murdock, *One Million Men: The Civil War Draft in the North* (Madison: State Historical Society of Wisconsin, 1971), 8, 27.

5 Conscription statute, § 12. Lincoln issued his proclamation for drafting 300,000 men on October 17, 1863. However, before that date, the president permitted, and the War Department conducted, actual drafts in places where enrollment had been completed: starting on July 7, 1863, in Rhode Island; July 11 in New York City; and July 17 in the remaining New England states. US Office of the Provost Marshal General, *Final Report Made to the Secretary of War, by the Provost Marshal General, of the operations of the Bureau of the Provost Marshal General of the US:* [...] *March 17, 1863 to March 17, 1866,* [...] (Washington, DC: Government Printing Office, 1866), pt. 2, 19–20, 26–27, and pt. 3, 27, 39 (hereafter, *PMG Final Report*). Also, because of "serious opposition to the enforcement of the law," the provost marshal general deemed it best to begin actual drafting surreptitiously, months before the presidential proclamation was issued. *PMG Final Report*, pt. 2, 27. For general information, see James W. Geary, *We Need Men: The Union Draft in the Civil War* (DeKalb: Northern Illinois University Press, 1991), 70.

6 James M. McPherson, *Battle Cry of Freedom* (New York: Oxford University Press, 1988), 601; Murdock, *One Million Men*, 62.

TABLE 2.1 Civil war draft calls by the union

Date of call	Number called
July–October 1863	300,000
March 1864	200,000
July 1864	234,327
December 1864	142,942

SOURCE: US OFFICE OF THE PROVOST MARSHAL GENERAL, *FINAL REPORT MADE TO THE SECRETARY OF WAR* (WASHINGTON, DC: GOVERNMENT PRINTING OFFICE, 1866), PT. 3, 41–45.

only sons of "aged or infirm" (again not defined) parents who were dependent on their sons, and fathers of motherless children younger than twelve.

The law specified that the draft would be organized by congressional districts. Each district had a chief conscription officer, the provost marshal, and the law created the national Bureau of the Provost Marshal General, to be led by a provost marshal general within the War Department to oversee the newly created national conscription system.

Section 9 of the law, which provided for the enrollment of draft-eligible men, contained the first of three highly problematic provisions in the statute. This section directed the appointment of one or more enrollment officers for each congressional district. Notably—and unlike each of the twentieth-century conscription statutes—the Civil War conscription statute placed no legal duty or obligation on men to enroll. Instead, perversely, the law placed the legal onus on the enrollment officer to find the eligible men in the district. In practice, this meant that enrollment officers traveled from town to town, knocking on doors to inquire at each residence whether any draft-age men lived at that address.[7]

The opportunities this presented for enrollment evasion were legion, subject only to the creativity of the individuals involved. Men failed to answer the door; a parent or sister who answered the door denied that any young men lived at the address; when intercepted, draft-age men lied, claiming to be underage or overage; and men gave false names, left town and hid in nearby woods, or fled to Canada or to the western frontier when enrollment officers came to town. In fact, these last two tactics became so common that both acquired nicknames: leaving town temporarily when the enrollment officers arrived was

7 Conscription statute, § 9; Murdock, *One Million Men*, 26.

called skulking (and those who did it were skulkers); fleeing to a distant area, never to return, was skedaddling (and those who did it were skeedaddlers).

The section that contained the actual conscription provision also provided that states would get credit for the number of military volunteers previously furnished by the state, thereby reducing the number of men drafted.[8] This system created an incentive for states to raise and provide cash bounties to volunteers to supplement the bounties already provided by the federal government.

The next section included two other major problem areas. It allowed any man drafted to provide a substitute instead of serving himself, and it allowed any man drafted to fully discharge his military obligation by paying a so-called commutation fee of $300.[9] The statute thus made explicit the rank economic and class discrimination that was more subtle or implicit in later US conscription laws. If drafted, wealthy men simply paid the commutation fee; only poor men found themselves on the front lines of the country's bloodiest conflict.

The law also provided that any man who was drafted and failed to serve (or provide a substitute or pay the commutation fee) would be "deemed a deserter" and tried by court-martial.[10] Desertion in time of war was a capital offense, and during the Civil War some draft refusers were, in fact, tried by court-martial, convicted, and shot. Treating draft evaders as deserters also obtained in the Confederacy during the Civil War (see chapter 3) as well as nationally during World War I (chapter 4) but not during World War II (chapter 5), the Korean War, or the Vietnam War.

The conscription statute made it a criminal offense to aid or counsel "any person to resist [the] draft."[11] Specifically, this section provided that

> if any person shall resist any draft of men enrolled under this act into the service of the United States, or shall counsel or aid any person to resist any such draft, or shall counsel or aid any person to resist any such draft, or shall assault or obstruct any officer in making such a draft ... such person shall be subject to summary arrest by the provost-marshal, and delivered to the civil authorities, and upon conviction thereof, be punished by a fine not exceeding five hundred dollars, or by imprisonment not exceeding two years, or by both said punishments.

---

8 Conscription statute, § 12.

9 Ibid., § 13.

10 Ibid.

11 Ibid., § 25. Interestingly, the statute uses that term: *resist* the draft. This is an early use of the term, and today it appears in much of the secondary literature specific to the Civil War draft; the term recurs in a variety of contexts in discussions of twentieth-century drafts.

This language was, in the most generous interpretation, ambiguous. Recall that the 1863 conscription statute contemplated a two-stage process: a registration, or enrollment, of draft-eligible men and the drafting of certain previously enrolled men. Though obviously related, the enrollment was separate and distinct from the draft. The express language of the law declared resisting the draft illegal; resisting enrollment was not illegal. This was a hole big enough to drive a covered wagon through. Some US attorneys, including the US attorney for New York, declined to prosecute enrollment resisters, in the belief that those people had not violated the statute.

On June 1, 1863, William Whiting, solicitor for the US War Department, told Robert Nugent, the acting assistant provost marshal general for New York, that he (Whiting) agreed that the statute did not bar enrollment resistance. Specifically, Nugent reported that Whiting had said "that in his opinion section 25 of the conscription act does not authorize the arrest of any person who refuses to give his name or otherwise embarrasses [hinders] the enrolling officer; that the penalties provided apply only to cases of men who resist the draft, and that the enrollment is not 'a service relating thereto,' i.e., to the draft."[12]

However, scarcely two weeks later, on June 16, 1863, the War Department issued its Circular No. 28, signed by Whiting, which informed draft officials that he now held the opposite opinion, that resisting enrollment was, in fact, resisting the draft under the statute.[13] However, a pronouncement by the War Department's lawyer, that the statute actually meant the opposite of what it plainly said, did not make it true. Congress fixed this problem in February 1864 with an amendment to the law that made resisting enrollment illegal.

The original 1863 conscription statute had one additional conspicuous omission: the law was completely silent on the subject of conscientious objectors (COs). This was not because government officials—including senators, representatives, Lincoln, Secretary of War Edwin Stanton, and other cabinet members—were unaware of the issue.

On February 23, 1863, while the conscription bill was pending in Congress, a group of Quakers met in Philadelphia and approved a memorial (a written request or petition) to the federal government seeking exemption from any draft for members of the pacifist denominations.[14] The memorial set forth the Quaker Peace Testimony, discussed its history and meaning in some detail, and explained why Quakers would be conscientiously unable to pay a

12 *OR*, ser. 3, 3:244. Note that many of the Civil War documents cited in this chapter and the next use the word *embarrass* in the dictionary sense of "obstruct, hinder, disconcert," not "cause to feel self-conscious."

13 *OR*, ser. 3, 3:368–69.

14 Quaker process approves such matters by consensus, not by vote.

commutation fee in lieu of personal military service if the commutation fee went to the military for military purposes (such as providing recruitment bounties to volunteers). The group chose four Philadelphia Quakers—Samuel Hilles, Charles Evans, James Emlen, and Charles Downing—to travel to Washington and present the memorial.[15]

Remarkably, the Quaker delegation was able to meet with Lincoln, Stanton, Attorney General Edward Bates, and other cabinet members. In addition, the delegation gave copies of the memorial to every member of the House of Representatives. On the Senate side, the delegation met with the chairman of the Senate Committee on Military Affairs, Henry Wilson, Republican of Massachusetts, and presented him with a copy of the memorial; at the request of the delegation, Wilson distributed copies of the Quaker memorial to every senator.[16]

Amendments to the conscription bill that would have exempted COs were proposed and debated in both houses of Congress. Ira Harris, Republican of New York, introduced the amendment on the Senate side. Harris's proposal would have added one exemption to the litany of exemptions found in section 2 of the bill for "persons who, being from scruples of conscience averse to bearing arms, are by the Constitution of any state exempt therefrom." Although Wilson supported the Harris amendment, it never came up for a vote.[17]

On the House side, Thaddeus Stevens, Republican of Pennsylvania, introduced an amendment that would have exempted any man who will "declare on oath or affirmation that he is conscientiously scrupulous against bearing arms, and that such is and has been his declared religion for more than [one] year."[18] Philip Johnson, Democrat of Pennsylvania, proposed a tweak to the Stevens amendment that would have changed the one-year requirement to three years, a revision that Stevens accepted. The House side defeated the amendment by a vote of 95 to 18.[19]

On March 17, 1863, Lincoln appointed James Barnet Fry to the post of provost marshal general; Fry thus became the overseer of the first US draft apparatus, a position he retained until the Bureau of the Provost Marshal General went out of business on August 28, 1866. Fry was a West Point graduate who happened

15 Edward Needles Wright, *Conscientious Objectors in the Civil War* (New York: A. S. Barnes, 1931), 58.

16 Ibid., 66–67.

17 Cong. Globe, 37th Cong., 3d Sess. 994 (1863); E. N. Wright, *Conscientious Objectors in the Civil War,* 62.

18 Cong. Globe, 37th Cong., 3d Sess. 1261 (1863). The original amendment, introduced on February 24, 1863, had a blank space where "[one]" appears in the text. Stevens had the number inserted in his proposed amendment the next day. Cong. Globe, 37th Cong., 3d Sess., pt. 2, 1291–92 (1863).

19 Cong. Globe, 37th Cong., 3d Sess. 1292 (1863).

to be in Virginia in the autumn of 1859 and participated in the military's capture of the abolitionist John Brown, whose raid on the US arsenal at Harper's Ferry was a failed attempt to spark a slave uprising.[20] On March 17, 1866, Fry presented the *Final Report by the Provost Marshal General*, which has formed the basis of much of the subsequent scholarship on the Civil War draft on the Union side.[21]

In 1940, Edward A. Fitzpatrick, who had been a draft official during World War I, published his widely respected *Conscription and America: A Study of Conscription in a Democracy*. The title of the section of Fitzpatrick's book dealing with the Civil War draft was succinct: "How Not to Draft Men." The Civil War scholar James M. McPherson agreed. Writing almost fifty years after Fitzpatrick, in his *Battle Cry of Freedom: The Civil War Era*, McPherson argued that the Civil War draft in the North involved "such inefficiency, corruption, and perceived injustice that it became one of the most divisive issues of the war and served as a model of how *not* to conduct a draft in future wars." Bruce Catton, in *The Civil War*, wrote, "The government could hardly have devised a worse law.... All in all, the conscription law was an atrocity."[22]

In 1955, the Selective Service System referred to the Civil War conscription statute as "the bungled Federal draft.... To say that the Federal draft law was unpopular and lacked public support would be putting it mildly.... As a matter of fact, there was much organized resistance to the draft in some parts of the country."[23] That resistance started with the enrollment process and then continued with the subsequent draft.

2 Enrollment Refusal

In his 1866 *Final Report*, Fry had this to say about the enrollment:

> Much difficulty was experienced in obtaining the services of reliable men to make the enrolment, this duty in some parts of the country being dangerous to life [because of] the disaffection of the inhabitants.... In some instances these [enrollment] officers were attacked and badly

---

20 Murdock, *One Million Men*, 15–16. For John Brown generally, see Tony Horwitz, *Midnight Rising: John Brown and the Raid That Sparked the Civil War* (New York: Picador, 2012).

21 Geary, *We Need Men*, xv. The PMG *Final Report* was a principal source for works by Murdock, Geary, and Levine. It was also a major source for this chapter.

22 McPherson, *Battle Cry of Freedom*, 600; Catton, *Civil War*, 208.

23 US Selective Service System, preface to *The Operation of Selective Service*, Special Monograph No. 17 (Washington, DC: Government Printing Office, 1955), 1:9–10.

> beaten, the sheets destroyed, and the work done before neutralized. Some of these officers were murdered, and others could do nothing unless protected by a strong military force. Many men, on the approach of an enrolling officer, left their homes, and their wives, mothers, or children, gave false names, or grossly misrepresented the age of the person to be enrolled.
>
> In the face of these difficulties it is not to be wondered that the original enrolment was very defective.[24]

The enormous failure of the entire Civil War enrollment process can be disaggregated into multiple component parts. Fry's 1866 assessment provided a reasonably good recitation of some of the major problems faced by the Bureau of the Provost Marshal General.

Fry was certainly correct about an initial problem: the difficulty of getting competent men to do the enrollment. Fry was also correct about the underlying reason for this difficulty: conscription was deeply unpopular. The draft was, as Murdock correctly observed, un-American.[25] As a result of these and other factors, little or no enrollment at all occurred in many places throughout the North.

Newark, New Jersey, in the state's Fifth Congressional District, was one such place. An enrollment officer was hired to enroll the eligible men of the district in June 1863 but was "compelled to leave the locality" because he did not have a sufficient military force backing him up. The Newark enrollment officer reported that men refused to provide names, women refused to answer doors, and that, even if drafted, no men would appear.[26]

Also in June 1863, in Ohio's Fourteenth Congressional District, the enrollment officer was chased out of one town by a stone-throwing mob. The local provost marshal was forced at gunpoint to release four enrollment refusers and requested federal troops to protect his men. Those troops were met by a thousand armed draft opponents, including enrollment refusers.[27] Again, part of what is important here is the large, local, grassroots support for the refusers.

Enrollment also proved difficult elsewhere in Ohio that June. In Morrow County, enrollment officers were "met with armed resistance" by one thousand organized, armed, antidraft men. In Crawford County, enrollment officers were

---

24 PMG *Final Report*, pt. 10, 143. Like the statute itself, the provost marshal general's *Final Report* sometimes spells *enrollment* with one *l*.

25 Murdock, *One Million Men*, 4.

26 E. N. Miller to R. C. Buchanan, June 22, 1863, *OR*, ser. 3, 3:400–1.

27 Robert E. Sterling, "Civil War Draft Resistance in the Middle West" (PhD diss., Northern Illinois University, 1974), 205.

mobbed and driven off. In Knox County, an enrollment officer was hired but resigned after being threatened. In Holmes County, enrollment officers were assaulted. The assailants were arrested, then freed from custody by an armed band of about seventy men.[28]

The Tenth Congressional District of Pennsylvania presented an interesting situation. In some parts of the district, enrollment officers had been appointed, "but they are so threatened that they do not dare to begin their work." However, in much more of the district no one would take the job because of the threats. The provost marshal for the district was sent fifty soldiers, but he reported that he could do nothing without a much larger force, because the resisters "are organized and armed with guns and revolvers ... they can summon 1,500 to 2,000 men at a call, and will do so to prevent an enrollment and draft."[29]

The provost marshal for this Pennsylvania district reported two separate incidents in June 1863 when his enrollment officers came under fire; some assailants were apprehended, while some others escaped. The damage, however, was done, and the local enrollment officer "was too much intimidated" and resigned in fear.[30]

Also in Pennsylvania in that month, an enrollment officer in Clearfield County was shot but survived.[31] In Bedford County, one enrollment officer had his sawmill burned down, another had his barn burned down, and a third was shot at. In Carbon County, a coal company executive was murdered because of a suspicion that he had provided the local enrollment officers with information about draft-eligible coal miners.[32]

In neighboring Delaware, enrollment officers fared no better. In June 1863, one was deterred by threats after shots were fired into his house.[33]

Little or no enrollment occurred in Kewaunee County, Wisconsin, or in portions of Manitowoc, Calumet, Outagamie, and Oconto Counties (Wisconsin); Camden, New Jersey; and Elk County, Pennsylvania.[34]

Indiana's Eleventh Congressional District never had enrollment officers because potential enrollers had "fear of personal injury or destruction of their property." Wisconsin's Fourth Congressional District lacked enrollment officers

---

28 E. A. Parrott to James Barnet Fry, June 12, 1863, *OR*, ser. 3, 3:349–50; Jack Franklin Leach, *Conscription in the United States: Some Background* (Rutland, VT: C. E. Tuttle, 1951), 258–59.

29 Murdock, *One Million Men*, 382.

30 C. Tower to Fry, June 10, 1863, *OR*, ser. 3, 3:330–31.

31 Robert M. Sandow, *Deserter Country: Civil War Opposition in the Pennsylvania Appalachians* (New York: Fordham University Press, 2009), 110–11.

32 Henry Ickes to George Eyster, June 5, 1863, *OR*, ser. 3, 3:324–25; Arnold M. Shankman, *The Pennsylvania Antiwar Movement, 1861–1865* (Rutherford, NJ: Fairleigh Dickinson University Press, 1980), 145–46, 148.

33 Edwin Wilmer to Fry, June 15, 1863, *OR*, ser. 3, 3:361–62.

34 Murdock, *One Million Men*, 30; Sandow, *Deserter Country*, 114.

because men there thought that acting as an enroller would be "hazardous to life." In Brown County, Wisconsin, it was said that no enrollment officer would be allowed to live. This was not tested; no one ever served as an enrollment officer in Brown County.[35]

On June 15, 1863, in Boone County, Indiana, "a mob of some twenty-five or thirty men surrounded the enrolling officer ... and caused the women of the town to assail him with eggs."[36] That same month, in Indiana's Fourth Congressional District, the deputy provost marshal and an enrollment officer were fired upon while on duty. The deputy was killed, the enrollment officer seriously wounded. Completing the enrollment in that district required two companies of infantry and an additional company of cavalry.[37] On June 18, in Indiana's Seventh Congressional District, the enrollment officer was shot and left to die on the road where he had been gunned down.[38]

The day before, enrollment officers in Rutland, Vermont, tried to enroll men, but the enrollment officers were met by an armed, well-organized band of five hundred resisters, the vanguard of a force of a thousand armed conscription resisters in Rutland. The enrollment officers said they were "mercilessly" attacked and driven out of town; both the provost marshal and the deputy sheriff were severely wounded.[39] The number of Rutland enrollment refusers suggests a widespread movement with broad local support, not the isolated acts of a few dissidents or malcontents.

In Illinois, men in many counties could be enrolled only with the heavy presence of troops; these counties included Clark, Coles, Fayette, Fulton, Morgan, and Scott. The counties saw armed conflagrations, with enrollment officers and military on one side and heavily armed enrollment resisters on the other; "men were killed and wounded on both sides." James Oakes, the acting assistant provost marshal general for Illinois, reported his view that "experience has demonstrated the comparative uselessness of [mere] infantry in dealing with roving bands" of resisters. "These gangs are usually well mounted, familiar with all the woods and swamps and byroads of the country, and can of course successfully elude the infantry and defy their pursuit."[40]

In July 1863, the provost marshal for Maryland's First Congressional District reported that "the intimidation is such that I am unable to procure the services

35 Sterling, "Civil War Draft Resistance," 185.

36 Conrad Baker to Fry, June 17, 1863, *OR*, ser. 3, 3:375.

37 Sterling, "Civil War Draft Resistance," 240.

38 R. W. Thompson to Baker, June 18, 1863, *OR*, ser. 3, 3:393–94; Sterling, "Civil War Draft Resistance," 244.

39 Murdock, *One Million Men*, 43.

40 James Oakes to Fry, August 9, 1863, *OR*, ser. 3, 5:837.

of enrolling officers." He reported: "Not only serious threats, but houses have actually been assaulted and fired into."[41]

In July in Bucks County, Pennsylvania, someone took a shot at an enrollment officer but he was not hurt. In adjoining Montgomery County, Pennsylvania, an enrollment officer was killed.[42]

As late as October 1863, seven months after the conscription statute was enacted, no enrollment had occurred in Morgan County, Kentucky, even though the enrollment officers were accompanied by one hundred soldiers from the Fortieth Kentucky Mounted Infantry. There were, however, armed skirmishes between the soldiers and armed resisters. Three resisters were killed and several soldiers were wounded.[43]

During the war, thirty-eight enrollment officers were killed and sixty more were wounded. These figures are an important indicator of just how widespread opposition to the Civil War draft was.[44]

Fry was also correct when he reported after the war that enrollment officers had had "the sheets destroyed, and the work done before neutralized."[45] A century after the Civil War ended, Vietnam era draft opponents engaged in the widespread destruction of draft files and believed that their tactic was dramatic, effective, and novel. They were correct about the drama and effectiveness (see chapter 7), but they were mistaken that their actions were an innovation; theft and destruction of enrollment records was widely engaged in during the Civil War.

In June 1863, in Fulton County, Illinois, an armed group of draft opponents, whom the provost marshal general characterized as desperadoes, took the enrollment officer hostage and relieved him of all of his enrollment books—after the enrollment for the entire county had been completed. It took "one full company of infantry" to enable him to redo his work.[46] At about the same time, the same thing happened in Owen and Clay Counties: antidraft activists stole and destroyed the completed enrollment records for those counties.[47]

On June 28, in Hancock County, Illinois, armed men raided the homes of several different enrollment officers, rousted them out of bed between 2:00

41 Fry to John Frazier, Jr., July 14, 1863, *OR*, ser. 3, 3:492.

42 Richard A. Sauers and Peter Tomasak, *The Fishing Creek Confederacy: A Story of Civil War Draft Resistance* (Columbia: University of Missouri Press, 2012), 28; Samuel Yohe to Fry, June 11, 1863, *OR*, ser. 3, 3:357.

43 W. C. Grier to W. H. Sidell, October 20, 1863, *OR*, ser. 3, 3:899–900.

44 *PMG Final Report*, pt. 1, 143, and pt. 2, 352. The same figure—an aggregate of ninety-eight enrollment officers killed or wounded—appears in Enoch Crowder, *The Spirit of Selective Service* (New York: Century, 1920), 84. Crowder served as the provost marshal general of the Selective Service System during World War I.

45 *PMG Final Report*, pt. 12, 143.

46 Baker to Fry, June 22, 1863, *OR*, ser. 3, 3:396–97.

47 Thompson to Baker, June 18, 1863, *OR*, ser. 3, 3:393–94.

and 3:00 a.m., and seized their enrollment papers, again at gunpoint. The enrollment officers had almost completed their enrollments, and the loss was nearly total. The local provost marshal had to send a company of cavalry to Hancock County as the enrollment officers started their effort all over again.[48]

In Pennsylvania, the enrollment evaders continued their mayhem. On July 20, the enrollment office in Columbia County was burglarized and the files destroyed. The same month, enrollment officers in Monroe and Pike Counties were assaulted and their files destroyed.[49]

Monroe County, Kentucky, also proved difficult to enroll, but that October enrollment officers were partially successful with the aid of one hundred infantrymen. But as soon as the partial enrollment was completed, a local draft opponent attacked the local provost marshal and burned his enrollment papers as he watched. The loss was total, and no further enrollment of the county was possible.[50]

On August 4, 1863, in Fulton County, Illinois, where two months earlier a group had taken the enrollment officer hostage and seized all his enrollment books, ten men intercepted the enrollment officer and relieved him of his enrollment book. Unlike groups of draft file destroyers during the Vietnam War, this group was heavily armed and took the enrollment book from him at gunpoint.[51]

Fry's assessment that the original enrollment was "very defective" rather understated the grim reality.

In his *Final Report*, Fry compares the number of men drafted during the Civil War with the number of men who failed to appear (broken out for each of the four separate draft calls). Fry starts with the number of men *called* in each of the four successive draft calls. Then he sets forth how many simply failed to report, as well as how many were exempted for reasons of disability, or because they paid the $300 commutation fee, or furnished substitutes and how many actually ended up in the army.[52]

These tabulations are valuable, because the percentage of drafted men who failed to report is remarkably large; thus, Fry's figures are a potent indicator of how unpopular the Civil War draft was. However, Fry's detailed and meticulous calculations suffered from a huge problem. By starting near the end of

48 Sterling, "Civil War Draft Resistance," 225–26.

49 Sauers and Tomasak, *Fishing Creek Confederacy*, 28; Yohe to Fry, June 11, 1863, *OR*, ser. 3, 3:357.

50 Ibid., 899–900.

51 Sterling, "Civil War Draft Resistance," 1–4.

52 *PMG Final Report*, pt. 12, 165–99, tables 5–10. Part 12 is an appendix.

the process, with the draft lottery, Fry skated past an antecedent problem of enormous dimensions: the vast number of men who were never enrolled.

Subsequent scholars of the Civil War draft, including Eugene C. Murdock, James W. Geary, and Peter Levine, follow directly in Fry's footsteps and make exactly the same mistake.[53] Levine goes so far as to say that the "'failed to report' figures ... represent the best available index" of illegal draft resistance.[54] But that is not accurate.

The Fry-Murdock-Geary-Levine recitations—comparing the number of men *called* with the number that *failed to report*—are not the "best available index" of illegal draft refusal because they omit all the men who were never enrolled. Fry's *Final Report* did not include any statistics pertaining to the numbers of men who were not enrolled. However, Fry's *Final Report* does set forth the number of men from each state who were actually enrolled. By comparing this figure to the 1860 US Census figures for men who were legally obligated to be enrolled, one can derive a reasonable estimate of the percentage of draft-age men in each state who simply never enrolled. Those figures are stunning.[55]

Of the twenty-two states in the Union (including the District of Columbia), four states had nonenrollment rates of *more than 50 percent*: Delaware (55.0 percent), Maine (56.8 percent), New Hampshire (52.4 percent), and Vermont (53.7 percent). Another five states had nonenrollment rates of 40 percent or more but below 50 percent: Connecticut (43.7 percent), Massachusetts (44.3 percent), Ohio (42.6 percent), Pennsylvania (40.0 percent), and Wisconsin (42.5 percent). Table 2.2 sets forth the nonenrollment numbers and percentages for each state in the Union. Previous scholars have never acknowledged or discussed these figures.

The enrollment that accompanied the nation's first federal conscription statute was not a success. Nearly a third of the men who were supposed to be enrolled never were. This figure provides important context for what followed: a highly unsuccessful draft.

53 Murdock, *One Million Men*, 350–56; Geary, *We Need Men*, 83, tables 2 and 3; Peter Levine, "Draft Evasion in the North during the Civil War, 1863–1865," *Journal of American History* 67, no. 4 (March 1981): 816.

54 Levine, "Draft Evasion in the North during the Civil War," 820.

55 US Census Bureau, *Population of the United States in 1860* (Washington, DC: Government Printing Office, 1864). The official date of the 1860 census was June 1, 1860. Note that draft enrollment commenced three years later, in the spring of 1863. See the note for table 2.2 for a more detailed discussion of the methodology used to account for men never enrolled.

TABLE 2.2 Enrollment refusers in the union (by state), 1863–65

State	Should have enrolled	Did enroll	Percentage of refusers
Connecticut	72,359	40,708	43.7
Delaware	15,584	7,011	55.0
Dist. of Columbia	9,915	10,725	0
Illinois	282,203	274,409	2.8
Indiana	224,771	153,238	31.8
Iowa	109,105	82,803	24.1
Kansas	22,923	23,022	0
Kentucky	159,043	104,082	34.6
Maine	106,746	46,121	56.8
Maryland	84,639	55,802	34.1
Massachusetts	189,656	105,650	44.3
Michigan	122,043	75,248	38.3
Minnesota	25,239	18,556	26.5
Missouri	186,843	133,501	28.5
New Hampshire	55,312	26,302	52.4
New Jersey	100,353	70,055	30.2
New York	551,426	431,462	21.8
Ohio	358,602	205,867	42.6
Pennsylvania	439,424	263,463	40.0
Rhode Island	25,924	16,256	37.3
Vermont	50,426	23,326	53.7
Wisconsin	95,178	54,737	42.5
Aggregate	**3,287,714**	**2,222,317**	**32.4%**

SOURCE: US CENSUS BUREAU, *POPULATION OF THE UNITED STATES IN 1860* (WASHINGTON, DC: GOVERNMENT PRINTING OFFICE, 1864) (FIRST COLUMN); US OFFICE OF THE PROVOST MARSHAL GENERAL, *FINAL REPORT MADE TO THE SECRETARY OF WAR*, PT. 2, 25 (SECOND COLUMN).

Note: See p. iv of the introduction to the census document for a tabulation of draft-age men in every state. The figures have been adjusted for age and alienage. Although the census considered draft age to be 18 to 45, the 1863 conscription statute considered draft age to be 20 to 45. Thus, the census figure for each state is reduced by 7.1% to account for the 18- and 19-year-olds. This assumes a proportional distribution of men across the different age groups counted. As for alienage, the introduction shows on p. xxix the number of "foreign residents" by state. Thus, the census figure for each state is reduced by the percentage of aliens in that state. This assumes that the proportion of draft-age men in a state who were aliens corresponded roughly to the percentage of all people in that state who were aliens. The provost marshal general's statement of population by state corresponds precisely with the census figures.

3 Draft Refusal

Between July and October 1863, as enrollments in different geographic areas were completed, the Bureau of the Provost Marshal General moved on to drafting.

Fry's description of the problems his bureau had encountered, quoted earlier, goes on to say: "The order for this draft was the signal for violent disturbance in many portions of the loyal States, and much blood was shed before these disturbances were quieted. In some parts of the country, particularly the city of New York, certain districts of Pennsylvania, Indiana, and Illinois, the draft was conducted under the protection of troops sent there … and in other districts the draft was deferred until troops could be furnished to protect officers."[56]

In New York City, the actual draft lottery began on Saturday, July 11, 1863, and continued on Monday morning, July 13. As is well known, the antidraft riot there consumed the city from July 13 through July 17.

A word of context is appropriate here. The Battle of Gettysburg took place July 1 to 3, 1863. On July 4 and 6, the *New York Times* ran casualty lists from the fighting on inside pages. On July 8 and 9, as well as on the eleventh, the very day the draft commenced in New York, much of the front page of the *Times*—column after column—was devoted to lists of war casualties identified by name and military unit and identifying those killed, wounded, or missing. Recall that in June 1969, at the height of the Vietnam War, peace advocates who were members of A Quaker Action Group stood, day after day, on the steps of the US Capitol, reading the names of American war dead from the *Congressional Record.* The pacifists also were arrested day after day for this nonviolent civil disobedience.[57] These Vietnam era activists understood that humanizing and personalizing the terrible human cost of war makes for a persuasive antiwar argument—so strong a propeace argument, in fact, that they were willing to be arrested and go to jail for simply reading those names aloud at the Capitol. The casualty reports that appeared in newspapers during the Civil War no doubt also had a powerful emotional impact on the public.

Different accounts provide different tallies of the number of people killed during the New York City antidraft riot. James McCague, writing in 1968, cited Thomas Acton, president of the New York City Police Commissioners at the time of the riots, as putting the death toll at 1,200. McCague himself believed that Acton's figure was too high and argued that 400 to 500 is more accurate. The Civil War scholar Bruce Catton said that the civilian death toll was 1,000 people. According to David M. Barnes, who covered the riot contemporaneously

56 PMG *Final Report*, pt. 12, 146.

57 "War Dead in Washington," *Resist* 31 (August 7, 1969): 3.

for the *New York Times* and published a book about the riots in September 1863, the death toll was 400 to 500 people. Barnet Schecter used the same figure in his book *The Devil's Own Work: The Civil War Draft Riots and the Fight to Reconstruct America* (2005).[58] By any of these measures, the New York City antidraft riot was the worst riot in American history.[59]

The New York City antidraft riot started as a general strike against the draft, with workers in multiple trades joining the strike. The strikers gathered early on the morning of Monday, July 13, in Central Park and paraded with antidraft placards to the provost marshal's office for New York's Ninth Congressional District, located at 677 Third Avenue, between Forty-Sixth and Forty-Seventh Streets. Along the way, strikers attacked hardware stores to steal axes and gun stores to steal firearms.[60]

Although several New York City draft offices were burned to the ground during the five-day riot, it is an exquisite irony that the first draft office to be razed was torched by the city's own fire department. When the draft lottery had begun the previous Saturday, the firefighters had been dismayed to learn that firefighters were not automatically exempted from the federal draft law, as they had previously been from New York State's militia draft. When the firefighters arrived at the Ninth District provost marshal's office, they set it aflame. As the draft office burned, several people, including Alderman Peter Masterson, gave antidraft speeches to the crowd, which cheered him wildly. When the flames from the burning draft office threatened to spread to adjacent tenements, the ever-dutiful local firefighters put out the fire they had set and thus protected nearby buildings.

---

58 James McCague, *The Second Rebellion: The New York City Draft Riots of 1863* (New York: Dial Press, 1968), 178; Catton, *Civil War*, 212; David M. Barnes, *The Draft Riots in New York: July 1863* (New York: Baker & Godwin, 1863), 6; Barnet Schecter, *The Devil's Own Work: The Civil War Draft Riots and the Fight to Reconstruct America* (New York: Walker, 2005), 251. Schecter explains how some contemporaneous observers determined that twelve hundred people were killed. However, Schecter says this figure may have been exaggerated. James H. Forest states, "At least 1,200 people were killed." Forest, "In Time of War," in *Delivered into Resistance* (New Haven, CT: Advocate Press, 1969), 3.

59 Compare, for example, the New York casualty count in July 1863 and the three hundred people killed in the notorious 1921 Tulsa Race Riot in Oklahoma. In the February 28, 2001, report *Tulsa Race Riot: A Report by the Oklahoma Commission to Study the Tulsa Race Riot of 1921*, the commissioners suggest that credible figures for the number of dead range from 75 to 300 victims. See also Maggie Astor, "What to Know about the Tulsa Greenwood Massacre," *New York Times*, June 20, 2020, in which Astor suggests, "The death toll may have been as high as 300."

60 This account of the antidraft riot in New York is drawn from Barnes, *Draft Riots in New York*; McCague, *Second Rebellion*; Schecter, *Devil's Own Work*; and Iver Bernstein, *The New York City Draft Riots: Their Significance for American Society and Politics in the Age of the Civil War* (Lincoln: University of Nebraska Press, 1990).

Later the same day, rioters raided an armory for weapons, burned down the provost marshal's office (at Broadway and Twenty-Ninth Street) for the Eighth Congressional District, and the Colored Orphans Asylum, at Fifth Avenue between Forty-Third and Forty-Fourth Streets. The motivation for many of the white rioters was virulent racism. They lynched many Blacks during the New York riots, some in exceptionally grisly ways. Multiple scholars ascribe the overtly racist sentiment of white rioters to Lincoln's promulgation of the Emancipation Proclamation, which had taken effect on January 1, transforming what had previously been a fight to maintain the Union against Confederate secession into a broader fight to abolish slavery. Many white New Yorkers were not eager to risk life and limb for the emancipation of their Black brethren.

The attack on the armory, located at Second Avenue and Twenty-First Street, was prolonged and bloody. Thousands of rioters outside made repeated assaults on the armory, besieging the police inside. In repulsing these assaults, the heavily armed police inside the armory shot and killed many attackers. Eventually, the rioters got the upper hand, all the police fled, and the demonstrators took over the armory and seized its cache of weapons. Later, police reinforcements returned and torched the armory. Many demonstrators who were trapped inside burned to death.

Starting on July 14, rioters erected barricades on many streets. Barricaded rioters and snipers hiding in structures and on rooftops fired on the police and soldiers trying to keep order. Police and military had to clear buildings of rioters, building by building. Soldiers brought in artillery to use on rioters. One day, Colonel Henry O'Brien used a howitzer to clear demonstrators on Second Avenue; the next day demonstrators took their revenge: they intercepted him outside his house, killed him, and left his body in the street.

Secretary of War Stanton sent five regiments—several of them straight from Gettysburg—to New York to restore order.

Antidraft riots also broke out in other cities, including Albany, Hartford, Boston, and Newark.

∴

On Wednesday, July 15, in response to the riots, New York City's Common Council, as the city council was then called, approved the sale of $2.5 million worth of interest-bearing bonds, to pay the full commutation fee for New Yorkers who could not afford $300. Mayor George Opdyke vetoed the bill, arguing that granting this money would reward rioters and encourage other municipalities to make similar grants, thereby vitiating the federal conscription law. In overriding Opdyke's veto, the council increased the amount appropriated to $3 million.

Opdyke was correct that other places would follow suit. The State of New Hampshire approved payment of $300 toward the purchase of a substitute for all New Hampshire draftees, and the State of Delaware approved $200 per man toward the commutation fee.[61] The Chicago Common Council voted $300,000 toward substitutes for indigent Chicagoans. In Milwaukee, a voter initiative created a $120,000 fund by increasing property taxes.[62] In each of these cases, the approval of these measures reflected the depths of popular antidraft sentiment.

∴

As dramatic as the New York City riots were, Fry was correct that the violence against conscription was geographically widespread.

On July 15, 1863, in Lancaster County, Pennsylvania, the local provost marshal was conducting the draft when a mob interrupted the proceedings. Drafting did not resume until troops arrived.[63]

A year later, on July 30, 1864, Lieutenant James Stewart Robinson and a posse of civilians went looking for local draft refusers in Luzerne County, Pennsylvania. The posse eventually laid an ambush for three fugitive draft evaders, the brothers Tommy and Miner Smith, and their friend Ellis Young. Unaware of the trap, the fugitives walked into the ambush. Many shots were exchanged. Robinson was shot, and his death was a lingering one.[64]

Once again, a notable feature of the episode was the widespread local support enjoyed by the draft refusers. Two days later, on August 1, the deputy provost marshal reported to the assistant provost marshal for Pennsylvania that Columbia County "is infested with deserters in large numbers who bid defiance to the government. They are fully armed and declare themselves ready to resist to the death.... The situation in this county can be determined by the fact that no union man living in the section of the county in which Robison [*sic*] was shot dare admit the dying man into his house." In response to a request from the deputy, a cavalry company arrived in the county to enable enrollment officers to do their job.[65]

Two months later, on October 3, 1864, Captain Eli McCarty was serving draft notices on those whose names had been called in the lottery in Daviess

61 The foregoing account of jurisdictions that opted to pay the commutation fees for their citizens is drawn from Geary, *We Need Men*, 113–14.

62 Sterling, "Civil War Draft Resistance," 486–87.

63 Shankman, *Pennsylvania Antiwar Movement*, 146.

64 Sauers and Tomasak, *Fishing Creek Confederacy*, 42–45.

65 Ibid., 45, 47.

County, Indiana, including Samuel Slicer. Slicer threatened McCarty, and later in the day Slicer and another man, Hillary Madden, went after McCarty. As McCarty was fleeing from Slicer and Madden, a third man, hiding in ambush, shot McCarty dead. The killers were never brought to justice.[66]

That same month enrollment officers in Poweshiek County, Iowa, were serving draft notices on men whose names had been called. Members of a local antiwar group shot them. One died instantly; the other lay in the road, wounded. The wounded officer fired at the attackers and wounded one. The wounded attacker took umbrage and bashed in the officer's head, killing him.[67]

On October 30, 1864, two enrollment officers went to arrest a draft refuser, a farmer, at his home near Clearfield, Pennsylvania. When the farmer came out of the house at dawn to milk his cows, the officers tried to arrest him. He fled inside and the officers followed. The farmer shot one officer, hitting him in the abdomen. The farmer fled to the second floor, and the wounded officer followed, then shot the farmer, who clubbed the officer unconscious with a rifle butt. The unconscious man tumbled down the stairs and soon died. The farmer's wound was superficial and he recovered. Local support for the farmer was so strong that his neighbors protected him, and he remained at large until 1875. At that time, he was arrested for the murder of the enrollment officer and acquitted by a jury.[68]

In December 1864, an enrollment officer was shot and killed in Bond County, Illinois, while searching for a hideout that he believed held several draft refusers. In January 1865, an enrollment officer in Jefferson County, Illinois, served a draft notice. The draftee was not pleased; he shot the officer dead.[69]

∴

As I have discussed, in his *Final Report*, Fry arrived at a number of draft refusers by comparing the number of men called in each of the four draft calls with the number of men who did not report. This produced the now-familiar figure that 161,244 men refused the Civil War draft in the North.[70] Fry's figures appear in the subsequent work of Murdock, Geary, and Levine.[71] However, although all four provide

66 Sterling, "Civil War Draft Resistance," 515–17.
67 Ibid., 505.
68 Sandow, *Deserter Country*, 124–26.
69 Sterling, "Civil War Draft Resistance," 536, 513–14.
70 Table 6, "Recapitulation, draft of July 1863," pt. 12, 174; table 8, "Recapitulation, draft of March 1864," pt. 12, 184; table 10, "Recapitulation, draft of July 1864," pt. 12, 198; and table 12, "Recapitulation, draft of December 1864," pt. 12, 211, all in PMG *Final Report*.
71 For further details, see PMG *Final Report*, pt. 12, 165–99, tables 5–10; Murdock, *One Million Men*, 350–56; Geary, *We Need Men*, 83, tables 2 and 3; and Levine, *Draft Evasion in the North*, 816.

readers with the raw numbers, none calculates the percentages of draft resistance that these numbers reveal. Table 2.3 provides both the absolute numbers and the previously omitted percentages derived from the raw numbers of draft refusers.

Of course, this accounts for only *draft* refusal, not enrollment refusal. Once we have the number of enrollment refusers (Table 2.2) and the number of draft refusers (Table 2.3), combining the figures to arrive at aggregate figures for overall refusal (that is, enrollment refusal and draft refusal combined) is a simple matter. See Table 2.4, which for the first time presents overall figures for refusal in the North.

TABLE 2.3 Draft refusers in the union (by state)

State	Number drafted	Did not report	Percentage of refusers
Connecticut	12,031	1,014	8.4
Delaware	8,635	1,443	16.7
Dist. of Columbia	14,338	5,594	41.5
Illinois	32,085	9,519	29.7
Indiana	41,158	6,235	15.1
Iowa	7,548	702	9.3
Kansas	1,420	419	29.5
Kentucky	29,421	9,503	32.3
Maine	27,324	3,760	13.8
Maryland	29,319	9,207	31.4
Massachusetts	41,582	5,167	12.4
Michigan	22,122	4,294	19.4
Minnesota	10,796	2,058	19.1
Missouri	21,519	9,444	43.8
New Hampshire	10,806	464	4.3
New Jersey	32,325	6,205	19.2
New York	151,488	31,745	21.0
Ohio	50,400	9,368	18.6
Pennsylvania	178,873	31,309	17.5
Rhode Island	4,321	249	5.8
Vermont	7,743	429	5.5
West Virginia	3,148	1,014	31.9
Wisconsin	38,395	11,742	30.6
Aggregate	776,829	161,244	20.8

SOURCE: US OFFICE OF THE PROVOST MARSHAL GENERAL, *FINAL REPORT MADE TO THE SECRETARY OF WAR*, PT. 12, 165–73, DOC. 6.

TABLE 2.4 Enrollment refusers and draft refusers in the union: Percentages by State

State	Percentage of enrollment refusers	Percentage of draft refusers	Aggregate percentage of refusers
Connecticut	43.7	8.4	52.1
Delaware	55.0	16.7	71.7
Dist. of Columbia	0	41.5	41.5
Illinois	2.8	29.7	32.5
Indiana	31.8	15.1	46.9
Iowa	24.1	9.3	33.4
Kansas	0	29.5	29.5
Kentucky	34.6	32.3	66.9
Maine	56.8	13.8	70.6
Maryland	34.1	31.4	65.5
Massachusetts	44.3	12.4	56.7
Michigan	38.3	19.4	57.7
Minnesota	26.5	19.1	45.6
Missouri	28.5	43.8	72.3
New Hampshire	52.4	4.3	56.7
New Jersey	30.2	19.2	49.4
New York	21.8	21.0	42.8
Ohio	42.6	18.6	61.2
Pennsylvania	40.0	17.5	57.5
Rhode Island	37.3	5.8	43.1
Vermont	53.7	5.5	59.2
West Virginia	–	31.9	31.9
Wisconsin	42.5	30.6	73.1
Aggregate	32.4	20.8	53.2

Note: The percentages of enrollment refusers are taken from Table 2.2; the percentages of drafter refusers come from Table 2.3.

The figures are staggering. The percentage of men in the Union who refused the draft (by either not enrolling or by evading induction) was more than 50 percent. Whether they were caught or not, all these men were committing felonies. In four geographically disparate states, the percentage of refusers exceeded 70 percent: Delaware (71.7 percent), Maine (70.6 percent), Missouri (72.3 percent), and Wisconsin (73.1 percent).

4 Amendments to the Conscription Statute of 1863

The manifest problems with the 1863 conscription statute led Congress to amend it three times: in February 1864, July 1864, and March 1865. However, the amendments were of limited benefit.

Lincoln signed the first set of amendments into law on February 24, 1864. This amendment was generally designed and intended to broaden the underlying conscription statute in ways that would result in more men entering the army.[72]

One section addressed the matter of substitutes. It provided that substitutes must not be liable to draft themselves.[73] The provision was meant to avoid double-dipping.

Another section addressed the controversial and sensitive issue of commutation.[74] The commutation provision in the original statute posed at least three separate problems. First, it was rank discrimination: rich men could (and did) buy their way out of the draft; poor men could not. Second, the provision was generally and deeply unpopular. For example, it was widely understood that the commutation provision had been a major contributing factor to the New York City antidraft riots.[75] Third, the availability of commutation substantially vitiated the underlying raison d'être of the conscription statute; so many men were paying the commutation fee that the draft wasn't working. In the first two drafts, while commutation was widely available, 133,000 men's names were called in the lottery; 85,000 of them, almost two-thirds, paid the commutation fee. In certain congressional districts the problem was even worse. In the March 1864 draft, 87 names were called in Madison County, New York; 86 men got out by paying the commutation fee. In Enfield, Connecticut, 53 men were called; 52 paid commutation. Fry, whose job was to run the system, strongly urged repeal of commutation. Congress was not yet willing to end the practice entirely, so there was a compromise. The amendment limited the benefit of paying the $300 commutation fee; going forward, payment of the fee exempted a draftee for only that draft call or for one year, whichever time was shorter.[76]

---

72 An Act for enrolling and calling out the National Forces, and for other Purposes, 13 Stat. 6 (1864) (hereafter, National Forces).

73 National Forces, § 4.

74 Ibid., § 5.

75 Schecter, *Devil's Own Work*, 113–15.

76 Murdock, *One Million Men*, 198, 200, 205.

A third section of the amendment repealed most of the exemptions that the original statute had provided: no longer would there be exemptions for only sons of widowed mothers dependent on the sons, only sons of "aged or infirm" parents who were dependent on the sons, and fathers of motherless children younger than twelve. The only exemptions that remained were for those "physically or mentally unfit for the service."[77]

Another section of the February 1864 amendment belatedly cleaned up an important omission in the original statute, the one that made it illegal to resist or obstruct the draft but did not bar resisting or obstructing enrollment. The amendment fixed this by making it illegal to "forcibly resist or oppose any *enrollment*" or to "incite, counsel, [or] encourage" others to resist enrollment. It also made it illegal to "assault, obstruct, hinder, impede or threaten any officer or other person" working on enrollment. The penalties for any such offense were increased to imprisonment for as long as five years, a fine of as much as $5,000, or both. If the obstruction led to the death of the enrollment officer, the penalty was death.[78]

The amendment also belatedly added a provision for COs:

> That members of religious denominations, who shall by oath or affirmation declare that they are conscientiously opposed to the bearing of arms, and who are prohibited from doing so by the rules and articles of faith and practice of said religious denominations, shall, when drafted in the military service, be considered non-combatants, and shall be assigned by the Secretary of War to duty in the hospitals, or to the care of freedmen, or shall pay the sum of three hundred dollars to such person as the Secretary of War shall designate to receive it, to be applied to the benefit of sick and wounded soldiers.[79]

That is, the amendment provided that COs had two choices if they were drafted. They could enter the military and become soldiers but not carry arms. In this case, they would still be under military control and would aid the broader military effort by tending to battle-wounded soldiers (or caring for freed slaves). In the alternative, they could pay the same $300 commutation fee that any other draftee could pay to completely evade service; however, in the case of COs the fee would go toward caring for soldiers wounded in battle. (For non-COs, commutation money often went toward federal bounties for volunteers.)

---

77 National Forces, § 10.
78 Ibid., § 12, emphasis added.
79 Ibid., § 17.

The provision regarding COs was the final outcome of a moderately lengthy dance between Stanton and multiple groups of Quakers. In November 1863, members of the Baltimore Friends Meeting had met with the secretary of war in Washington and urged him to support a statutory amendment that would exempt COs from both the draft and the commutation fee. Until then, the official government view had been that Quakers should have no complaints about conscription; after all, they could pay the $300 commutation fee and thereby avoid military service, just like everyone else. In the November meeting, the Quakers explained to Stanton that most Quakers had religious objections to paying the commutation fee since the Provost Marshal General's Bureau used the money to obtain a substitute.[80]

Stanton presented a counterproposal that he asked the delegation to share with Quakers throughout the North: Quakers would pay the commutation fee of $300, but the War Department would segregate the funds so received into a separate account that would not be used to procure substitutes or for other military purposes. In fact, Stanton even offered that the money would be spent only at the direction of Quakers themselves.[81]

On December 7, representatives from Quaker Yearly Meetings in New England, New York, Ohio, Indiana, and elsewhere met in Baltimore to consider Stanton's offer. The consensus of the group was to reject Stanton's proposal. Their view was that, because the $300 was seen "as equivalent for or in lieu of" military service, Quakers could not in good conscience accept the proposal.[82]

A second, three-member committee consisting of Francis King of Baltimore Yearly Meeting, Charles Coffin of Indiana Yearly Meeting, and Samuel Tobey of New England Yearly Meeting met with Stanton in Washington later in December to convey the response. Stanton tried but failed to persuade the Quakers to accept his proposal.[83]

Thus, the section of the February 1864 amendment addressing the commutation fee substantially reflected Stanton's proposal, which the Quakers had rejected. The new provisions met with a mixed response from COs.

The February 1864 amendment also addressed conscription of Black men. Recall that the original 1863 conscription statute applied only to white men. The amendment did several things. It required that all Black men (free and enslaved) aged twenty to forty-five be enrolled. After enrollment, of course,

80 E. N. Wright, *Conscientious Objectors in the Civil War*, 72–73; Rufus Jones, introduction to Cyrus Pringle, *The Record of a Quaker Conscience: Cyrus Pringle's Diary* (New York: Macmillan, 1918).

81 E. N. Wright, *Conscientious Objectors in the Civil War*, 72–74.

82 Ibid., 73–74.

83 Ibid.

these men would be liable for the draft, just as whites had been. The amendment also provided that for any enslaved man who was drafted, the usual Union bounty of $100 would be paid—*to the slave's owner*. In addition, it provided that for each enslaved man who volunteered, "just compensation" of up to $300 would be paid—again, not to the enslaved but to the man's owner. In both cases, if the enslaved man survived military service, he would be freed.[84] Whatever else might be said about the two compensation provisions, they comported well with the prevailing law at the time (even if not with ethics or morality). By law, slaves were chattel property. If they were drafted, or were compelled by their owners to "volunteer," the owner was losing valuable property and needed to be compensated.

Clearly, several of these provisions were intended to increase the number of soldiers produced by the conscription statute: the sections concerning substitutes, limiting the benefit of commutation, repealing all exemptions based on hardship, and expanding the draft to include Blacks.[85]

∴

The second of the three amendments to the 1863 conscription statute was signed into law on July 4, 1864, just four months after the first set of amendments. This amendment was a short, terse document.[86] It restored the federal bounties of $100 for volunteers that Lincoln had suspended on April 1.[87] It also repealed commutation entirely (except for COs). And it allowed Northern governors to recruit volunteers in the Confederate states (except Arkansas, Tennessee, and Louisiana) and have the men so recruited count against draft calls in their Northern state.[88]

The last amendment to the 1863 conscription statute was signed into law on March 3, 1865.[89] It had little real-world effect because General Robert E. Lee surrendered at Appomattox on April 9, and on April 13, Stanton ended conscription in the Union. However, this detailed and lengthy set of amendments was a highly revealing document for what it said about systemic problems plaguing the entire draft apparatus that these amendments were designed to address.

84 National Forces, § 24.
85 Ibid., §§ 4, 5, 10, and 24.
86 An Act further to regulate and provide for the enrolling and calling out the National Forces, and for other Purposes, 13 Stat. 379 (1864).
87 Geary, *We Need Men*, 132.
88 An Act further to regulate, §§ 1–3.
89 An Act to amend the several Acts heretofore passed to provide for the Enrolling and Calling out the National Forces, and for other Purposes, 13 Stat. 487 (1865).

The March 1865 amendment required (rather belatedly) that the War Department honor promises it had made to Blacks pertaining to "pay and allowances." It took an act of Congress to get the War Department to keep these promises.

The amendment also created a residency requirement in order for volunteer enlistments to be credited against a (subsequent) draft call of a city or town.[90] This was necessary because municipalities would sometimes get into bidding wars over which could attract more volunteers by paying higher bounties. Towns and counties had an interest in doing this because those volunteers would be credited against the draft call of the municipality that got the volunteer—even if that volunteer lived and worked somewhere else. For example, in December 1863, Oneonta, New York, which paid a $300 bounty to volunteers, was outbid by neighboring Oswego, which paid $600. Consequently, men from Oneonta volunteered in Oswego. Oneonta then had both a higher draft call to fill than it otherwise would have had (if the Oneonta residents had volunteered in Oneonta) *and* fewer men available to fill that higher call. The March 1865 amendment was designed to fix this widespread problem.[91]

Another section was designed to rein in brokers who had gone into business because of the bounties paid to volunteers and the fees paid to substitutes. Brokers were, for the most part, a bunch of scoundrels, crooks, fraudsters, and con men. This provision made it illegal for brokers to profit by providing underage children, insane people, criminals, people already in the military, and deserters from the military as either volunteers or substitutes. It also made it illegal for these scam artists to defraud or cheat volunteers out of their bounty money by acting as sham intermediaries between the volunteer and the provost marshal. The amendment set penalties for these practices at three months to two years in prison and/or a fine of $200 to $1,000.[92] These remedial steps were necessary because these practices were ubiquitous. However, coming only days before the draft ended, the amendment had little or no effect.

The March 1865 amendment also was designed to ameliorate a similar set of problems associated with substitutes. It made it illegal to evade military service by knowingly providing a substitute who was underage, insane, a criminal, already in the military, a deserter from the military, or ineligible to serve by reason of any other mental or physical infirmity.[93] Again, Congress deemed this provision absolutely necessary because of the pervasive nature of precisely these abuses; again, the section did no good because it came at the end of the war.

---

90 Ibid., § 5.
91 Murdock, *One Million Men*, 174–75.
92 An Act to amend the several Acts, § 17.
93 Ibid., § 19.

Finally, the third amendment repealed the section of the July 1864 amendment that allowed Northern governors to recruit in most of the Confederate states.[94]

5 Bounties

In addition to the bounties paid by the federal government, states, counties, and municipalities also paid supplemental bounties to draftees. As noted earlier, competition among neighboring jurisdictions was common. Depending on where they were located, enterprising men could piece together federal, state, and local bounties amounting to $1,000 or more.[95] For context, at the time the Union army paid soldiers a salary of $13 per month. Local governments would raise bounty money through fundraising drives and patriotic speeches at public meetings. In some cities, local newspapers attempted to shame noncontributors by publishing the names of people who had contributed to local bounty funds—along with the amounts that each person had given.[96]

A great many soldiers procured by the bounty program were merely gaming the system and had not the slightest intention of serving. Many deserted as soon as they had their money in hand.[97] Sometimes, when guards shot at fleeing bounty men, the bounty men escaped safely, but civilian bystanders were killed. This became common enough that there were public scandals about these civilian deaths in both Philadelphia and Albany. There were even mass bounty jumps, in which large groups of newly enrolled bounty men would flee together.[98] Some bounty men who had found it impossible to desert earlier promptly surrendered to Confederate forces when they arrived at the front.[99]

Perhaps the most egregious aspect of the bounty system was the substantial number of men who made a career out of bounty jumping multiple times. Men would enlist in one town, collect the bounty, desert, and go to another town, repeating the pattern six, twelve, or more times. In states bordering Canada, Canadian citizens went into business crossing the border, enlisting in the US Army for the bounty, then fleeing home to Canada. Repeatedly.[100]

The record for the largest number of bounty jumps is believed to have been held by one John O'Connor of New York; he received thirty-two bounties by

94 Ibid., § 22.

95 McPherson, *Battle Cry of Freedom*, 605–6; Catton, *Civil War*, 206.

96 Murdock, *One Million Men*, 154–62.

97 Catton, *Civil War*, 206–8.

98 Murdock, *One Million Men*, 218–22, 233.

99 McPherson, *Battle Cry of Freedom*, 606.

100 Murdock, *One Million Men*, 228.

enlisting thirty-two times. Close behind O'Connor was Thomas Ryan, who collected thirty bounties. The list could go on almost endlessly. Punishments for bounty jumpers were notoriously inconsistent. The record holder, O'Connor, received a four-year prison sentence when caught. But in April 1864, Charles McCarty and John Riley were executed at Fort Warren, Massachusetts, for a single jump each. James Devlin was executed on February 3, 1865, for two jumps.[101]

In some cases, bounty jumpers were tortured before being executed. On August 29, 1863, for example, the Army of the Potomac executed five men—George Kuhn, John Ferlane, George Reinese, Charles Walters, and Emile Lai—for bounty jumping. The lengthy execution process was made into a spectacle for twenty-five thousand observers, mostly troops who, one supposes, were meant to derive a salutary lesson from the experience. The condemned men were paraded up and down to martial music in front of the assembled troops, each man directly behind his own coffin. Eventually, the parade arrived at five open graves. Each man was ceremonially shot into his open coffin, which was then lowered into the appropriate grave. *Harper's Weekly* published an eyewitness account of the spectacle.[102]

However, the deterrent value of whatever punishments were meted out to bounty jumpers was relatively small because of both the infrequency of bounty jumpers' being caught and the complete unpredictability of any sentence.

Toward the end of the war, as other methods of fundraising ran dry, municipalities turned to assessing real estate taxes to raise bounty money. This did not always end well. Stow, Ohio, resorted to such a real estate tax in early 1865. A local farmer, William Graham, did not fancy paying this tax. On March 2, a delegation of twenty townspeople paid a call on Graham to explain that he owed the tax money. Graham shot two of the townspeople dead before the rest fled.[103] The record does not reveal whether he was prosecuted.

6 Conscientious Objectors

The responses of both individual COs and of Quaker meetings varied widely after the conscription law was amended to give COs the option of noncombatant military service or a $300 commutation fee that would go to nonmilitary purposes. Many Quaker meetings took strong positions against payment of the commutation fee because it was inextricably linked to the military draft.

101 Ibid., 226, 247–49.
102 Ibid., 247–48.
103 Ibid., 3–4.

Some men who considered themselves COs chose noncombatant duty; some paid the commutation fee. Some COs, who had scruples about paying the commutation fee, ended up having the fee paid for them; it is likely that in some of these cases the payment was made over the objection of the CO and in other cases with active complicity of the CO. In yet other cases, COs refused both options; these objectors sometimes ended up in military prisons for varying lengths of time.[104]

The treatment of the COs in the last category varied widely depending on where they had been drafted and where they ended up in military prison. In his diary of his Civil War imprisonment as a CO, the Quaker Cyrus Pringle records that one of his CO prisonmates was threatened with being summarily shot. Pringle himself was mistreated by being tied, spread-eagle, to the ground for many hours a day in all types of weather. Yet other COs were treated well.

Pringle was drafted in July 1863, before the February 1864 amendment was enacted, so he could not have paid the commutation fee and have the money go to nonmilitary purposes. Still, his diary reflects a view toward the commutation fee widely shared by Quakers, even after the February 1864 amendment: "We were urged by our acquaintances to pay our commutation money; by some through well-meant kindness and sympathy; … and by others through a belief [that] it was our duty. But we confess a higher duty than to county; and, by asking no military protection of our government and grateful for none, deny any obligation to support so unlawful a system as we hold war to be even when waged in opposition to an evil and oppressive power and ostensibly in defence of liberty, virtue, and free institutions."[105]

COs who were unable conscientiously to either accept noncombatant military service or pay the commutation fee were liable to end up in military prisons. However, there was a sharp difference between official government policy and what actually happened to these men in almost all cases. Government policy was that the law was the law; men needed to avail themselves of one of the two prongs of the CO provision or wind up in military prison. In reality, most COs, including Pringle, were "paroled until called for" after a short prison stay. This was the norm after July 1864, although the practice certainly existed earlier.[106]

There is no accurate count of the number of men who were COs during the Civil War, because neither the Provost Marshal General's Bureau nor the religious denominations kept records. The Philadelphia Yearly Meeting (Quakers),

104 E. N. Wright, *Conscientious Objectors in the Civil War*, 180–219.

105 Murdock, *One Million Men*, 2.

106 E. N. Wright, *Conscientious Objectors in the Civil War*, 86–87; Murdock, *One Million Men*, 211.

however, did keep good records. Extrapolating from those records, it seems that hundreds of COs entered the military as noncombatants; hundreds of others paid the commutation money or had it paid for them; and still hundreds more did neither, were arrested, and "paroled until called for" after longer or shorter periods of time.[107]

The account in the wartime diary of Jesse Macy, nineteen at the start of the war, exemplifies the widely differing approaches taken by different Quakers. Before the Emancipation Proclamation, Macy and many of his friends were firmly against service: "Our army was engaged in upholding slavery and in returning escaped slaves to their masters ... the war was in no sense waged for the liberation of the slave. I could not support it." However, immediately after the Emancipation Proclamation, Macy records that a meeting was held at the Quaker Academy at Oskaloosa where he was a student. "The Academy was soon after depleted of nearly all its men suitable for military service. I was one of the few who did not volunteer."

After the February 1864 amendments were enacted, Macy enlisted as a noncombatant. He served under several commanding officers, yet not one honored his noncombatant status. Macy refused to don a uniform or carry a weapon. He was variously ordered to act as a cook, a military freight train attendant, a telegrapher, and other war-related roles; he refused all such service. Officers threatened Macy with a variety of punishments, including death, but Macy was steadfast in his refusal. "After the war was over I was told by a member of Company K that a very large proportion of the soldiers sympathized with me ... and a number of them had agreed to stand together in forcible resistance in case extreme measures were instituted against me."[108]

7 Legal Cases Pertaining to the Conscription Statute

The constitutionality of the 1863 conscription statute was never considered by the US Supreme Court. The constitutionality was, however, challenged in the Pennsylvania Supreme Court. Three men—Henry S. Kneedler, William F. Nickels, and Francis B. Smith—had been enrolled and then drafted. They were

---

107 Wright, *Conscientious Objectors in the Civil War*, 182–84.

108 Jesse Macy, *Jesse Macy: An Autobiography*, ed. Katherine Macy Noyes (Springfield, IL: Charles C. Thomas, 1933), 36, 41–51, 61, 76–78; "A Reluctant Conscientious Objector," in Peter Brock, ed., *Liberty and Conscience: A Documentary History of the Experiences of Conscientious Objectors in America through the Civil War* (New York: Oxford University Press, 2002), 166–72.

ordered to report for military duty or "be deemed as deserters, and subject to the penalty prescribed therefor by the rules and articles of war." The men sued the officers of their local enrollment board, filing their lawsuits directly with the Pennsylvania Supreme Court, something that was procedurally acceptable at the time. The men sought a declaration from the court that the conscription statute was unconstitutional and an injunction (a court order) directed to the enrollment board enjoining (preventing or stopping) the enrollment board from any further proceedings against the three men. By a vote of 3 to 2, the court ruled that the conscription statute was unconstitutional and granted Kneedler, Nickels, and Smith the injunction against the enrollment board.[109]

Chief Justice Walter H. Lowrie delivered the opinion of the court. The heart of Lowrie's opinion was based on the enumerated powers doctrine (although he did not call it that). Lowrie explained that the federal government has limited powers and that the only powers held by the federal government are those expressly conferred by the Constitution.

Lowrie distinguished between the two military-related powers conferred on the federal government in article 1, section 8, of the Constitution: the clause 15 power to second the state militias into federal service, and the clause 12 power to raise and support armies. The militias, Lowrie said, are, and always have been, the *state* militias. That is why, according to Lowrie, the Militia Act of 1862 (in which the federal government had "called forth" the state militias) had been constitutional: calling forth the militia is a power the Constitution expressly confers on the federal government.

However, the clause 12 power to raise and support armies refers to the federal government's *standing* army. The federal government certainly has the right to recruit a standing army under clause 12, the chief justice said, but it has no power to *conscript* such an army.

Lowrie wrote that if imposing conscription could be considered a necessary and proper way for Congress "to raise and support armies," then the government could surely also "take their houses for offices and courts; their ships and steamboat for the navy; their land for its fortresses; [and] their mechanics and workshops for the different branches of business that are needed for army supplies." But Congress can do none of these things legally, Lowrie said; instead, Congress should raise its standing army "in the ordinary way, by voluntary enlistments."

109 Kneedler v. Lane, 45 Pa. 238 (1863).

Associate Justices William Strong and Joseph Reed dissented in *Kneedler*. Their argument was direct. For context, they started by observing that the Preamble to the Constitution explains that the Constitution was adopted, in part, to "provide for the common defense." In the circumstances of the Civil War, they said, providing for the common defense meant conscription.

Strong and Reed's view was that the provisions of article 1, section 8, fully support conscription. Clause 12 allows Congress to "raise and support armies." Clause 18 confers on Congress the power "to make all laws which shall be necessary and proper for carrying forth the foregoing powers."[110]

But the case was not over. A month after the *Kneedler* decision, Lowrie's term on the Pennsylvania Supreme Court expired, and he was replaced by Daniel Agnew. With the new justice seated, the US attorney promptly approached one of the (previously) dissenting justices, Strong, and asked for a do-over. Strong agreed, and on January 16, 1864, the Pennsylvania Supreme Court issued an entirely different opinion in the same case. It took no new evidence, held no new hearing, and heard no new argument. The newly reconstituted Supreme Court of Pennsylvania simply issued a new decision in the already-decided case.

This time, Agnew, the new justice, joined Strong and Reed to form a new majority that declared the conscription statute to be constitutional. To further complicate matters, in today's Pennsylvania law books, the two diametrically opposite decisions in the same case carry the same identification number: 45 Pennsylvania 238.[111]

∴

Soon after Lincoln signed the conscription statute into law, both federal and state courts started receiving applications for writs of habeas corpus. When a judge issues a writ of habeas corpus, the judge is ordering the person holding the prisoner to physically bring the prisoner into court. When the prisoner is in court, standing before the judge, the judge will determine whether the prisoner should continue to be detained or be set free. That is, the writ of habeas corpus is a way for a prisoner to obtain a hearing in order to have a court determine whether a detention (or punishment) is lawful. Today, prisoners typically seek writs of habeas corpus against the warden of the prison in which the prisoner is being held; during the Civil War, conscripted men often sought the writ against the local provost marshal who wanted to conscript them.

110 Ibid., 277.

111 J. L. Bernstein, "Conscription and the Constitution: The Amazing Case of *Kneedler v. Lane*," ABA *Journal* 53, no. 709 (1967): 711–12.

The writ of habeas corpus has for centuries held a hallowed place in Anglo-American jurisprudence, and it is the only common law writ to be specifically mentioned in the US Constitution. In 1830, Chief Justice John Marshall said that "the great object" of the writ is "liberation of those who may be imprisoned without sufficient cause." In 1969, the US Supreme Court said that the writ of habeas corpus "is the fundamental instrument for safeguarding individual freedom against arbitrary and lawless state action."[112]

Some of the habeas corpus cases dealing with the Civil War conscription statute were brought in federal courts.[113] Some were brought in state courts.[114] In both sets of cases, issues pertaining to the courts' jurisdiction to hear the cases were prominent. *Jurisdiction* refers to a court's right or authority to hear (and rule upon) a particular case. Controversies involving a court's jurisdiction can be complicated and arcane, but in the context of conscription cases, it is important to understand these jurisdictional issues.

In the federal cases, the jurisdictional question hinged on the meaning of section 14 of the 1863 conscription statute, which said that "all persons drafted and claiming exemption from military duty on account of disability, or any other cause, shall present their claims to be exempted to the [local enrollment] board, whose decision shall be final." The reference to disability in section 14 relates to section 2, which vaguely exempted men who were "physically or mentally unfit for the service" without defining or explaining what that meant.

Provost marshals, understandably, took the position that this language precluded men from appealing to any court the local enrollment board decisions that they disagreed with because—as the law said—decisions of the board were final. Draftees, and some judges, disagreed. They said that for centuries the Great Writ, as habeas corpus is known, had been a last resort for those wrongfully imprisoned and that Congress could not override the centuries of Anglo-American jurisprudence since medieval times by enacting conscription.

The underlying jurisdictional issue that these cases presented continued into the twentieth century and was extensively litigated during World War I, World War II, and the Vietnam War. Yet even during the Civil War, the question of whether draft board decisions were (or were not) judicially reviewable was central to these court cases, and none settled the matter once and for all.

---

112 *Ex parte* Watkins, 28 U.S. 193, 202 (1830); Harris v. Nelson, 394 U.S. 286, 290–91 (1969).

113 *In re* Blum, 3 F. Cas. 752 (D. Mass. July 1863); *ex parte* Burke, 4 F. Cas. 731 (E.D. Pa. 1863); and Antrim's Case, 1 F. Cas. 1062 (E.D. Pa. 1863).

114 *In re* Spangler, 11 Mich. 298 (1863); *in re* Hopson, 40 Barb. 34 (N.Y. 1863); *in re* Barrett, 25 How. Pr. 380, 42 Barb. 479 (N.Y. 1863).

For example, the habeas corpus case of a young man named Burke was brought in federal court in Pennsylvania and decided in August 1863. The soldier's mother brought the case; she claimed that her son, who had volunteered, was underage. In what must have been deeply alarming to the Provost Marshal General's Bureau in Washington, the judge in that case ignored the threshold jurisdictional issue, just assumed that the court had jurisdiction, and went straight to the merits of the case. In an opinion that was all of three sentences long, the judge ruled in Mrs. Burke's favor; her son was released from the army and set free.[115]

In the state court cases, the jurisdictional issue was a bit different. The question there was whether state courts could properly exercise jurisdiction over men who were (at least arguably) subject to the federal draft under the federal conscription statute—or whether such cases must necessarily be heard in a federal court. This, too, is an important question, because it inevitably implicates the dual sovereignty nature of the United States. The federal Constitution contemplates the side-by-side coexistence of state sovereigns and the federal sovereign. The US Constitution gives some powers to the federal government, while the states continue to hold other powers. (To make matters even more complicated, some matters can be addressed concurrently by both the federal and state governments.)

In Civil War habeas corpus cases at the state court level, the decisions provided no consistency on the jurisdictional question. Indeed, in the same month—August 1863—two different New York State courts came to diametrically opposite conclusions on the identical question of whether a state court has jurisdiction to issue a writ of habeas corpus in a case arising under the federal conscription statute. One court answered this question with a resounding yes and issued the writ over the objection of the local provost marshal, while another court held that it had no jurisdiction over this federal matter.[116]

On September 9, 1863, Judge John Cadwalader of the US District Court for the Eastern District of Pennsylvania decided *Antrim's Case*.[117] This was a habeas corpus action brought against the local provost marshal by a draftee named Antrim, who claimed that he was the only son of a widow dependent on him for support and that he therefore was entitled to a draft exemption under the conscription statute.[118] The judge discussed the provision of the conscription statute that purported to make enrollment board determinations final but

115 *Ex parte* Burke.

116 *In re* Barrett; *In re* Hopson.

117 Antrim's Case, 1 F. Cas. 1062, 5 Phila. 278 (E.D. Pa. 1863).

118 Conscription statute, § 2.

ruled that a mere statute of Congress could not trump the Great Writ: "Powers of government must be administered constitutionally, and their execution must be regulated in due subservience to judicial authority.... A condition as would abrogate, or abridge, the effect of a constitutional provision as to the judicial power cannot be [allowed]."[119] Cadwalader issued the writ of habeas corpus that Antrim had requested.

Fry was understandably alarmed, fearing that the entire conscription apparatus could be effectively nullified by a flood of habeas corpus petitions from disgruntled draftees. Fry wrote to Secretary of War Stanton, saying that "the necessity for having no appeal from the decision of the Board of Enrollment in questions of exemptions is of first importance to the military service." Stanton took the matter to Lincoln, and on September 15, 1863—a mere six days after Cadwalader's decision was issued—Lincoln issued his Proclamation 104, which suspended the writ of habeas corpus for the remainder of the Civil War in all cases of men "resisting the draft" (Lincoln's words).[120]

∴

The ubiquity of city and town ordinances that would pay all or part of local men's commutation fees gave rise to an unusual and interesting set of lawsuits raising a variety of issues under these ordinances. Here are some examples.

On July 1, 1863, citizens at a town meeting in Camden, Maine, voted to pay the full $300 commutation fee for any citizen of Camden who was drafted. The townspeople also voted to issue bonds to raise the money for this. Simon S. Barbour was a citizen of Camden, he was drafted, and he sought the $300 from the town, which, for some reason not stated in the case, refused to pay. The case went to the Maine Supreme Court and Barbour lost. According to the court, the "town had no legal authority to assess taxes or raise money to pay the commutation of one, who had been drafted in pursuance of the Act of Congress of March 3, 1863.... The government of the United States were [*sic*] in need of soldiers."[121]

In New Jersey, a local taxpayer, Henry Beckett, sued Bordentown. In 1865, the town had imposed a property tax to raise money to pay the commutation fees of local men. Beckett's argument was that, since he was not a US citizen, he would not be subject to the draft; thus, he would not directly benefit from

119 Ibid., § 14; Antrim's Case, 1 F. Cas. at 1067.
120 Sterling, "Civil War Draft Resistance," 620–21.
121 Barbour v. Camden, 51 Me. 608 (1865).

money raised to pay commutation fees; therefore the tax should not apply to him. Beckett lost the case.[122]

Similar cases did not always lead to similar results. The Township of Delaware, New Jersey, voted to impose a tax to raise $31,000 to pay the commutation fee of every town resident who might be drafted. Issachar Hoagland and other resident taxpayers sued, saying that the tax was illegal and unconstitutional. They lost; the court upheld the tax in 1864. But when the same tax levied by the same township for the same purpose was challenged again, the court three years later held the tax to be illegal because it "was calculated to defeat or obstruct the operation of [an] act of congress … [of] raising an army to suppress the rebellion."[123]

On August 13, 1863, in the immediate aftermath of the New York City draft riots, the selectmen of Woodbury, Connecticut, voted to pay the commutation fees of town residents and to float a bond issue in the amount of $6,400 to raise the necessary funds. On November 13, 1863, the Connecticut legislature passed a bill saying that Connecticut municipalities have the power to do that. Charles Booth, a local taxpayer who did not want to be taxed to pay for other men's commutation fees, sued the town, saying that the imposition of the tax was unjust and was designed to undermine the conscription law enacted by the federal government. (On that second point, Booth was certainly correct.) The Supreme Court of Connecticut ruled that the town ordinance was legal.[124]

Even the end of the Civil War did not stop the lawsuits pertaining to municipality payment of commutation fees. In 1864, Franklin, Minnesota, issued bonds to pay the commutation fee of men from the town who were drafted. Louis Kunkle bought a Franklin bond but wanted his money back early, so he brought a lawsuit against the town saying that it did not have legal authority to sell bonds to raise money for the commutation fee. In 1868, the Minnesota Supreme Court disagreed with Kunkle and ruled in favor of the town.[125]

8 Conclusion

The title of the last chapter of Eugene Murdock's comprehensive study of the Civil War draft in the North is "But It Worked." Murdock argues that, although "it had weaknesses," the draft was successful. James Geary agrees with

122 State *ex rel.* Beckett v. Collector of Bordentown, 32 N.J.L. 192 (1867).

123 Hoagland v. Delaware, 17 N.J. Eq. 106 (1864); State v. Jackson, 33 N.J.L. 450 (1867).

124 Booth v. Woodbury, 32 Conn. 118 (1864).

125 Murdock, *One Million Men*, 333–49, esp. 334.

Murdock; in the conclusion of his book on the Civil War draft, Geary says "it worked despite its flaws."[126]

The most obvious rejoinder to the contention that the Civil War draft worked is to look at the raw statistics. Between March 3, 1863, when the conscription statute was signed into law and the end of the war, 776,829 men had their names called in draft lotteries. (Bear in mind the substantial number of men whose names never got into the lotteries because they skulked or skedaddled.) Of those 776,829 men whose names were called, only 46,347 men (5.96 percent) were conscripted.[127] Nearly twice that number (86,724 men) evaded the draft by paying the $300 commutation fee. A mere 3.67 percent of the Union army were conscripts.[128] (For purposes of comparison, approximately 78.8 percent of the US Army was conscripted during World War I.)[129] Just judging by the raw numbers, the Civil War draft was not successful.

Murdock was not unaware of these figures but had a response ready: Don't look at only the raw figures; look at the broader context: "The Civil War draft law was not really a draft law at all.... The idea was simply to use the threat of the draft to spur voluntary enlistments."[130] That is, the specter of conscription encouraged men to seek out and provide substitutes, and it encouraged states and municipalities to provide bounties for volunteers, because the number of volunteers from the county or municipality were deducted from local draft calls. Thus, Murdock argues that, in the end, through substitutions, bounties, and volunteers, many more men were induced to serve than it might seem at first.

But if Murdock is correct that it is important to look at the wider context, then there are several additional aspects of that wider context that also need to be addressed. First, the figures used by Fry (and by Murdock, Geary, Levine, and others) mistakenly start with men actually drafted. But more than 50 percent of the men who were supposed to be enrolled either refused enrollment or refused to be drafted. That is a stunningly high figure. In Delaware, Maine, Missouri, and Wisconsin the percentage was more than 70 percent. A law

126 Ibid.; Geary, *We Need Men*, 167–74.

127 *PMG Final Report*, pt. 12, 165–99, tables 5–10.

128 Geary, *We Need Men*, 173.

129 On November 11, 1918, the day of the armistice ending the First World War, approximately 3.5 million men were serving in the US armed forces, 2,758,542 of whom had been conscripted. These figures are described in the US Office of the Provost Marshal General, *Final report of the Provost Marshal General to the Secretary of War on the operations of the Selective Service System to July 15, 1919* (Washington, DC: Government Printing Office, 1919), 14.

130 Murdock, *One Million Men*, 335.

violated by more than half of the cohort affected by the law (in this case, men aged twenty to forty-five) is not a law that can be said to be working.

Second, if the purpose of law is, in part, as the Preamble to the US Constitution has it, to "insure domestic tranquility," then the Civil War conscription statute was surely a failure. Civil War conscription led to the worst outbreak of urban rioting in the history of the United States up to that time, a record that remains unbroken today. In addition to riots all over the country, the draft led directly to many other outbreaks of violence, including the murders of many enrollment officers charged with carrying out the statute. The Civil War draft actively disrupted domestic tranquility in multiple important ways.

Third, if the purpose of law is, in part, as the Preamble says, to "establish justice," then the Civil War draft was a colossal failure. Any man could buy his way out of the draft for $300—but this was a huge sum for a working person at the time. Murdock actually argues that the commutation provision was a gift to the poor, because it put an effective cap on the cost of buying a substitute. That is, prospective substitutes could hardly charge more than $300, because every prospective principal knew that he could commute for $300. Even if the commutation fee did work as Murdock claimed—and the evidence on the point is distinctly mixed—his argument is wholly irrelevant, because as long as the commutation fee existed, all rich men could avail themselves of it, regardless of what the going rate for substitutes was. The short of it was that rich men bought their way out of the draft, while poor men went to the front. This was not just.[131]

Fourth, the bounty system had at least two important problems associated with it. At the purely practical level, it was grossly inefficient. It encouraged competition between neighboring counties or municipalities that drove up bounty prices in some places while stripping other places of sufficient men to satisfy draft quotas. This hindered the ability of the government to acquire conscripts. At the philosophical level, the bounty system actively encouraged bounty jumping. If, with Thomas Aquinas, one believes that "the purpose of law is to lead men to virtue," then the Civil War conscription statute was a bad law indeed because it provided a strong economic incentive to break the law and to act wickedly.[132] We know this because, according to one scholar, "those states which paid the largest bounties produced the largest number of deserters."[133] That is, the payment of bounties actively encouraged men to enlist merely to

131 Ibid., 7, 340–41.

132 Thomas Aquinas, *Summa theologica*, trans. Fathers of the English Dominican Province (New York: Benziger Brothers, 1911–25), pts. 1–2, question 13; and pt. 2, reply to question 2.

133 Ella Lonn, *Desertion during the Civil War* (Lincoln: University of Nebraska Press, 1998), 142.

receive the bounty under false pretenses, with absolutely no intention whatsoever of actually serving.

One reason that Murdock believed the Civil War draft worked well is that it "did raise the army that won the war." But Murdock confused association with causation. The North did have a draft and did win the war, but the Confederacy also had a draft—adopted earlier than the Northern draft—but the Confederacy did not win the war.[134]

---

134 Murdock, *One Million Men*, 334.

CHAPTER 3

The Civil War: the Confederacy

On April 16, 1862, the Confederate States of America (CSA) enacted a broad conscription statute. For obvious reasons, some scholars refer to the April 1862 Confederate conscription statute as the first conscription in American history: it predated the draft in the North by almost a year.[1] For equally obvious reasons, other scholars refer to the March 3, 1863, Union conscription statute as the first conscription in American history: when the Confederacy enacted conscription, the Confederacy had seceded from the United States and considered itself to be a separate, independent country.[2] There is no need to address that issue here. The people who lived in the Confederate states during the Civil War had been US citizens only a short time earlier, and they would again be US citizens just a short while later. Thus, it is appropriate to consider here conscription in the Confederacy.

Passage of any statute in the Confederacy immediately raises several questions for the contemporary observer. When we speak of a measure's being enacted in the North, the modern reader will immediately understand that both houses of Congress passed a bill that the president then signed into law. But what kind of government did the Confederacy have? How were laws in the Confederacy made?

Although the Confederacy started out with a provisional constitution in 1861, by the time the first Confederate conscription statute was enacted in April 1862, the Confederacy had a permanent constitution in place. That Constitution, which governed the enactment of Confederate conscription, will look familiar to American readers who remember their high school civics

1 See, for example, James M. McPherson, *Battle Cry of Freedom* (New York: Oxford University Press, 1988), 430, and William L. Shaw, "The Confederate Conscription and Exemption Acts," *American Journal of Legal History* 6, no. 4 (October 1962): 374. McPherson calls Confederate conscription "the first national conscription act in American history" (430). See also John F. Reiger, "Deprivation, Disaffection and Desertion in Confederate Florida," *Florida Historical Quarterly* 48, no. 3 (January 1970): 279, 286; and David Carlson, "The 'Loanly Runagee': Draft Evaders in Southern Georgia," *Georgia Historical Quarterly*, 84, no. 4 (Winter 2000): 589, 593.

2 See, for example, Jack Franklin Leach, *Conscription in the United States: Historical Background* (Rutland, VT: Charles E. Tuttle, 1952), 386, which refers to the Union draft as "the first national draft in American history"; Timothy J. Perri, "The Evolution of Military Conscription in the United States," *Independent Review* 17, no. 3 (2013): 429, 430; Eugene C. Murdock, *One Million Men: The Civil War Draft in the North* (Madison: State Historical Society of Wisconsin, 1971), 5–6.

 | DOI:10.1163/9789004546684_005

classes because, with only a few salient exceptions, the Confederate Constitution tracked the US Constitution more or less exactly.[3]

Exactly like the US Constitution, article 1 of the Confederate Constitution vested legislative authority in a bicameral legislature consisting of the House of Representatives and Senate. Exactly like the US Constitution, the Confederate Constitution provided that members of the House of Representatives were elected for two-year terms by direct popular vote and had to be at least twenty-five years old.[4] Exactly like the US Constitution, the Confederate Constitution provided that members of the Senate were elected for six-year terms by the respective state legislatures and had to be at least thirty years old.[5] The Confederate Constitution conferred on the Confederate Congress pretty much the same powers that the US Constitution conferred on the federal Congress and in pretty much the same order, too: to lay and collect taxes, borrow money, coin money, establish post offices, declare war, raise and support armies, provide for calling forth the militia, and "make all laws which shall be necessary and proper for carrying into execution the foregoing powers." (One of the few differences between the US and CSA Constitutions is that the Confederate Constitution had several provisions related to slavery not found in the US Constitution.)[6]

The parallels continued: article 2 of the Confederate Constitution vested executive authority in a president who was elected by an electoral college, not by direct popular vote. The Confederate Constitution provided that if the president died in office or resigned, the vice president would become president. The Confederate Constitution made the Confederate president the commander in chief of the armed forces, allowed him to nominate judges and other officials with the advice and consent of the Senate, and allowed him to make treaties subject to ratification by two-thirds of the Senate. (Another difference between the US and CSA Constitutions is that the Confederate Constitution established a single six-year term of office for the president.)

Exactly like the US Constitution, article 3 of the Confederate Constitution vested the judicial power of the Confederacy in "one Supreme Court and in such

3 The Constitution for the Provisional Government of the Confederate States of America came into effect on February 8, 1861, until a permanent constitution was passed on February 22, 1862. See "Constitution of the Confederate States; March 11, 1861," Yale Law School: The Avalon Project, https://avalon.law.yale.edu/19th_century/csa_csa.asp#prea, last accessed April 26, 2022 (the Confederate Constitution is hereafter Conf. Const.).

4 Conf. Const. art. 1, §§ 1 and 2.

5 The popular election of US senators did not occur until ratification of the Seventeenth Amendment in 1913.

6 Conf. Const. art. 1, §§ 8 and 9.

inferior courts as the Congress may, from time to time, ordain and establish." The Confederate Constitution also granted judges life tenure. The Confederate Supreme Court never came into existence; as a result, state courts, usually state supreme courts, decided all the reported Confederate cases pertaining to conscription.

1 The First Confederate Conscription Statutes

The Confederacy adopted conscription in 1862 for reasons similar to those of the Union the following year. Confederate forces had suffered military setbacks in a significant battle in early February 1862 at Fort Henry on the Tennessee-Kentucky border and a week later in the nearby battle at Fort Donelson, in eastern Tennessee.[7] The Union general George McClellan had started his Peninsular Campaign in March, and his troops were approaching the Confederate capital of Richmond. And 148 regiments of twelve-month volunteers who had entered the Confederate army after Fort Sumter in April 1861 were coming to the end of their enlistment.[8]

On March 28, 1862, Confederate President Jefferson Davis sent a message to the CSA Congress urging it to adopt conscription. His message was terse. The first paragraph of the message read, "The operation of the various laws now in force for raising armies has exhibited the necessity for reform. The frequent changes and amendments which have been made have rendered the system so complicated as to make it often quite difficult to determine what the law really is, and to what extent prior enactments are modified by more recent legislation."[9]

Davis was absolutely correct. In the preceding year, the Confederate Congress had passed no fewer than eleven partially overlapping, sometimes contradictory, statutes pertaining to military recruitment: on March 6, May 6 and 8, August 8, and December 11, 1861, and on January 22, 23, 27, 29, and February 17, 1862.[10] The system was a shambles, and the introduction of conscription did little to fix the problem. For the remaining years of its existence,

7 McPherson, *Battle Cry of Freedom*, 428.

8 Albert Burton Moore, *Conscription and Conflict in the Confederacy* (1924; Columbia: University of South Carolina Press, 1996), 14.

9 *Journal of the House of Representatives of the First Congress of the Confederate States of America, 1861–1865* (Washington, DC: Government Printing Office, 1904), 5:156–57. Ironically, this seven-volume set was published by the US government between 1904 and 1905.

10 A. B. Moore, *Conscription and Conflict in the Confederacy*, 1–8.

the Confederate Congress hastily passed multiple ill-considered conscription bills, only to immediately revise (and sometimes re-revise) them in relatively short order. Only the most important ones, with the furthest-reaching effects, are addressed here.

The April 16, 1862, Confederate conscription statute contained two main provisions. It declared that "all white men who are residents of the Confederate States between the ages of eighteen and thirty-five" are subject to conscription for "three years, unless the war shall have been sooner ended." In addition, it provided that the enlistment terms of all the twelve-month men who had enlisted voluntarily during the first rush of war fever in the spring of 1861 would—involuntarily—have their terms extended for three years from the date of their enlistment (unless the war ended sooner).[11]

The conscription statute also allowed the Confederate president, with the consent of state governors, to second state staff for conscription purposes. Should the president fail to obtain the necessary gubernatorial consent, the statute empowered him to use Confederate officers for that purpose.[12]

The law also provided that draftees who evaded service were to be treated as deserters "and punished as such." The statute further provided that men could evade service entirely by providing a substitute.[13]

One section appears to be peculiar by contemporary standards, but it must be borne in mind that, at the outset of the Civil War, the Confederacy was severely lacking in industrial capacity in general and the capability to produce weapons in particular. Thus, this section encouraged a conscript to bring his own "musket, shot-gun, rifle or carbine," and, if he did so, the Confederacy would pay him the value of the weapon.[14] The consequence of this arrangement was that Confederate forces had no standard-issue weapons and therefore no standard-issue ammunition.

---

11 An Act to further provide for the public defense (hereafter public defense law), April 16, 1862, § 1, in James M. Matthews, ed., *Public Laws of the Confederate States of America, passed at the first session of the First Congress, 1862, Carefully collated with the originals at Richmond* (Richmond, VA: R. M. Smith, 1862–64), 29–32. Statutes enacted by the Confederate States of America are available today from several sources—but no single comprehensive source. The sources include Matthews, *Public Laws of the Confederate States of America*, and US War Department, *The War of the Rebellion: A Compilation of the Official Records of the Union and Confederate Armies*, 128 vols. (Washington, DC: Government Printing Office, 1880–1901 (hereafter, OR). In every case, my notes cite to the actual source I used for that statute.

12 Public defense law, § 3.

13 Ibid., § 9.

14 Ibid., §§ 6 and 8.

The law made no mention whatever of COs, and, unlike the federal conscription statute enacted a year later, the Confederacy's measure had no commutation provision.

Before the ink had dried on the April 16 conscription statute, the Confederate Congress realized it had made something of an error: if *all* men aged eighteen to thirty-five were in the army, not enough men would be left behind to grow food (including to feed the army), run industry (including to produce weapons), and generally keep the economy afloat. Thus, a mere five days later, on April 21, the Confederate Congress passed another statute enumerating whole categories of people who were to be exempt from the just-enacted conscription statute.[15] These categorical exemptions included employees of the Confederacy; members of the legislative, executive, and judicial branches of the Confederate and state governments; all other state employees; mail carriers; ferrymen on mail routes; telegraph operators; ministers; printers; college professors; teachers; nurses at hospitals; druggists; and staff at wool and cotton factories. As was utterly predictable, thousands of men suddenly became druggists in shops that sold no drugs, teachers without students, or mail carriers on nonexistent routes.[16]

Most of the Quakers and Mennonites who might have been eligible for CO status, had there been one, were concentrated in North Carolina and Virginia. After the April statutes had been passed, these groups began petitioning the Confederate government, asking that some accommodation be made for COs. North Carolina Quakers appointed a committee of four Friends to communicate their concerns to Secretary of War James Seddon. In August, a somewhat larger group of North Carolina Quakers met with Davis and members of the Confederate Congress to urge that provisions be made in the conscription statute for COs. Virginia Mennonites and Dunkards, an Anabaptist denomination, presented similar appeals to their members of Congress.[17]

On October 11, 1862, the Confederate Congress further amended its conscription statute. The new statute contained several important sets of changes.[18]

---

15 An Act to exempt certain persons from enrollment for service in the armies of the Confederate States (hereafter exemption law), April 21, 1862, in Matthews, *Public Laws of the Confederate States of America*, 1st Cong., 1st sess., 51–52.

16 A. B. Moore, *Conscription and Conflict in the Confederacy*, 53–56.

17 Edward Needles Wright, *Conscientious Objectors in the Civil War* (New York: A. S. Barnes, 1931), 93, 101–4.

18 An Act to amend an act entitled 'An act to further provide for the public defense' approved 16th April 1862 and the act to amend the same, approved September 27th, 1862 (hereafter third amendment), *OR*, ser. 4, 2:160.

First, it added yet additional categories of people exempted from the draft: railroad presidents, conductors, engineers, station agents, and track hands; public printers; shoemakers; tanners; blacksmiths; wagon makers; millers of grain and all their employees; all the staff and attendants at hospitals (not just the nurses); everyone employed in wool and cotton factories; all postal employees (not just mail carriers); all employees involved in manufacturing arms, ordnance, or munitions of war; all those involved in manufacturing saddles or harnesses; all those with government contracts for arms, ordnance, or munitions of war; all people engaged in the construction of ships or engines; everyone involved in burning coke, smelting iron, mining coal, or making charcoal; and everyone involved in making salt.

The resulting situation—an underlying conscription statute that purported to draft *everyone* aged eighteen to thirty-five with two follow-ups that provided total exemptions for so many professions—had predictable results. On July 25, 1863, a group of fourteen Confederate generals, including Braxton Bragg, wrote ruefully to the Conscription Bureau about the problem. They complained that suddenly the numbers of "shoemakers, blacksmiths, tanners, wagon makers, millers and their engineers, millwrights ... railroad and telegraph" operators, post office employees, and printing press operators were multiplying way beyond the needs of any civil society.[19]

Second, the October 11 Confederate conscription statute included a limited provision for COs: it exempted Quakers, Nazarenes, Mennonites, and Dunkards from the draft if they provided a substitute (this was not new; the April 16 statute exempted *all* men if they provided a substitute) or paid a commutation fee of $500 (this was new).

In January 1863, a young man named Levi M. Stringer brought a case testing the newly enacted CO provision. Stringer sought a writ of habeas corpus, claiming that as a member of a "Christian Church," he had conscientious scruples against bearing arms and should be allowed to be classified as a CO. The Supreme Court of Alabama ruled against him because "conscientious scruples against bearing arms, unless the party entertaining them belongs to one of the religious sects mentioned in the statute, presents to the courts of the country no legal ground for declaring the petitioner exempt from military duty."[20] As a purely legal matter, the Alabama court was correct.

Because neither the government nor the religious denominations kept accurate records, it is impossible to determine how many men declined to serve in the Confederate army because of conscientious scruples—whether

---

19 Generals H. D. Clinton et al. to General S. Cooper, July 25, 1863, *OR*, ser. 4, 2:670–71.

20 *Ex parte* Stringer, 38 Ala. 457 (1863).

they satisfied the requirements of the October 1862 amendment and paid the commutation fee, refused to pay the fee despite being eligible to do so, or considered themselves to be COs but did not meet the statutory definition. After the October 1862 amendment became law, the North Carolina Yearly Meeting (of Friends) approved a minute (decision) opposing payment of commutation by Quakers: While "we do pay all taxes imposed on us as citizens and property holders, in common with other citizens ... yet we can not conscientiously pay the specified tax, it being imposed on us on account of our principles—being the price enacted of us for religious liberty."[21] During the Civil War, some Quakers followed this advice and declined to pay the required commutation fee.

"The position of Friends [during the Civil War] was a peculiar one," wrote North Carolina Quakers.[22] They were not merely opposed to the war—they were also well-known abolitionists. This combination often brought down upon the heads of Quaker COs a double measure of opprobrium from their neighbors. As a result, Southern Quakers felt acutely that they suffered for their beliefs more than their Northern coreligionists did.

The North Carolina Yearly Meeting published a report after the war about the problems North Carolina Quakers had faced during the Civil War. It is a detailed, often grisly, account of the treatment of Quaker COs in the Confederacy: hung by their thumbs, stabbed repeatedly with bayonets, subjected to mock executions, and starved for extended periods. The report includes explicit descriptions of a particular punishment known as bucking, "in which the person is placed in a stooping position, the wrists firmly tied and brought in front of the knees, thus keeping the body in a painful and totally helpless position." It recounts one case in which two Quaker brothers died of their abuse, one after being sent to the front even though he would not bear arms, the other starved to death in a Confederate prison in Richmond.[23]

The third change in the law wrought by the October 11, 1862, statute included what the literature reports variously referred to as the "twenty-slave law," the "twenty-negro [*sic*] law," and the "twenty-nigger law."[24] This change provided for the exemption of one white "owner or overseer on each plantation of

---

21 *An Account of the Sufferings of Friends of North Carolina Yearly Meeting, In Support of Their Testimony Against War, From 1861 to 1865* (New Vienna, NC: Peace Association of Friends in America, 1872), 15–16. *Minute* is the Quaker term for a formal decision, statement, or declaration.

22 *An Account of the Sufferings of Friends*, 3–5.

23 Ibid., 9, 13.

24 David Williams, *A People's History of the Civil War: Struggles for the Meaning of Freedom* (New York: New Press, 2005), 77; John M. Sacher, "Twenty-Negro,' or Overseer Law: A Reconsideration," *Journal of the Civil War Era* 7, no. 2 (June 2017): 269; A. B. Moore,

twenty negroes, on which there is no [other] white male."[25] The exemption also applied where one white male was overseeing twenty or more enslaved people on two or more plantations within five miles of each other.

Multiple scholars—Albert Burton Moore (1924), Georgia Lee Tatum (1934), and John M. Sacher (2017)—quote the same portion of an 1862 letter from Senator James Phelan of Mississippi to Jefferson Davis about the despicable nature of the twenty Negro law. Phelan's much-cited denunciation of the statute is this: "Never did a law meet with more universal odium than the exemption of slave owners. Its injustice, gross injustice, is denounced even by men whose position enables them to take advantage of its privileges. Its influence upon the poor is most calamitous, and it has awakened a spirit and elicited a discussion of which we may safely predicate the most unfortunate results."[26]

Writing in 1924, Moore, still considered by many to be the preeminent authority on Confederate conscription, explained why the twenty slave law was regarded as absolutely essential: "Of course agriculture and the lives of families could not be entrusted to slaves unrestrained by overseers. They had as a rule been managed by overseers and had neither the will nor the knowledge of their own to farm successfully; and without the awe-inspiring presence of the overseer the social and economic order of the Confederacy would be endangered."[27]

It may be tempting to dismiss Moore's view—that slaves would be too stupid or lazy to farm without the supervision of their white masters—as merely an unfortunate example of racism from a benighted earlier era, but, amazingly, justifications for the twenty slave law persist into the twenty-first century. Recently, John M. Sacher published a scholarly article on the subject. Sacher recites at some length the views of modern historians who criticize the law and then provides another explanation of why the statute was needed: "These historians overlooked the necessity of such an exemption for slave control." Lest a reader miss the racist sexual implication, Sacher clarifies that the concern was "about violence on a home front devoid of white men" and "the potential for the rape of white women." Contemporary critics of the twenty slave law simply do not appreciate the matter of sexual propriety, Sacher tells his readers: the

---

Conscription and Conflict in the Confederacy, 70; W. L. Shaw, "Confederate Conscription and Exemption Acts," 368.

25 An Act to amend an act, § 1.

26 A. B. Moore, *Conscription and Conflict in the Confederacy*, 71; Georgia Lee Tatum, *Disloyalty in the Confederacy* (Chapel Hill: University of North Carolina Press, 1934), 89; Sacher, "'Twenty-Negro,' or Overseer Law," 269–70; James Phelan to Jefferson Davis, December 9, 1862, *OR*, ser. 1, vol. 17, pt. 2, 790.

27 A. B. Moore, *Conscription and Conflict in the Confederacy*, 70.

"guarantee of ... safety, especially for white southern women and girls, was ... the key that most scholars who overestimate [the law's] pernicious impact ... miss."[28]

∴

The Confederacy did not keep anything approaching the North's meticulous and detailed records on conscription.[29] There is, however, enough of a record to discern that conscription in the Confederacy could best be described as slow to be organized, chaotic and ineffective when operating, and deeply and widely opposed by the populace. Perhaps reflecting the views of the public, governors, other state officials, and even judges sometimes worked assiduously to interfere with the conscription system. The system was run by personnel who were often incompetent, corrupt, or both. Throughout the South evasion of conscription was widespread. At best, conscription in the Confederacy was only partially successful in raising troops. With its provisions to allow drafted men to buy substitutes and the twenty slave law, conscription in the Confederacy also had a strong class bias. As one historian wrote of the latter provision, "the infamous twenty-slave law ... virtually excused planters outright from the draft."[30]

But perhaps the most startling feature of Confederate conscription was not the evasion, official opposition, ineptness, corruption, or class bias. To the modern observer the most startling feature of Confederate conscription may be that it gave rise, throughout the Confederacy, to roving bands of heavily armed men—10, 20, 50, 100, or 300 men to a band—who evaded enrollment officers by hiding out in hills, mountains, and swamps and sometimes living by means of banditry, preying on local townspeople. These conscription evaders often merged with similar bands of deserters from the Confederate army, who frequently absconded with significant quantities of arms and ammunition. Indeed, because draft evaders were legally classified as deserters, it is sometimes difficult to distinguish in the historical records what admixture of conscription evaders and army deserters were in these outlaw bands.

2 General Responses to Confederate Conscription

In his authoritative work on Confederate conscription, Moore concludes that "the enrolling officers were ignorant, incompetent, and corrupt." The Civil War archives certainly provide ample evidence to support that conclusion.

28 Sacher, "'Twenty-Negro,' or Overseer Law," 217, 273, 287.

29 A. B. Moore, *Conscription and Conflict in the Confederacy*, 127, 355.

30 Williams, *A People's History of the Civil War*, 77.

In January 1863, Colonel John L. Preston, superintendent of the Conscription Bureau, wrote to a member of Congress and referred to Confederate enrollment officers as "ignorant" officers who were "selected and sent out because they are useless." Preston continued, noting that "many of them can scarcely write their own names." The Confederate conscription system "is much worse than nothing," and so "utterly fruitless," he declared, that it would be "better [to] repeal the whole system and start anew."[31]

In Preston's view, the Bureau of Conscription did not become more effective as the war dragged on. A year and a half later, on November 29, 1864, Preston wrote to Seddon, saying, "The agencies employed in the conscript service are extremely defective in proper qualification as well for this as other grave duties, and I have found the utmost difficulty in getting the duties … performed even by the highest officers in the service."[32] These were the words of the superintendent of the Conscription Bureau, writing to the secretary of war, based upon significant firsthand experience.

Another Confederate officer wrote of the conscription system: "Seven months' experience has proved to me conclusively that out of the whole number employed in the conscription service in the State there are not twenty who possess one qualification for the office.… So long … as the present system of appointing disabled, stupid, and illiterate men to perform duty requiring activity, intelligence, and education, the result will be failure.… I have alluded only to the incompetency of the enrolling officers. I have not mentioned the frauds and corruption which mark their track everywhere."[33]

In December 1864, five months before the end of the war, H. W. Walter, the assistant adjutant general of the Conscription Bureau, said, point-blank: "I regard the conscript department in Georgia, Alabama, and Mississippi as almost worthless.… The department might be abolished without material injury to the service."[34]

The near-total ineffectiveness of the Confederate conscription apparatus is apparent in a letter from Preston to Seddon dated August 17, 1863, more than a year after the Confederate Congress passed the first conscription statute. In it, Preston makes three startling assertions:

- Fully fifteen months after the statute took effect, the Conscription Bureau, charged with carrying out the law, was not yet fully established *in any one of the eleven Confederate states*. Preston was blunt: "The agencies prescribed by

31 A. B. Moore, *Conscription and Conflict in the Confederacy*, 355; John L. Preston to William Porcher Miles, January 8, 1863, *OR*, ser. 4, 2:307.

32 Jno. S. Preston to James A. Seddon, November 29, 1863, *OR*, ser. 4, 2:866–67.

33 Wm. M. Browne to Preston, November 24, 1864, *OR*, ser. 4, 2:1049–50.

34 H. W. Walter to J. W. C. Watson, December 29, 1864, *OR*, ser. 4, 3:976.

law have not yet been brought into action." In many places, Preston reported, no one had appointed enrolling officers. Perhaps even worse, Preston stated that "even if successful in procuring these valuable adjuncts" (that is, enrolling officers), that would be "not sufficient for the ends proposed" (that is, conscription); he did not say why he believed that to be the case.

– The two exemption statutes that immediately followed enactment of the April 16, 1862, conscription statute had severely limited conscription. In fact, Preston said bitterly, "Had the exemption act been entitled an act to aid the enemy in diminishing the number of men in the Army … it could not more thoroughly have effected its purpose." Preston claimed that the exemption acts required the Conscription Bureau "to let drop ninety-nine through the expanded and broken meshes of the law while it takes in secure hold of one and lands him in the army." Even allowing for some exaggeration, this was a remarkable claim.
– The Conscription Bureau was meeting deeply felt and widespread popular opposition to the draft. Preston knew the origin of that opposition: conscription works "a minute interference with every civil, political, and domestic right of the citizen" in opposition to the most cherished beliefs of the people. Preston advised Seddon that putting conscription into effect required that troops be placed in every congressional district in the Confederacy.[35]

Evasion of Confederate conscription was so widespread that in many areas of the South evasion was the rule rather than the exception. On September 23, 1862, Florida governor John Milton wrote to Davis to report that conscription was failing in Florida: enrollment officers had told Milton that more men eligible for conscription were evading service than were being enrolled. Even the men being drafted were resisting and could not be brought to training camps without military force.[36]

On November 6, 1862, a Virginia enrollment officer wrote to Adjutant General and Inspector General Samuel S. Cooper to try to explain away "the discrepancy between the total number enrolled as liable to duty and the total number actually received into the [army]." This was the result of two major problems: "A great many of the men fail to report at the time and place appointed." And many of those who did report deserted en route to their training camp: "Four-fifths of these [deserters] have been substitutes deserting within twenty-four hours of their having been received as such."[37]

35 Preston to Seddon, August 17, 1863, *OR*, ser. 4, 2:723–25.
36 John Milton to Davis, September 23, 1862, *OR*, ser. 4, 2:92–93
37 J. C. Shields to Cooper, November 6, 1862, *OR*, ser. 4, 2:171–72.

Nineteen months later, in correspondence to Preston dated May 24, 1864, the inspector general of the Conscription Bureau, R. H. Chilton, reported an estimate of "quite 10,000 men liable to military duty absent from their commands and evading" conscription just from the district that included most of Mississippi.[38]

In December 1864, Assistant Adjutant General H. W. Walter summed up many of the problems with Confederate conscription: "First, the number of deserters is alarmingly large." Walter cited a recent group of 537 men conscripted in Mississippi and reported that 302 had deserted almost immediately. Second, "the number of skulkers … is also very large." And the "medical boards have in many instances done a disservice to the cause. The Board at Brookhaven, Mississippi, from April to October [1864] examined 1,125 men, and of these discharged 807 (over 70 per cent) as unfit for any duty." Walter describes one such medical evader whom he met: the man "looks strong enough to brain an ox with his knuckle and eat him afterward." Finally, the entire Conscription Bureau was larded with the "large numbers of supernumerary and unnecessary officers [that] are found everywhere." The net result of all of these problems, Walter said, is that "there are this day in Mississippi alone a sufficient number of deserters, skulkers, idle officers … and useless exempts to give victory to any army to which they are sent."[39]

∴

As a purely descriptive matter, the provision in the statute that allowed for substitution created rank economic discrimination in the Confederacy (just as it had in the Union). However, as a normative matter, the opinions of scholars about substitution vary considerably. To David Williams, author of *A People's History of the Civil* War (2005), allowing the rich to evade service while forcing the poor to the front was one of "the most offensive provisions of conscription.... At least fifty thousand southern gentlemen politely sidestepped military service by hiring substitutes. So few men of means served that soldiers in Richmond jokingly said it was easier for a camel to go through the eye of a needle than for a rich man to enter Camp Lee."[40]

Moore, who was writing eighty years earlier, however, was more understanding. The law, which was designed "to reserve skill and talent for service in the essential industries[,] … was not generally understood and accepted." The

---

38 R. H. Chilton to Preston, May 24, 1864, *OR*, ser. 4, vol. 3, sec. 1, 445–46.
39 Walter to Watson, December 29, 1864, *OR*, ser. 4, vol. 3, pt. 2, 976–79.
40 Williams, *A People's History of the Civil War*, 76.

problem, according to Moore, was that appreciating the substitute law required a "keener analysis" of the Confederacy's wartime needs than many poor people were capable of understanding. To Moore there was sense and logic to exempting the rich and financially successful; in fact, their very riches tended to show their merit: "Substitution would give an advantage, of course, to the individual who had accumulated wealth, but likely he had succeeded through his enterprise and efficiency and was just the sort of person whose services were needed at home."[41]

Regardless of whether substitution constituted excellent public policy (to serve essential industries) or terrible public policy (because of class bias), in practice, there were three broad, real-world problems with it.

Most of the men who served as substitutes did so only for the money they were paid and soon deserted. Assistant Secretary of War John A. Campbell estimated that two-thirds of all substitutes deserted. That is an astonishingly high figure—but it is dwarfed by the estimate of Inspector General Cooper, who estimated that 99 percent of substitutes deserted. The fourteen Confederate generals, writing together, agreed with Cooper. Their view was that more than 150,000 men had procured substitutes and 99 percent of them had simply deserted.[42]

Another major problem with substitution had to do with the certificates of exemption that a principal received when the army accepted his substitute. Thousands of forged exemption papers circulated, and thousands of draft-eligible men were able to evade service by means of these forged papers.[43] In other cases, wealthy men had procured exemption papers by bribing the officers involved, not by procuring a substitute.[44]

The third major problem with substitution was that—just as in the North—a group of dishonest brokers came into existence to help men evade conscription. These brokers were guilty of "habitual frauds" and "grossly criminal and mischievous" behavior. General George Lay complained to Seddon that the Conscription Bureau had been completely unable to address the problem or prosecute any of the corrupt brokers.[45]

---

41 A. B. Moore, *Conscription and Conflict in the Confederacy*, 27.

42 J. A. Campbell to E. Barksdale, July 21, 1863, *OR*, ser. 4, 2:655–56; Cooper to Braxton Bragg, August 6, 1863, *OR*, ser. 4, 2:695–96; H. D. Clayton et al. to Cooper, July 25, 1863, *OR*, ser. 4, 2:670–71.

43 T. P. August to S. W. Melton, November 7, 1863, *OR*, ser. 4, 2:939–40; Melton to Seddon, November 11, 1863, *OR*, ser. 4, 2:944–52; P. N. Paige, [Circular], September 14, 1863, *OR*, ser. 4, 2:808.

44 Gid. J. Pillow to Falconer, March 19, 1863, *OR*, ser. 4, 2:443.

45 George Lay to Seddon, June 9, 1863, *OR*, ser. 4, 2:582–83; Geo. W. Randolph to David, August 12, 1862, *OR*, ser. 4, 2:42–49.

3 New Conscription Statutes, December 1863–February 1864

By late 1863 it was clear that the Confederacy needed more soldiers. On November 26, 1863, Seddon proposed to Davis that the conscription law be rewritten to eliminate both the provision that allowed substitution and the categorical exemptions for trades and professions. Seddon was clear-eyed about the problems with substitution, which, he said, "has done more than any single measure to excite discontent ... among the soldiers." In addition to the general worthlessness of the substitutes themselves, Seddon acknowledged the class issue: "The fact that the wealthy could ... purchase liberation ... naturally produced among the less fortunate and poorer classes ... discontent."[46]

Less than two weeks later, Davis sent a message to Congress urging complete abolition of substitution and significant scaling back of categorical exemptions to conscription.[47] On December 28, 1863, Congress enacted a statute, all of one sentence long, abolishing substitution. In its haste, however, the members of Congress forgot one small matter: they had not made liable for the draft the *principals who had previously provided substitutes*. So, a week later, on January 5, 1864, Congress fixed its mistake by enacting another single-sentence statute, this time making principals liable for service.[48]

Five weeks later, Congress completed its rewrite of the Confederate conscription statute with a more comprehensive amendment. The statute of February 17, 1864, expanded the Confederate draft to all men aged seventeen to fifty.[49] It also provided that all soldiers aged eighteen to forty-five who were currently serving in the army would be retained for the duration of the war. The new draft law also made explicit that the previous, quite modest, provisions for COs would be retained but repealed most, but not all, categorical exemptions.[50] One exemption that was retained was that for state employees, a matter that the governors did not overlook.

---

46 Seddon report to Davis, November 26, 1863, *OR*, ser. 4, 2:996–97.

47 Davis to Congress, December 7, 1863, *OR*, ser. 4, 2:1024–49. For the section on conscription, see 1041.

48 An Act to prevent the enlistment or enrollment of substitutes in the military service of the Confederate States., December 28, 1863, in Matthews, *Public Laws of the Confederate States of America*, 1st Cong., 4th sess., 172; An Act to put an end to the exemption from military service of those who have heretofore furnished substitutes, January 5, 1864, in Matthews, *Public Laws of the Confederate States of America*, 1st Cong., 4th sess., 172.

49 This remains the most sweeping conscription statute ever enacted in the United States (quite aside from the question of whether the Confederacy was part of the United States).

50 An act to organize forces to serve during the war, February 17, 1864, §§ 1, 2, 4, and 10, in Matthews, *Public Laws of the Confederate States of America*, 1st Cong., 4th sess., 211–15.

The December 1863 statute abolishing substitution and, especially, the January 1864 statute, which made principals who had previously provided substitutes liable to be drafted, naturally met with opposition. Many of those principals who were now liable to be drafted had paid their substitutes quite handsomely to take their places at the front. These principals were quick to go to court to challenge the constitutionality of the post hoc revocation of what the principals, at least, had considered to be a settled matter. The Georgia Supreme Court decided one such early case in *Daly v. Harris*. Like many of the reported Confederacy conscription cases, this case was filed by several draftees seeking writs of habeas corpus. Daly was one of the draftees; Harris was the commandant holding them. The Georgia Supreme Court ruled against the conscripts, holding that substitution was "a gratuitous privilege" granted by Congress, which could repeal it at will.[51]

In addition to arguing that making principals liable for the draft retroactively was unconstitutional, the three draftees in *Daly* made a contract argument. They argued that they had, in effect, entered into a contract with the Confederate government—they had paid valuable consideration for the contracted-for military exemption in the form of payment to the substitute, and the government could not retroactively undo the contract. The court rejected the contract argument as well.

The North Carolina Supreme Court considered the issue in *Gaitlin v. Walton*. The court ruled, by a 2–1 majority, that Congress's post hoc revocation of substitution was constitutional. Justice William Horn Battle delivered the opinion of the court. The court relied on the duty of self-preservation that all governments share. Battle also cited early US Supreme Court cases, Alexander Hamilton in the *Federalist* No. 23, and James Madison in *Federalist* No. 41. Then Battle reached back to English common law and cited Sir William Blackstone and Sir Edward Coke for the following proposition: "The power and jurisdiction of Parliament ... are so transcendent and absolute, that it cannot be confined, either for causes or persons within any bounds. It hath sovereign and uncontrollable authority in making, confirming, enlarging, restraining, abrogating, repealing, reviving and expounding of laws, concerning matters of all possible denominations, ecclesiastical and temporal, civil, military, maritime or criminal."[52]

This was a peculiar argument for a Confederate justice to make. The citation to Blackstone and Coke is fair. However, an important reason that the thirteen colonies, including North Carolina, broke away from England in 1776 was that

51 Daly v. Harris, 33 Ga. Supp. 50 (1864).

52 Gatlin v. Walton, 60 N.C. 333, 4012 (1864).

the British Parliament had powers that the American colonists considered to be too vast. That is why, when they wrote the US Constitution in 1787, the framers were at pains to limit the powers of the US Congress. In addition, a principal reason that the Confederate States of America, including North Carolina, seceded from the United States in 1861 was that the US Congress had powers that the Confederate states considered to be too vast. In this context, it is strange indeed to read Battle's opinion citing Blackstone and Coke and making approving reference to the legislature's "uncontrollable authority."

4 Official State Exemptions

To a remarkable degree, Confederate governors actively impeded conscription by inventing thousands of bogus jobs for "state officials" so that the men in those jobs could illicitly evade the draft. Records of the Bureau of Conscription show that Governor Zebulon Baird Vance of North Carolina certified 14,675 men as being state officers, Governor Joseph Emerson Brown of Georgia certified 8,229 men, and Governor Charles Clark of Mississippi certified 4,000 men. Based on 1860 US Census figures for the draft-eligible male populations of these states, Vance designated fully 12.7 percent of North Carolina's draft-eligible population as conscription-exempt state officials; Brown so designated 7.4 percent of Georgia's draft-eligible population; and Clark so designated 5.6 percent of Mississippi's draft-eligible population.[53]

However, Bureau of Conscription records were haphazardly kept and incomplete, so the actual figures for these supposed state officials were almost certainly higher. For example, on April 9, 1864, Georgia's Governor Brown issued an official proclamation in which he described at some length a new set of categories of state officials that would be draft exempt. The list was, to say the least, comprehensive. Among the categories exempted were judges, court clerks, court reporters, deputy court reporters, sheriffs, unlimited numbers of deputy sheriffs, county clerks, unlimited numbers of deputy county clerks, jailers, deputy jailers, tax collectors, tax receivers, county surveyors, county coroners, constables, all agents involved in cotton export, all agents involved in importation of clothing, all staff and cadets at the Georgia Military Academy, all state militia officers, all employees of the state road department, all mayors, city council members, town council members, unlimited numbers of justices

53 A. B. Moore, *Conscription and Conflict in the Confederacy*, 94–95, 250–51; US Census Bureau, *Population of the United States in 1860* (Washington, DC: Government Printing Office, 1864), xvii.

of the peace, statehouse officers, statehouse clerks, and statehouse secretaries. It takes little imagination to understand what happened next in Georgia.[54]

5 Armed Resistance

Perhaps the most remarkable aspect of Confederate conscription was the widespread presence, throughout the war, of armed bands of conscription refusers. What follows is a review of some notable episodes involving these bands, as reflected in the Civil War archives. In considering the following examples, readers ought to keep four things in mind:

- The presence of these bands was not restricted to only a few geographical areas; they were present throughout the Confederacy.
- Many bands were large, well organized, and heavily armed.
- Many bands were comprised of a combination of conscription refusers and armed army deserters. (Note that the documents refer to conscription refusers variously as resisters, skulkers, tory conscripts, fugitive conscripts, outlying conscripts, and stragglers.) These resisters were not reluctant to defend themselves with weapons.
- The bands were often popular. One reason they were often hard to bring to justice was that they enjoyed broad public support. This is not difficult to understand: the Civil War was the bloodiest war in American history, and these men were often hiding out in areas populated with their relatives, neighbors, and friends.

On October 17, 1862, Major General Samuel Jones in the military Department of East Tennessee wrote to then–Secretary of War George W. Randolph urging the suspension of conscription in eastern Tennessee. Conscription, Jones said, was unenforceable because of widespread opposition. He wrote: "Many men have deserted their homes and gone to the mountains ... to avoid conscription.... Some of them, I am sorry to say, are reported as banding themselves together to resist authority.... I have sent out several detachments to kill, capture, or destroy these bands. They generally scatter, however, and conceal themselves in the mountains, where it is difficult to find them."[55]

On November 28, 1862, the enrollment officer for Austin County, Texas, wrote to his superior about a group of thirty-two local draft refusers and requested that troops be sent to aid in capturing them. Notably, he asked for regular army

54 *OR*, ser. 4, vol. 3, pt. 1, 345–47.

55 Jones's letter reflected that he had previously written to Jefferson Davis with the same recommendation. See Samuel Jones to George W. Randolph *OR*, ser. 1, vol. 16, pt. 2, 954–55.

troops, because the members of the local militia supported the conscription refusers. This reflected the local popularity of the draft refusers.[56]

A few weeks later, on January 3, 1863, the same enrollment officer wrote to his superior again, stressing the popularity of the resisters and the need for Confederate forces to address the issue: the local population "of my district and of the adjoining counties are in a state of open rebellion to our Government. They are holding meetings almost every day, and held a large meeting, consisting of about 600 persons, on the 31st of December, 1862.... The drafted men have continued to refuse to be sworn in ... [the enrollment officer] was assaulted and driven from the place appointed for him for said purpose; also a friend of his was actually mobbed, being beaten with sticks, iron bars, &c."[57]

On January 5, 1863, Governor Vance of North Carolina wrote to Seddon to report on resistance to conscription throughout his state. Conscription evaders, said Vance, were joining armed army deserters. Vance believed that he would be able to make arrests and prevent armed bands of draft evaders from forming in the central and eastern portions of North Carolina. "But in the mountains of the west [of North Carolina] the case is different. The enforcement of the conscript law in East Tennessee has filled the mountains with disaffected desperadoes of the worst character, who joining with deserters from our army, form very formidable bands of outlaws, who hide in the fastnesses, waylay the passes, rob, steal, and destroy at pleasure. The evil has become so great that travel has been almost suspended through the mountains."[58]

Throughout the Civil War, the so-called Quaker Belt of central Piedmont North Carolina—consisting of fifteen counties, including Yadkin, Guilford, Randolph, and Chatham—was a relative hotbed of anticonscription sentiment. Quakers and Moravians, members of traditional peace churches, had settled the area in the mid-eighteenth century; thus, both antiwar and antislavery sentiment were strong the area.[59]

Correspondence from Yadkin County tells of a shoot-out between a squad of army officers and a well-armed group of draft evaders. The letter, from Lieutenant Colonel W. A. Joyce to Vance, was dated February 16, 1863:

> It is my painful duty to inform you of a most melancholy affair which occurred in this county last Thursday. I sent out a squad of 14 men in

56 A. J. Bell to J. P. Lewellan, November 28, 1862, *OR*, ser. 1, 15:887.

57 Bell to Lewellan, January 3, 1863, *OR*, ser. 1, 15:925.

58 Zebulon Baird Vance to Seddon, January 5, 1863, *OR*, ser. 1, 18:821–22.

59 William T. Auman, *Civil War in the North Carolina Quaker Belt: The Confederate Campaign against Peace Agitators, Deserters, and Draft Dodgers* (Jefferson, NC: McFarland, 2014), 1–10.

> search of conscripts. They found a number of them (now supposed to be between 20 and 30) lodged in a school-house and attacked them. The conscripts were well armed and resisted most stubbornly. They shot through the windows and cracks at our men, and killed two of the best citizens in this county—James West, a magistrate, and John Williams, a most excellent man. These men were shot dead in their tracks. Two of the conscripts were killed and the balance escaped, though two of them are seriously wounded.… If they do not leave the country they will rally the crowd (as I understand some of them are trying to do) and fight again. They are much better armed than the militia, having been preparing for it for months.[60]

Three days later, on February 19, Robert F. Armfield (the records do not identify his position or status) wrote about the same event. One thing that stands out in this account—aside from the violence of the draft refusers—is the description of the broad support refusers had among residents of the county:

> We have had a startling occurrence in this county.… I mean the murder of two of our best citizens, magistrates of the county, by a band of deserters and fugitive conscripts. The circumstances are these: There has been a strong feeling against the conscript law … ever since its passage … [and] when it came time for them to go perhaps nearly 100 in this county took to the woods, lying out day and night to avoid arrest; and although the militia officers exerted themselves with great zeal, yet these skulkers have always had many more active friends than they [the officers] had and could always get timely information of every movement to arrest them and so avoid it. The militia officers have been able to arrest very few of them.

So that his point about broad public support for the draft evaders could not be missed, Armfield repeated it: "The section of the country in which they lurk is so disloyal (I grieve to say it), and the people so readily conceal the murderers and convey intelligence to them, that it will be exceedingly difficult to find them."[61]

On May 11, 1863, Vance issued a proclamation aimed at "the many persons in [North Carolina] who incite and encourage these deserters and harbor and conceal these misguided men at home." Vance understood that the armed bands of conscription resisters could not and would not have been as

---

60 W. A. Joyce to Vance, February 16, 1863, *OR*, ser. 1, 18:880–81.

61 R. F. Armfield to "Dear Sir," February 19, 1863, *OR*, ser. 1, 18:886–87.

successful in evading the law had it not been for the widespread support they received from the citizenry. Thus, Vance's proclamation targeted these aiders and abettors, instructing them "to desist from such base, cowardly, and treasonable conduct" lest they "subject themselves to indictment and punishment in the civil courts … [and] the everlasting contempt and detestation of all good and honorable men."[62]

On June 24, General Lay of the Conscription Bureau wrote to Cooper about the evaders of conscription who were banding together with armed deserters: "The evil of desertion from the Army, with the determination to avoid and even resist future service, appears of late to be on the increase, and it is more difficult to deal with that [because] the soldiers now bring with them Government arms and ammunition, banding among themselves and with evaders of conscription. The enrolling officers are sometimes shot by them and the community kept in terror. This state of things exists more or less in each of our Atlantic states."[63]

The day before, Colonel J. J. Siebels had written to Assistant Secretary of War John Campbell about the situation in Alabama: "In several counties of this State there are, unfortunately, quite a considerable number of troops, or rather men fit for troops, hiding and skulking in the hills and mountains, where it would require a force nearly equal to themselves to ferret them out and arrest them."[64]

On July 14, 1863, General Gideon J. Pillow, who, for part of the Civil War operated a kind of separate conscription apparatus in competition with the official Bureau of Conscription's, wrote to Brigadier General William MacKall about draft refusal in the mountains of northern Alabama. What was striking in Pillow's account was the degree to which he believed his own men were unable to rein in the forces of the draft evaders: There "has been great increase of stragglers, who now infest the mountains of North Alabama, and who are so formidable in numbers and so thoroughly armed and organized as to hold the citizen population in terror of their displeasure. They rob, burn, and murder the unarmed population of the country with impunity. As a general thing, they have been joined by the tory population, who were liable under the conscription law. They thus constitute a force with which the officers of this bureau can do nothing unaided by a supporting force of cavalry."[65]

62 Ibid., vol. 51, pt. 2, 706–8.
63 George Lay to Cooper, June 24, 1863, *OR*, ser. 4, 2:607–8.
64 J. J. Siebels to John Campbell, June 23, 1863, *OR*, ser. 4, 2:603–4.
65 Gideon J. Pillow to William MacKall, July 14, 1863, *OR*, ser. 4, 2:638.

On August 4, Governor John Shorter of Alabama wrote to General Howell Cobb after the governor sent troops to capture draft evaders:

> For some time past a band of deserters and outlying conscripts have been infesting the lower part of Henry County, upon the border of Florida, having for the place of their concealment the swamps of the Chipola River and its tributaries.... I ordered Captain Armstrong [to use] such force as he might think necessary, from his and Captain Chisholm's company ... in making arrests.
>
> Some 6 or 7 men, liable to Confederate service, were captured and sent back under escort for safe-keeping, when the escort was attacked by a superior force in ambush, and the prisoners rescued and 1 of the escort seriously wounded.[66]

Less than a week later, D. C. Pearson, the enrollment officer for North Carolina's Tenth Congressional District, wrote to his superior about a force of draft evaders and deserters in Cherokee and Henderson counties that he estimated to be between three hundred and five hundred men: "It will be impossible for me to execute the conscription in certain counties without the aid of some military force."[67]

Three of the now-familiar themes were present in the August 7 report of John D. Ashmore, the chief enrollment officer for the Fifth Congressional District of South Carolina: the combination of draft evaders and deserters, the armed nature of the combined gangs, and the general support the gangs enjoyed from the surrounding population. The report is addressed to Major C. D. Melton, the commandant of conscripts for South Carolina:

> The conscription of persons liable to military duty ... has been nearly perfected, but a large number ... particularly from Spartanburg and Pickens, have not reported for duty and are now declared deserters by law. In addition to these a very large number of deserters from different commands in the armies of the Confederate States have been reported for arrest, making in the aggregate 502 deserters.... This large body of deserters are spread over a large frontier border of over 150 miles, every foot of which is a mountain country and much of it almost inaccessible. They are banded together in tens, twenties, and thirties, are bold, defiant, and even threatening.... They have spies and signals to indicate the approach of strang-

66 Jno. Gill Shorter to H. Cobb, August 4, 1863, *OR*, ser. 1, vol. 28, pt. 2, 273–74.

67 D. C. Pearson to J. C. McRae, August 10, 1863, *OR*, ser. 4, 2:734.

> ers, and are armed, many of them with Government arms with which they deserted, and are all with weapons such as fowling pieces, revolvers, &c.… They are unfortunately sustained in their conduct by many persons who have been heretofore regarded as good and loyal citizens.[68]

On August 25, 1863, Melton wrote to Preston, head of the Conscription Bureau, about the same mountainous area of Pickens, Spartanburg, and Greenville Counties. Melton described how the conscription evaders were combining with armed deserters and explained that the resistance was well organized. He was especially explicit about the reasons that these armed bands of resisters were so successful in evading the law—everyone supported the resisters: "There can now be found few families which have not a husband, a son, a brother, or kinsman, a deserter in the mountains.… All are ready to encourage and aid the efforts of those who are avoiding duty, and to refuse information and thwart and even resist those who seek to make arrests."[69]

On September 2, 1863, Lay wrote to Preston, noting that the Confederacy's original notion of capturing "deserters and skulkers by the use of force was doubtless based on the supposition that such characters would be found lurking about singly, unarmed, acting in no concert, and supported by no public opinion." However, Lay continued, evasion had taken "a very different and more formidable shape and development than could have been anticipated"; indeed, every one of those earlier assumptions had been wrong:

> In Cherokee County a large body of deserters (with whom I class also those in resistance to conscription) have assumed a sort of military occupation, taking a town, and … in Wilkes County they are organized, drilling regularly.… These men are not only determined to kill in avoiding apprehension (having just put to death yet another of our enrolling officers), but their esprit de corps extends to killing in revenge as well as in prevention of the capture of each other. So far they seem to have had no trouble for subsistence. While the disaffected feed them from sympathy, the loyal do so from fear. The latter class (and the militia) are afraid to aid the conscript service lest they draw revenge upon themselves and their property.[70]

On September 18, 1863, Brigadier General H. E. McCulloch wrote from Bonham, Texas, that "deserters and conscripts … have gone into the brush in large

68 John D. Ashmore to C. D. Melton, August 7, 1853, *OR*, ser. 4, 2:771–73.
69 C. D. Melton to Preston, *OR*, ser. 4, 2:769–70.
70 Lay to Preston, September 2, 1863, *OR*, ser. 4, 2:783–85.

numbers." A month later, McCulloch, still in Bonham, wrote to General John B. Magruder: "There are not less than 1,000 deserters, from the army, conscription, and the militia, in the woods, ready to take to the brush... . They keep every road picketed [watched by guards or sentries] that goes into their vicinity so perfectly that not a man, woman, or child goes near them that they don't know it; they have sympathizers all through the country."[71]

On March 4, 1864, from his headquarters in west Florida, Brigadier General Alexander Asboth wrote of a band of one hundred deserters and conscription refugees that "entered Pikesville, Alabama, on the 11th of February, killing, wounding, and capturing several of the Confederate garrison." He stated that "Col. W. Miller, of the rebel conscript bureau for the Southern District of Florida and Alabama, intends to make a raid on East Bay with 300 cavalry, for the purpose of gathering up all deserters and refugees secreted in the woods and abandoned farms."[72]

6 The March 1865 Statute

By January 1865, the exigencies of the war had forced Southern military and civilian leaders to contemplate a truly radical notion: arming slaves to fight for the Confederacy. One of the proponents of this step was General Robert E. Lee. He discussed this possibility in a January 1865 letter to Andrew Hunter. (Just before the war, in October 1859, Hunter had been the lead prosecutor of John Brown and his men for the raid on the Harper's Ferry arsenal.) Lee explained to Hunter that while he (Lee) certainly did not *want* to arm slaves, he believed that there could be no choice if the Confederacy was to survive.[73]

Not all of Lee's colleagues agreed with him. General Patton Anderson viewed the proposal as a "monstrous proposition." He wrote, in a letter marked "Confidential," "I will not attempt to describe my feelings on being confronted by a project so startling in its character—may I say so revolting to Southern sentiment, Southern pride, Southern honor."[74]

Cobb, writing to Seddon, was aghast. While acknowledging that Confederate conscription had not seen much success ("there are many liable to conscription who have not been reached, and ... I fear never will be reached"), Cobb

71 H. E. McCulloch to Edmund P. Turner, September 18, 1863, *OR*, ser. 1, vol. 26, pt. 2, 236; McCulloch to John B. Magruder, October 21, 1863, *OR*, ser. 1, vol. 26, pt. 2, 344–45.

72 Alexander Asboth to Charles P. Stone, March 4, 1864, *OR*, ser. 1, vol. 35, pt. 2, 4–6.

73 Tony Horwitz, *Midnight Rising: John Brown and the Raid That Sparked the Civil War* (New York: Picador, 2011), 201; Robert E. Lee to Andrew Hunter, January 11, 1865, *OR*, ser. 4, 3:1012–13.

74 Patton Anderson to L. Polk, January 14, 1864, *OR*, ser. 1, vol. 52, pt. 2, 598–99.

was still not ready to entertain the notion of arming the enslaved: "I think the proposition to make soldiers of our slaves is the most pernicious idea that has been suggested since the war began.... You can't keep white and black troops together, and you can't trust negroes by themselves.... Use all the negroes you can get, for all the purposes for which you need them, but don't arm them."[75]

The skeptics did not prevail. On March 13, 1865, about a month before the Civil War ended, the Confederate Congress passed a statute permitting the arming of the enslaved to fight in the Confederate army. The statute provided that the new soldiers be organized into companies, battalions, regiments, and brigades. In an outburst of truly eleemosynary spirit, the statute provided: "That while employed in the service the said troops shall receive the same rations, clothing, and compensation as are allowed to other troops in the same branch of the service." Lest anyone get the wrong idea that military service might lead to eventual freedom for the enslaved or for their families, another section made clear "that nothing in this act shall be construed to authorize a change in relation which the said slaves shall bear toward their owners."[76]

The political, ideological, and cultural magnitude of this step cannot be overstated. For centuries, the concept of an armed slave had been a nightmare, an anathema, to Southern whites. John Brown was hanged because he had sought to arm slaves—indeed, this was the very reason he and his comrades had raided the Harper's Ferry arsenal.[77]

Criminal laws throughout the South made the arming of slaves a felony, as is seen in many reported cases in the antebellum period. For example, *State v. Hannibal* was a criminal case brought against an enslaved man who had been armed by his owner to protect the owner's property. In convicting the enslaved man, the court emphasized that arming slaves is prohibited in any circumstance: "The prohibition is expressed in the strongest and broadest terms, and rendered emphatic by the concluding words, *upon any pretense whatsoever*."[78]

Southern civil law reflected the same abhorrence of armed slaves. *Laparouse v. Rice* was a civil case in which Laparouse sued Rice for monetary damages because Rice had killed Laparouse's slave; Laparouse wanted to recover the dollar value of his slave property. But Laparouse lost the case, because his slave was armed when he was killed, and it was illegal for a slave to be armed and legal to kill an armed slave.[79]

---

75 Cobb to Seddon, January 8, 1865, *OR*, ser. 4, vol. 3, pt. 2, 1009–10.

76 An Act to increase the military force of the Confederate States, March 13, 1865, §§ 2, 3, and 5, *OR*, ser. 4, vol. 3, pt. 2, 1161.

77 Horwitz, *Midnight Rising*, 70–72.

78 State v. Hannibal, 51 N.C. (6 Jones) 57 (1858), emphasis in the original.

79 Laparouse v. Rice, 13 La. Ann. 567 (1858).

However, coming as it did just weeks before the end of the Civil War, the March 1865 conscription statute had little real-world effect.

7 Legal Cases Pertaining to the Conscription Statute

Since the Supreme Court of the Confederacy never came into existence, such constitutional challenges to conscription as there were in the South were decided by several state supreme courts. In relatively short order, the state supreme courts of Texas, Georgia, and Alabama ruled that Confederate conscription was constitutional; the Supreme Court of Virginia agreed but did not rule until later in the war.

The first Confederate state supreme court decision on the constitutionality of the conscription statute was the Texas case, *Ex parte Coupland.* Like the other cases discussed in this section, *Coupland* came to the Texas court on a writ of habeas corpus from an unwilling conscript who claimed that the conscription statute was unconstitutional. A two-justice majority (Chief Justice Royall T. Wheeler and Associate Justice George F. Moore) found the Confederate conscription statute to be constitutional; Associate Justice James H. Bell filed a dissenting opinion. Perhaps because this was the first state supreme court case addressing the constitutionality of the conscription statute, both sides presented detailed arguments for their opposing conclusions. Their arguments rehearsed the familiar discussion of the delegated powers doctrine.[80]

Moore delivered the opinion of the court. He started out by stating the issue before the court: the obligation of "the Confederate government to show that the authority [to conscript] assumed by the Confederate States is sanctioned by an expressly delegated power; or … is necessary and proper for the carrying into effect an expressly delegated power." The court quickly concluded that, "the power … to raise and support armies is expressly granted by the constitution to the congress of the Confederate States." Next, the court acknowledged that Coupland's argument was that Congress "can only do this by voluntary enlistments" and not by involuntary conscription. However, the court disagreed: "The grant of power to make war carries with it by necessary implication … the right to demand compulsory military service from the citizens."[81]

Like the majority, Bell, the dissenter, identified the question as one hinging on the delegated powers doctrine and specifically cited the US Supreme Court case of *McCulloch v. Maryland* for the proposition: "The government is one of limited powers, and whoever asserts the constitutionality of a particular power must show it. Is the power expressed in the constitution?" Bell acknowledged

80 *Ex parte* Coupland, 26 Tex. 386 (1862).

81 Ibid., 392–93.

that "the majority of the court say, the power to raise and support armies is expressly granted. This, of course, is not denied." However, he goes on to say, "The question is not whether or not the power to raise and support armies is expressly granted; the question is, whether or not the power to raise armies *by conscription* is expressly granted." To this latter question, Bell answers a resounding no: "It cannot be contended that conscription is a means necessary to carry into effect the power to raise armies. Such a proposition, besides being contrary to reason, would find its contradiction in the facts of history, and in the events which are transpiring before our eyes. The government of the United States has always kept an army on foot, maintained an honorable contest with Great Britain in 1812, and planted her banners on the walls of the capital of Mexico in 1847, but has never raised troops by conscription."[82]

Shortly after the Texas court handed down the decision in *Coupland*, the Georgia Supreme Court upheld conscription in *Jeffers v. Fair*. The court's reasoning was familiar: "The Congress shall have the power to raise armies.... Language could not express a broader, more general grant of a specific power. We look in vain for the limitation to voluntary enlistment as a means." And this: "Compulsory enrollment is a proper incident of the power to raise armies."[83]

The Alabama Supreme Court upheld the constitutionality of conscription in *Ex parte Hill*. The Virginia Supreme Court did not rule on the constitutionality of conscription until 1864, in *Burroughs v. Peyton*. By that time, the Confederate Congress had abolished substitutes and made principals liable for service, so the Virginia Supreme Court addressed in the same case both the constitutionality of the original April 1862 conscription statute and the drafting of principals who had previously provided substitutes. The court found that conscription itself was constitutional because of the express power to raise and support armies. The court also found that making principals liable for service—even after they had provided substitutes—was not unconstitutional, again on the familiar ground that Congress was not violating a binding contract that it had made with a conscript.[84]

Different state courts sometimes came to different legal conclusions on the same question, a problem recognized by Confederate officials at the time. In December 1864, Seddon wrote: "Unfortunately the Confederate States have not organized their judiciary so as that questions arising under the Constitution

82 Ibid., 417–19, emphasis his.
83 Jeffers v. Fair, 33 Ga. 347 (1862).
84 *Ex parte* Hill, 38 Ala. 429 (Jan. 1863); Burroughs v. Peyton, 57 Va. (16 Gratt) 470 (1864).

and laws of the Confederacy shall be finally settled by their own tribunals.... This has compelled the [War] Department to submit to decisions by [state] courts that are not under the [Confederate] Constitution ... [and] in different States diverse opinions upon the same questions have been more than once expressed by their highest courts."[85]

Not only did different courts sometimes give inconsistent decisions on the same issue; sometimes the same court gave inconsistent decisions on the same issue. In January 1863, the Georgia Supreme Court heard two cases that presented the identical issue in every regard. (The cases are confusing, because both bear the same name, *Ex parte Hill*, because the cases were named for the local conscription officer, L. H. Hill. However, these cases were brought to the court by two conscripts, Asa Willis and W. B. Armistead.)[86]

Both cases were brought as requests for writs of habeas corpus; the same Georgia Supreme Court decided both cases in January 1863. In both cases the court addressed the same issue: Does a state supreme court in the Confederacy have jurisdiction to hear and decide a habeas corpus case brought by a would-be conscript? In one case, the Georgia Supreme Court decided that it *did* have jurisdiction in Confederate conscription cases.[87] In the other case, the court came to the opposite conclusion: it did *not* have jurisdiction.[88] In neither case did the justices acknowledge the obvious inconsistency, much less try to address or resolve it.

Since the 1970s, a central tenet of scholars of critical legal studies has been that the law is indeterminate. By *indeterminate*, these scholars mean that the rules of law taught to students in law schools don't really exist in the real world. Thus, the outcomes of legal cases are random, and predicting case outcomes by reference to those rules of law is impossible. In that sense, the Georgia Supreme Court's pair of *Ex parte Hill* rulings is a poster child for the critical legal studies view of the law as indeterminate: two diametrically opposite holdings by the same court at the same time on substantially identical facts.[89]

∴

85 Seddon to William L. Brandon, December 25, 1864, *OR*, ser. 4, vol. 3, pt. 2, 965–66.

86 The case brought by Armistead was *Ex parte Hill*, 38 Ala. 429 (1863). Asa Willis was joined by E. P. Johnson and Calvin Reynolds in *Ex parte Hill*, 38 Ala. 458 (1863)

87 *Ex parte Hill*, 38 Ala. 429 (1863).

88 *Ex parte Hill*, 38 Ala. 458 (1863).

89 See, for example, John Batt, "American Legal Populism: A Jurisprudential and Historical Narrative, Including Reflections on Critical Legal Studies," *Northern Kentucky Law Review* 22, no. 48 (1995): 651–762. Batt explains that "Critical Legal Studies stresses the law is largely indeterminate.... Indeterminacy in this context refers to our inability to predict case outcomes by reference to a particular rule of law. This idea is subversive to traditional views of the utility of 'legal reasoning'" (658).

Immediately after the Confederate Congress abolished substitution and made principals liable for service, Edward S. Walton made a request for a writ of habeas corpus to the North Carolina Supreme Court; he claimed that what Congress had done was unconstitutional. Because the North Carolina Supreme Court was not sitting at the time, Chief Justice Richmond M. Pearson heard the case alone in chambers. Pearson agreed with Walton and ruled that the change in law was unconstitutional.[90]

The Confederate government duly noted Pearson's decision in the Walton case. On February 3, 1864, Jefferson Davis sent a message to Congress urging it to suspend the writ of habeas corpus. Less than two weeks later, Congress did just that. One of the specific circumstances for which the writ was suspended was for all cases involving "attempts to avoid military service."[91] The suspension of the writ remained in force until July 1, 1864. However, Congress's suspension of the writ did not deter Pearson. While the writ was suspended, Pearson granted a writ of habeas corpus to a would-be conscript in the case *In re Roseman.* Pearson used the following, rather tortured, logic to evade the clear meaning of the statute suspending the writ. He reasoned that the word *avoid* really meant to improperly evade; since Roseman was claiming to be *entitled* to an exemption, if correct, he would not be improperly evading conscription. Hence, Pearson reasoned, the suspension of the writ did not apply to Roseman's situation.[92]

Pearson did the same thing in several other habeas corpus cases. In each of the following cases the chief justice issued the writ after Congress had suspended it: *Russell v. Whiting, In re Wyrick, In re Bradshaw,* and *In re Cain.*[93] And Pearson was not alone. The Georgia Supreme Court granted writs of habeas corpus in multiple conscription cases during the period in 1864 when the

90 Pearson *ex parte* Walton, 60 N.C. (Win.) 425 (1864). Note, however, that as we saw earlier, when the full North Carolina Supreme Court had the opportunity to review Chief Justice Pearson's ruling, it came to the opposite conclusion and ruled that the post hoc revocation of exemption by reason of substitution was perfectly constitutional.

91 "Message of Jefferson Davis to House and Senate, February 3, 1864," in *Journal of the … Congress of the Confederate States,* 1st Cong., 4th sess., 699–701; An Act to suspend the writ of habeas corpus in certain cases, in Matthews, *Public Laws of the Confederate States of America,* 1st Cong., 4th sess., 187–89, and § 5, 188.

92 *In re* Roseman, 60 N.C. (Win.) 443, 450, 454, 463, 466 (1864).

93 Russell v. Whiting, 60 N.C. 463 (1864); *in re* Bradshaw, 60 N.C. 454 (1864); *in re* Wyrick, 60 N.C. 450 (1864). Chief Justice Pearson references *in re Cain* on page 466 of his opinion in *Russell.*

Confederate Congress had suspended the writ: *Gates v. McManus*, *Connell v. Leonard*, and *Scott v. Lazenby*.[94]

The Alabama Supreme Court granted the writ in several cases while the writ was suspended, including *Ex parte Cain* and *Ex parte Barton*. In *Ex parte Lockhart*, the Alabama Supreme Court ordered the release of a soldier from the Confederate army because he claimed that he had just become a letter carrier, a then-exempted position. The Alabama Supreme Court granted the writ in *Lockhart* even though the Confederate Congress had suspended the writ. The case report contains was no mention of that inconvenient little fact.[95]

Pearson was, however, sui generis. Soon after Confederate conscription had been enacted in April 1862, he developed a reputation throughout the Confederacy for his opposition to conscription and his propensity to rule in favor of conscripts and soldiers who came before him on writs of habeas corpus. One contemporaneous writer referred to Pearson "as the friend and protector of deserters and conscripts" and said that "the impression has gone abroad" that Pearson meant to discharge every man who appeared before him on a writ of habeas corpus.[96] As a result, conscripts and soldiers from throughout the Confederacy would make their way to Pearson in North Carolina in an effort to escape military service. By June 1863, Justice Pearson had heard no fewer than 27 such habeas corpus petitions, and by early the following year the number had grown to 115 to 120.[97]

8 Conclusion

Using some of the same criteria discussed at the end of chapter 2, conscription in the Confederacy was not a success. Although the Civil War was a time of great turmoil in both the North and the South, Confederate conscription

94 Gates v. McManus, 33 Ga. Supp. 67 (1864); Connell v. Leonard, 33 Ga. Supp. 58 (1864); Scott v. Lazenby, 33 Ga. Supp. 134 (1864).

95 *Ex parte* Cain, 39 Ala. 440 (1864); *ex parte* Barton, 39 Ala. 452 (1864); *Ex parte* Lockhart, 39 Ala. 450 (1864).

96 Memory F. Mitchell, *Legal Aspects of Conscription and Exemption in North Carolina, 1861–1865* (Chapel Hill: University of North Carolina Press, 1965), 46. Mitchell cites Daniel G. Fowle to Richard M. Pearson, May 22, 1863, in Richmond Pearson Papers, Southern Historical Collection, Wilson Library, University of North Carolina at Chapel Hill.

97 Jennifer Van Zandt, "Confederate Conscription and the North Carolina Supreme Court," *North Carolina Historical Review* 72, no. 1 (January 1995): 54, 58; A. B. Moore, *Conscription and Conflict in the Confederacy*, 46, 71; Ths. Bragg to Seddon, March 29, 1864, OR, ser. 4, vol. 3, pt. 1, 256.

statutes did the opposite of "insur[ing] domestic tranquility." Instead, they triggered, stimulated, and encouraged deep and widespread opposition, including armed opposition.

Confederate conscription also did not "establish justice." Substitution ensured that the rich need never be drafted and that only the poor would die at the front. The twenty slave law not only further entrenched the class bias that resulted from substitution but added to it a large dose of overt racism.

The quality of Confederate conscription officers was obvious both to the men who tried to run the system at the time and to later scholars. William M. Browne, commandant for Georgia, called enrollment officers ignorant, stupid, and lazy and then went on to add, "I have not mentioned the frauds and corruption which mark their track everywhere." Writing in 1924 and echoing Browne, Albert Burton Moore concluded that "enrolling officers were ignorant, incompetent, and corrupt."[98]

In the Confederacy, as in the Union, conscription was deeply unpopular, and many people viewed it as un-American. Moore concluded that conscription was "contrary to the spirit of the people ... [and] there was always strong opposition to it."[99] (The hypocrisy of white Southerners was patent: they fussed that conscription was enforced service or involuntary servitude even as they sought to build a nation and culture predicated on chattel slavery.)

A broad point here should not be missed. Based on colonial era and postindependence state constitutions, on the debates attending the ratification of the Constitution, and on state and federal statutes from both the colonial era and the period after independence, it is clear that—whatever jingoism may be present in the American character—antiwar and antidraft sentiment also runs deep in the American psyche. Just as it had in the North, that antidraft sentiment manifested itself in many ways throughout the South for the entire period that the Confederacy had conscription in effect. And, critically, the responses of individuals to conscription (widespread evasion) and of the community (opposition to the draft and general support for the evaders) was similar in the South and the North.

Some aspects of Confederate opposition to conscription have absolutely no modern parallel. Large bands of heavily armed draft resisters were present everywhere; they not only evaded service themselves but waged active war on enrollment officers, and it took companies or even battalions of soldiers to oppose them. And governors contrived to undermine conscription, as did some judges.

---

98 Browne to Preston; A. B. Moore, *Conscription and Conflict in the Confederacy*, 355.

99 A. B. Moore, *Conscription and Conflict in the Confederacy*, 354.

CHAPTER 4

World War I

On April 2, 1917, President Woodrow Wilson addressed a joint session of Congress to seek a declaration of war against imperial Germany. The thrust of Wilson's speech was that the United States was already at war with Germany; Germany's policy of waging submarine warfare against international shipping, including both civilian vessels and ships of the US Navy, made this true. Wilson told Congress that a formal declaration of war would simply be recognizing in law a reality that already existed. In a notable sentence midway through his address, Wilson explained that, in addition "to the armed forces of the United States already provided for by law in case of war," he was seeking "at least 500,000 men who should, in my opinion, be chosen upon the principle of *universal liability to service*, and also the authorization of subsequent additional increments of equal force so soon as they may be needed and can be handled in the training."[1] It was widely understood at the time that Wilson was referring to conscription.

Four days later, on April 6, 1917, Congress declared war on Germany. On that date, the US Army stood at 127,588 men.[2] This was 8.7 times the size of the US Army at the start of the Civil War, although the population of the United States had grown to about 3.6 times its earlier size during the intervening fifty years (to 93.4 million people in the 1910 census).[3]

The next day, Secretary of War Newton D. Baker submitted the administration's conscription bill to Congress.[4] The military had drafted the bill the

1 "President Calls for War Declaration, Seeks Navy, New Army of 500,000 Men," *New York Times*, April 3, 1917, emphasis added.

2 This figure appears in US Office of the Provost Marshal General, *Second Report of the Provost Marshal General to the Secretary of War on the Operation of the Selective Service System to December 20, 1918* (Washington, DC: Government Printing Office, 1919), 227, table 80 (hereafter, *PMG 2d*).

3 US Census Bureau, *Sixteenth Census of the United States. Vol. 1: Population, General Report and Analysis* (Washington, DC: Government Printing Office, 1913), 21.

4 On the World War I draft, see John Whiteclay Chambers II, *To Raise an Army: The Draft Comes to Modern America* (New York: Free Press, 1987); Norman Thomas, *The Conscientious Objector in America* (New York: B. W. Huebsch, 1923); and Christopher Capozzola, *Uncle Sam Wants YOU: World War I and the Making of the Modern American Citizen* (Oxford: Oxford University Press, 2008). On the World War I draft in the context of the broader peace movement, see John Whiteclay Chambers II, ed., *The Eagle and the Dove: The American Peace Movement and United States Foreign Policy, 1900–1922* (Syracuse, NY: Syracuse University Press, 1991);

 | DOI:10.1163/9789004546684_006

previous week under the direction of General Enoch H. Crowder.[5] The same day, the House Military Affairs Committee began ten days of secret hearings on the bill during which the members heard from Baker, Crowder, and the army's general staff. Yet so hastily had the bill been prepared that, on the first day of the hearings, no printed copies of the proposed bill were available to committee members, and even the few typewritten copies contained major errors, with one section omitted entirely and another section appearing twice.[6]

Under the new statute, Crowder went on to become the only provost marshal general of the Selective Service during World War I (just as General James Barnet Fry had been the only provost marshal general during the Civil War). Crowder produced three book-length annual reports on the work and functioning of the Selective Service. In addition, in 1920, Crowder published an unofficial, and rather more personal, memoir about the draft, *The Spirit of Selective Service*. Like Fry's post–Civil War report, Crowder's multiple volumes are enormous compendiums of highly detailed information that have formed the basis of much subsequent scholarship concerning the World War I draft apparatus.[7]

1 The Statute

On May 18, 1917, Congress passed An Act To authorize the president to increase temporarily the Military Establishment of the United States; this was the nation's World War I draft law. War had been raging in Europe for almost two

David M. Kennedy, *Over Here: The First World War and American Society* (Oxford: Oxford University Press, 2004); Scott H. Bennett and Charles F. Howlett, eds., *Antiwar Dissent and Peace Activism in World War I America* (Lincoln: University of Nebraska Press, 2014); Gilbert C. Fite and H. C. Peterson, *Opponents of War: 1917–1918* (Seattle: University of Washington Press, 1957); Michael Kazin, *War against War: The American Fight for Peace, 1914–1918* (New York: Simon & Schuster, 2017); Charles Chatfield, *For Peace and Justice: Pacifism in America, 1914–1940* (Knoxville: University of Tennessee Press, 1971); and Joshua Kastenberg, *To Raise and Discipline an Army: Major General Enoch Crowder, the Judge Advocate General's Office, and the Realignment of Civil and Military Relations in World War I* (DeKalb: Northern Illinois University Press, 2017). Frederick C. Giffin, *Six Who Protested: Radical Opposition to the First World War* (Port Washington, NY: Kennikat Press, 1977), includes profiles of Eugene Debs, Emma Goldman, "Big" Bill Haywood of the Industrial Workers of the World, and others.

5 Chambers, *To Raise an Army*, 153; Enoch H. Crowder, *The Spirit of Selective Service* (New York: Century, 1920), 122.

6 *Increase of Military Establishment of 1917: Hearings Before the Committee on Military Affairs, House of Representatives*, 65th Cong., 1st sess., 25–26, 79–81 (1917).

7 Crowder, *Spirit of Selective Service*, 122–23; D. Kennedy, *Over Here*; Chambers, *To Raise an Army*; Capozzola, *Uncle Sam Wants YOU*.

years, and the United States had formally entered the war the previous month. The eight-page conscription statute fleshed out admirably well the single sentence about conscription in Wilson's April 2 speech.[8]

Section 1 of the conscription statute allowed the government "immediately to raise" the full increment of forces that the National Defense Act of 1916 had provided.[9] That earlier statute increased the strength of the US Army to 175,000 men in five equal annual incremental increases in peacetime but as many as 298,000 men in the event of war.[10] This is what Wilson had meant when he had spoken in his speech of increases "to the armed forces of the United States already provided for by law in case of war."

Section 1 also allowed the government "to draft into the military service of the United States ... all members of the National Guard and of the National Guard Reserves." This paragraph made a specific reference to section 111 of the National Defense Act of 1916. This section of the law requires a bit of explanation.

The National Guard is today, and was at the time of World War I, the state militias that are referenced in article 1 of the US Constitution. The federal government can second state militias into federal service but only for one of three domestic purposes: to repel invasion, to suppress insurrection, or to execute federal laws. In no case can these state militias be used in foreign wars. That's why, during the War of 1812, the governors of Massachusetts and Connecticut refused to allow their state militias to be seconded into federal service and why thousands of state militiamen refused to obey unlawful orders to invade Canada. That's also why, during the Mexican-American War, President James Polk had to fight exclusively with volunteers.

This long-standing, constitutionally based requirement that seconded state militias could not be used outside the United States underwent a gradual metamorphosis in the early years of the twentieth century. The change proceeded in three steps, each step reflected in a separate statute.[11]

The first was the Militia Act of 1903, also called the First Dick Act, after then-representative Charles Dick (R-Ohio). The First Dick Act changed the name of state militias to National Guard. It allowed federal secondment of state

8 An Act To authorize the President to increase temporarily the Military Establishment of the United States, May 18, 1917, Pub. L. No. 65-12, 40 Stat. 76 (1917) (hereafter conscription statute of 1917).

9 Ibid., § 1.

10 National Defense Act of 1916, 64 Pub. L. 64-85, 39 Stat. 166 (1916), § 2; Chambers, *To Raise an Army*, 117.

11 William M. Shaw, "The Interrelationship of the United States Army and the National Guard," *Military Law Review* 31, no. 39 (January 1966): 39–84.

militias for no more than nine months—but, importantly, the law retained the constitutional limitation that any such secondments could be made only in cases of invasion, insurrection, or the need to enforce federal laws.[12]

The second step in the change came in the Militia Act of 1908, also called the Second Dick Act, also named for Dick, who had been elected to the Senate. This lifted the nine-month time limit on secondments, and, crucially, the statute purported to completely eliminate the constitutional restriction on sending seconded militiamen on foreign military adventures by allowing such service "either within or without the territory of the United States."[13]

But merely passing the Second Dick Act did not make this radical change constitutional. In 1911, the US Army judge advocate general issued an opinion saying that the use of seconded militiamen outside the United States was unconstitutional.[14] In 1912, in response to a query from Secretary of War Henry Stimson, US Attorney General George Wickersham issued a formal opinion to the same effect; section 5 of the Second Dick Act was plainly unconstitutional.[15]

Congress was not pleased with these opinions and so attempted to address the situation with the third incremental statute in this gradual evolution. Section 111 of the National Defense Act of 1916 stated that in time of war the president may draft the National Guard into the US Army. In such cases, these state militiamen would no longer be called the National Guard; instead, they would be called the US Army. In the conscription statute that Congress passed in May 1917, the section that drafts the National Guard into the army recites that this was being done "in accordance with the provisions of section one hundred and eleven of said national defense Act." No one ever challenged this provision of the conscription statute specifically on these grounds; even if it had been challenged, it is not clear that any court would have found it unconstitutional. Nevertheless, changing the nomenclature that we apply to state

12 An Act To promote the efficiency of the militia, Pub. L. No. 57-33, 32 Stat. 775 (1903).

13 An Act To further Amend the Act entitled "An Act To promote the efficiency of the militia," Pub. L. No. 60-145, 35 Stat. 399 (1908).

14 Michael J. Golden, "The Dormant Second Amendment: Exploring the Rise, Fall, and Potential Resurrection of Independent State Militias," *William & Mary Bill of Rights Journal* 21, no. 4 (May 2013): 1036.

15 US Department of Justice, Authority of President To Send Militia Into a Foreign Country, 29 Op. Att'y Gen. 322 (1912); US War Department, *1912 Report of the Chief of Staff US Army to the Secretary of War. Annual Reports, War Department Fiscal Year Ended June 30, 1912* (Washington, DC: Government Printing Office, 1912), 1:147.

militiamen does not change the substance of what they are, and section 111 of the National Defense Act of 1916 was probably unconstitutional.[16]

The May 1917 conscription statute also called for the draft of an additional 500,000 men. In his speech, Wilson had spoken of raising exactly this number of additional men. The statute also authorized the president to "raise and begin training" 500,000 more men beyond the 500,000 to be drafted immediately and beyond the 298,000 provided for in the National Defense Act of 1916.[17]

The 1917 statute also provided that, with two relatively narrow exceptions, these forces "shall be raised and maintained by selective draft exclusively." As I will show, the decision to rely almost exclusively on conscripts rather than on volunteers was a critical one. The law also provided that draft quotas would be allocated to the states in proportion to population, a seemingly reasonable provision that was, however, changed by statutory amendment later in the war.[18]

The 1917 conscription law corrected two of the three major logistical (and moral) problems that had made the Civil War draft such a disaster: it prohibited enlistment bounties and substitutes.[19]

Further, the law exempted from the draft (but not from the registration requirement) legislative, executive, and judicial officers of both the federal government and the states, ordained clergy and divinity students, county officials, and mail carriers. Importantly, it also authorized the president to provide draft exemptions or deferments to men working in armories, arsenals, navy yards, defense industries, and agriculture "necessary to the maintenance of the Military Establishment or the effective operation of the military forces." These critical provisions melded with the ban on voluntary enlistments to effectively give the federal government the legal authority to decide how and where *all* men would be employed while ostensibly drafting only *some* men into the army.[20]

The statute also addressed conscientious objectors (COs): "Nothing in this Act ... shall be construed to require or compel any person to serve in any of the forces herein provided who is found to be a member of any well-recognized religious sect or organization at present organized and existing and whose existing creed and principles forbid its members to participate in war in any form ... but no person so exempted shall be exempted from service in any

16 An Act for making further and more effectual provision for the national defense, and for other Purposes, June 3, 1916, Pub. L. No. 64-85, § 111, 39 Stat. 166 (1916).

17 Conscription statute of 1917, § 1.

18 Ibid., § 2.

19 Ibid., § 3.

20 Ibid., § 4.

capacity that the President shall declare to be noncombatant."[21] That is, the CO provision required that COs be *religious* objectors, not secular or political objectors, and required that they must be absolute pacifists—opposed to participation in all wars in any form—not merely conscientiously opposed to a particularly obnoxious or pointless war.

Perhaps the biggest problem with the CO provision was the obligation to serve "in any capacity that the President shall declare to be noncombatant." This required that the president actually make such a declaration as to what constituted noncombatant duty. In addition, as the statute was written, such noncombatant duty need not even be civilian; the president could require all COs to enter the army and don uniforms; indeed, as the statute was written, the president could declare that loading shells into mortars is noncombatant so long as the CO has another soldier actually pull the trigger. In actual practice, nearly all COs during World War I were required to serve as soldiers in the army.

Next, the law provided for the creation of local civilian draft boards in every county, with additional draft boards in larger cities, one draft board for every thirty thousand people. Each draft board would "consist of three or more members, none of whom shall be connected with the Military Establishment." These draft boards would be made up of well-respected local citizens; service on draft boards was not compensated. Local draft boards were to be the basic working units that carried out all the primary aspects of the draft and operated nearly all portions of the conscription apparatus: they conducted registrations, processed forms from registrants, decided who would be exempted or deferred under the statute, and sent out draft notices.[22]

The idea behind having local civilian draft boards was that they would put a friendly local nonmilitary face on the conscription machinery. As Crowder put it in his postwar memoir, having local civilian draft boards "put the administration of the draft into the hands of the friends and neighbors of the men to be affected; it would operate through familiar, well-known instrumentalities.... It was the enunciation of the true democratic doctrine of local self-government ... the very embodiment of American ideals."[23]

In the event, 4,648 local draft boards were created, and 14,416 men served on them while they were in existence.[24] The conceit was that local draft board members would be drawn from all walks of life; however, 8,604 of them—more than 63 percent of the total—came from three professions: they were

21 Ibid., § 5.
22 Ibid., § 4.
23 Crowder, *Spirit of Selective Service*, 120–21.
24 *PMG 2d*, 276.

doctors, lawyers, and public office holders. Only 121 were laborers (fewer than 1 percent), and 142 (slightly more than 1 percent) were teachers.[25] Nevertheless, it is worth noting that the Socialist Party leader Norman Thomas—who actively opposed both US entry into the war and the draft—generally agreed with Crowder on the role that local draft boards played in fostering public acceptance of conscription: "On the whole the country accepted conscription docilely.... One reason for this was doubtless the fact that the selective service law ... left power with local boards.... No effort was spared to make it appear that the drafted man was honored by being chosen to serve his country ... by a committee of his fellow citizens."[26]

The conscription statute also created so-called district boards to hear appeals from local boards; these were effectively administrative appellate tribunals.[27] One hundred fifty-five of these district boards were created, and most had five members. The statute also provided for one further level of appeal, from the district board to the president, but these presidential appeals were rare.[28]

Importantly, the statute conferred rule-making authority on the president in order to flesh out the provisions of the statute and to guide the work of the enormous conscription machinery that would be embodied in the Selective Service System, which was an administrative agency. "Administrative agency" has a technical legal meaning, and understanding it is crucial to understanding conscription in the United States both during World War I and for the remainder of the twentieth century.

Although American students are taught that the Constitution created a government consisting of three branches, since the early twentieth century the nation has had a de facto fourth branch of government—the administrative agencies. An administrative agency is created by Congress but technically is part of the executive branch. To take some contemporary examples, the Environmental Protection Agency, the Federal Aviation Administration (FAA), the Food and Drug Administration, and the Occupational Safety and Health Administration (OSHA) all are administrative agencies. When Congress creates an administrative agency, it delegates to the agency the legal authority to control or regulate certain area(s) of commerce or life. A central part of this delegation is that Congress wants the administrative agency to decide exactly *how* to effect that regulation (although always within the broad parameters set by Congress). Part of the public policy (ideological) underpinning of having

---

25 Ibid., 277, table 95.

26 Thomas, *Conscientious Objector in America*, 82.

27 Conscription statute of 1917, § 4.

28 *PMG 2d*, 268.

administrative agencies is that they are supposed to be staffed with nonpartisan experts in the particular field that is being regulated.[29]

A noncontroversial example is the FAA. Congress created it to regulate airline safety. Congress does want air travel to be safe, but members of Congress do not want to be bothered with the minutiæ of what makes air travel safe. The FAA promulgates the actual safety rules, for example, when planes take off, they must have enough fuel to reach their destination; passengers may not threaten flight crew members. The FAA may hire experts in its field—say, pilots or engineers—to carry out its tasks; members of Congress need not be pilots or engineers.

Administrative agencies are sometimes referred to as a fourth branch of government because their roles do not fit neatly into the trifurcated structure contemplated by the Constitution. Administrative agencies have some powers that are similar to legislative powers: the rules issued by administrative agencies have the force of law. If you violate a rule issued by an administrative agency (e.g., the FAA rule not to threaten flight attendants), you can be prosecuted criminally, just as you can be prosecuted if you violate a criminal statute. Administrative agencies also have some powers similar to the powers of courts in that administrative agencies determine the outcome of certain disputes and controversies. An OSHA official can decide whether Joe Smith's auto shop meets OSHA safety guidelines for auto shops (that is, if Smith says yes but OSHA says no).

Administrative agencies saw rapid growth in both number and authority during Franklin Roosevelt's New Deal. Some New Deal era agencies, such as the Works Progress Administration and the Civilian Conservation Corps, no longer exist. But many administrative agencies created during the New Deal not only exist today but have grown in both importance and power, for example, the Federal Deposit Insurance Corporation, Federal Communications Commission, National Labor Relations Board, and Securities and Exchange Commission.

The Selective Service System created during World War I was a classic administrative agency. Indeed, according to one scholar, "one of the chief drivers of the wartime expansion of the American administrative state was the passage of the Selective Service Act of 1917."[30] Congress created the Selective Service to

---

29 As one court decision put it, "An Administrative agency has been realistically characterized as a fourth branch of government." American Beauty Homes Corp. v. Louisville & Jefferson City, 379 S.W.2d 450, 454 n.4 (Ky. 1964).

30 Jeremy K. Kessler, "The Administrative Origins of Modern Civil Liberties Law," *Columbia Law Review* 114, no. 5 (June 2014): 1083, 1087.

carry out a specific function (conscription). Congress provided the broad outlines of how it was to function (for example, workers crucial to defense industries were to be exempt). The agency was given rule-making authority (similar to legislation), and the rules thus promulgated had the force of law. Within the agency, a quasi-judicial appeal procedure was created.

The World War I conscription statute contained a registration requirement. Crucially—and unlike the Civil War statute—the 1917 conscription statute placed the legal obligation to register on the men to be registered, not on the government. "Willful" failure to register was a misdemeanor, triable in a civilian (federal) court, and punishable by up to a year in prison.[31]

Violating *any* provision of the statute, or urging someone else to do so, was a misdemeanor punishable by up to a year in prison.[32] Thus, urging young men not to register or to refuse induction was made a criminal offense.

An amendment to the draft law was enacted the following year, on August 31, 1918. This extended the age of men liable for registration and conscription to all those aged 18 (rather than 21) to 45 (rather than 30).[33] (The so-called Furlough Law of March 16, 1918, is discussed in the section dealing with COs.)

2 The Espionage Act of 1917 and Sedition Act of 1918

On June 15, 1917, less than a month after the conscription statute was enacted, Congress passed the Espionage Act of 1917. In addition to outlawing conventional spying, the Espionage Act did three things relevant to the discussion of conscription during World War I.[34]

- The Espionage Act made it illegal to interfere with the draft by publishing an antidraft essay, magazine, or newsletter or by making an antidraft speech. Such acts were punishable by a prison sentence of as long as twenty years,

31 Conscription statute of 1917, § 5.

32 Ibid., § 6.

33 An Act Amending the Act entitled "An Act to authorize the President to increase temporarily the Military Establishment of the United States," Pub. L. No. 65-210, 40 Stat. 955 (1918).

34 An Act To punish acts of interference with the foreign relations, the neutrality, and the foreign commerce of the United States, to punish espionage, and better to enforce the criminal laws of the United States, Pub. L. No. 65-24, 40 Stat. 217 (1917) (hereafter, Espionage Act of 1917).

a fine of as much as $10,000, or both.[35] This provision overlapped with section 6 of the conscription statute (though the penalties in the Espionage Act were far more draconian). During the war, people who were prosecuted criminally for speaking or publishing against the draft were sometimes indicted, for the very same conduct, for violations of both the Selective Service statute and the Espionage Act. (On May 16, 1918, Congress passed the Sedition Act as an amendment to the Espionage Act; this significantly broadened the reach of the original statute. Now it was illegal to say or do anything to hinder the sale of war bonds; write, print, utter, or publish anything critical of the military, the flag, the army uniform, or the Constitution; or to "incite, provoke, or encourage resistance to the United States.")[36]

- The Espionage Act allowed the issuance of warrants in cases where a person may have violated any other provision of the Espionage Act or, indeed, may have committed any felony at all—such as publishing a magazine opposed to the draft. These warrants were exceptionally broad and allowed not merely the search of homes, offices, or other premises but also the seizure of any and all materials that violated the law. If the police suspected someone of, say, having antidraft leaflets in her possession, they could get a warrant to search her home or office and seize the offending material.[37] The government used this provision liberally.
- The Espionage Act made it illegal to send through the mails anything that violated any other part of the act, such as an article or magazine that opposed the draft or anything that advocated resistance to any federal law whatever. These materials were not merely to be banned from the US mails; the sender could face a prison sentence as long as five years, a fine of as much as $5,000, or both.

In both conception and execution, the Espionage Act was very much of a whole with the conscription statute passed the previous month. The government intended to, and did, make extensive use of the Espionage Act to crush

35 Espionage Act of 1917, Title I, §§ 1 and 2. Title I, § 4, of the Espionage Act was a broad conspiracy provision. If two or more people conspired to speak or write against the draft, every member of the conspiracy became liable for all the overt actions of every other member of the conspiracy.

36 An Act To amend section three, title one, of the Act entitled "An Act to punish acts of interference with the foreign relations, the neutrality, and the foreign commerce of the United States, to punish espionage, and better to enforce the criminal laws of the United States, and for other purposes," Pub. L. No. 65-150, 40 Stat. 553 (1918).

37 Espionage Act of 1917, Title XI, §§ 1 and 2.

opposition to the conscription statute. And the Espionage Act was only too effective. After the war, Norman Thomas wrote, "Nearly all organized agitation in favor of conscientious objectors disappeared under threat of the espionage law and the general pressure of public opinion."[38]

The Espionage Act is still on the books today and has not fallen into desuetude.[39] On June 28, 1971, the United States used the Espionage Act of 1917 to indict Daniel Ellsberg and Anthony Russo in the Pentagon Papers case, for giving a secret government study about US involvement in Vietnam to the *New York Times*, the *Washington Post*, and other publications. On June 21, 2013, the United States used the Espionage Act to indict Edward Snowden for giving military secrets to unauthorized persons.[40]

3 Registration and Opposition to It

As I have noted, the World War I conscription statute wisely corrected one of the major shortcomings of the Civil War conscription statute by placing the onus for registration on the registrant, not on the government. Nevertheless, on May 18, 1917, when Wilson signed the conscription bill into law, there remained the challenge of when and how the government would register the millions of men between the ages of twenty-one and thirty. Wilson wasted no time. On the same day he signed the conscription bill into law, he issued a presidential proclamation setting June 5 as the date all men in the relevant age cohort would be legally obligated to register for the draft.

The next day the *New York Times* printed the proclamation on its front page. Wilson's proclamation recited that it was a criminal offense not to register for the draft, punishable by a year in prison, and then went on to say that the draft "is in no sense a conscription of the unwilling. It is, rather, a selection from a nation which has volunteered in mass."[41] That last statement may have been

38 Thomas, *Conscientious Objector in America*, 81.

39 Today the Espionage Act of 1917 is codified as Gathering, transmitting, or losing defense information, 18 U.S.C. §§ 793–99 (2022).

40 *In re* Ellsberg, 446 F.2d 954, 954 (1st Cir. 1971); Peter Finn and Sari Horwitz, "U.S. Charges Snowden with Espionage," *Washington Post*, June 22, 2013.

41 "A Proclamation by the President of the United States," *New York Times*, May 19, 1917. This may be Wilson's most famous and widely quoted statement concerning World War I conscription, and not all those who cite this quotation do so ironically, that is, because it turned out to be so spectacularly wrong. Crowder plagiarized the line, without attribu-

wishful thinking on Wilson's part; if the nation had, in fact, volunteered en masse, there would have been no need for conscription at all.

Also on May 18, the president federalized the entire National Guard, the state militias of the forty-eight states, pursuant to the National Defense Act of 1916.

Opponents of conscription also wasted no time. Almost immediately, the anarchists Emma Goldman and her sometime lover Alexander Berkman started the No-Conscription League (for a copy of the league's manifesto, see Figure 4.1).[42] On the evening of May 18, Goldman delivered an antidraft speech in which she urged men of conscription age not to register. According to the police stenographic transcript of Goldman's speech at New York's Harlem River Casino, she boasted that there would be perhaps fifty thousand men who would refuse to go if they were conscripted. (Police stenographers were present at Goldman's various public antidraft speeches that spring, and the transcripts were introduced as evidence in the trial that summer when Goldman and Berkman were criminally prosecuted for their antidraft activities.) Goldman also pledged to support those who resisted conscription.[43]

A branch of the No-Conscription League in Seattle, Washington, produced a flyer entitled "No Conscription, No Involuntary Servitude, No Slavery." The flyer began with the text of the Thirteenth Amendment and then continued: "We, signing this, are native-born citizens, within the age limit set for the compulsory draft.... Resist! Refuse! Don't yield the first step toward conscription. Better to be imprisoned than to renounce your freedom of conscience.... Seek out those who are subject to the draft! Tell them that we are refusing to register or to be conscripted and to stand with us like men." The flyer included a

tion to Wilson, in his 1920 memoir, *The Spirit of Selective Service*, 125, as a straightforward account of what Crowder believed had occurred. Later scholars, however, do cite the quotation ironically. See, for example, Fite and Peterson, *Opponents of War*, 21, 23; and Chambers, *To Raise an Army*, 170.

42 Emma Goldman (1869–1940) and Alexander Berkman (1870–1936) were communist anarchist organizers, public speakers, and prolific writers. Goldman's two-volume autobiography is *Living My Life* (New York: Dover, 1970). *Anarchism and Other Essays* (New York: Dover, 1969) is a collection of her essays. For a double biography of Goldman and Berkman, see Paul Avrich and Karen Avrich, *Sasha and Emma: The Anarchist Odyssey of Alexander Berkman and Emma Goldman* (Cambridge, MA: Harvard University Press, 2012). Berkman is the author of *The ABC of Anarchism* (New York: Dover, 1972) and *Prison Memoirs of an Anarchist* (New York: New York Review of Books, 1999). Multiple essays by Goldman, Berkman, and others are reprinted in Peter Glassgold, ed., *Anarchy! An Anthology of Emma Goldman's* Mother Earth (Washington, DC: Counterpoint, 2001).

43 Emma Goldman, "We Don't Believe in Conscription" speech at Harlem River Casino, New York, New York, May 18, 1917, available online at Emma Goldman Archive, University of California, Berkeley, https://www.lib.berkeley.edu/goldman/pdfs/Speeches-WeDon't BelieveinConscription.pdf, last accessed April 26, 2022.

NO CONSCRIPTION!

CONSCRIPTION has now become a fact in this country. It took England fully 18 months after she engaged in the war to impose compulsory military service on her people. It was left for "free" America to pass a conscription bill six weeks after she declared war against Germany.

What becomes of the patriotic boast of America to have entered the European war in behalf of the principle of democracy? But that is not all. Every country in Europe has recognized the right of conscientious objectors—of men who refuse to engage in war on the ground that they are opposed to taking life. Yet this democratic country makes no such provision for those who will not commit murder at the behest of the war profiteers. Thus the "land of the free and the home of the brave" is ready to coerce free men into the military yoke.

No one to whom the fundamental principle of liberty and justice is more than an idle phrase, can help realize that the patriotic clap-trap now shouted by press, pulpit and the authorities, betrays a desperate effort of the ruling class in this country to throw sand in the eyes of the masses and to blind them to the real issue confronting them. That issue is the Prussianizing of America so as to destroy whatever few liberties the people have achieved through an incessant struggle of many years.

Already all labor protective laws have been abrogated, which means that while husbands, fathers and sons are butchered on the battlefield, the women and children will be exploited in our industrial bastiles to the heart's content of the American patriots for gain and power.

Freedom of speech, of press and assembly is about to be thrown upon the dungheap of political guarantees. But crime of all crimes, the flower of the country is to be forced into murder whether or not they believe in war or in the efficacy of saving democracy in Europe by the destruction of democracy at home.

Liberty of conscience is the most fundamental of all human rights, the pivot of all progress. No man may be deprived of it without losing every vestige of freedom of thought and action. In these days when every principle and conception of democracy and individual liberty is being cast overboard under the pretext of democratizing Germany, it behooves every liberty-loving man and woman to insist on his or her right of individual choice in the ordering of his life and actions.

The NO-CONSCRIPTION LEAGUE has been formed for the purpose of encouraging conscientious objectors to affirm their liberty of conscience and to make their objection to human slaughter effective by refusing to participate in the killing of their fellow men. The NO-CONSCRIPTION LEAGUE is to be the voice of protest against the coercion of conscientious objectors to participate in the war. Our platform may be summarized as follows:

> We oppose conscription because we are internationalists, anti-militarists, and opposed to all wars waged by capitalistic governments.
>
> We will fight for what we choose to fight for; we will never fight simply because we are ordered to fight.
>
> We believe that the militarization of America is an evil that far outweighs, in its anti-social and anti-libertarian effects, any good that may come from America's participation in the war.
>
> We will resist conscription by every means in our power, and we will sustain those who, for similar reasons, refuse to be conscripted.

We are not unmindful of the difficulties in our way. But we have resolved to go ahead and spare no effort to make the voice of protest a moral force in the life of this country. The initial efforts of the conscientious objectors in England were fraught with many hardships and danger, but finally the government of Great Britain was forced to give heed to the steadily increasing volume of public protest against the coercion of conscientious objectors. So we, too, in America, will doubtless meet the full severity of the government and the condemnation of the war-mad jingoes, but we are nevertheless determined to go ahead. We feel confident in arousing thousands of people who are conscientious objectors to the murder of their fellowmen and to whom a principle represents the most vital thing in life.

Resist conscription. Organize meetings. Join our League. Send us money. Help us to give assistance to those who come in conflict with the government. Help us to publish literature against militarism and against conscription.

We consider this campaign of the utmost importance at the present time. Amid hateful, cowardly silence, a powerful voice and an all-embracing love are necessary to make the living dead shiver.

NO-CONSCRIPTION LEAGUE
20 East 125th St., New York.

FIGURE 4.1 The No-Conscription League's manifesto was printed in New York and distributed nationally.
USED BY PERMISSION OF THE EMMA GOLDMAN PAPERS

paragraph from Daniel Webster's antidraft speech in 1814. The group printed twenty thousand copies of the flyer and distributed them door to door.[44]

In May 1917, immediately after the Selective Service Act was signed into law, several people in West Virginia, including Edward Firth, the state secretary of the Socialist Party, wrote and circulated an antidraft pamphlet entitled *Are We Facing a Militarized America?* in which they exhorted men not to register and to refuse induction. Firth and his codefendants were sentenced to six months in prison for circulating the antidraft pamphlet.[45]

44 Wells v. United States, 257 F. 605 (9th Cir. 1919).
45 Firth v. United States, 253 F. 36 (4th Cir. 1918); Fite and Peterson, *Opponents of War*, 30.

Different activists took different positions on how draft-age men ought to respond to the new draft law. The American Union Against Militarism issued a flyer on May 22 that urged men to follow the law and to register as the law required: "You can only make your protest effective by registering. See that the clerk puts down your claim to exemption from service as a 'conscientious objector to war.'... We therefore urge all conscientious objectors to register."[46] In contrast, the anarchists' No-Conscription League urged men not to register or cooperate with the conscription machinery in any way. Its manifesto, printed in New York and distributed nationally, said: "We oppose conscription because we are internationalists, anti-militarists, and opposed to all wars waged by capitalistic governments. We will fight for what we choose to fight for; we will never fight simply because we are ordered to fight.... Resist conscription. Organize meetings. Join our league."[47]

On May 22, eight men were arrested in Snyder and Abilene, Texas, as part of an anticonscription conspiracy that the government believed involved more than fifty men. The conspirators had bought high-powered rifles and fled from their homes to a nearby canyon in order to resist conscription officers.[48] Also on May 22, the California Supreme Court rejected a lawsuit that would have enjoined (legally prevented by means of a court order) Oakland officials from assisting with registration on the ground that the draft law was an unconstitutional form of involuntary servitude in violation of the Thirteenth Amendment.[49]

During the last week of May 1917, there were organized antidraft protests in Texas, Virginia, West Virginia, Chicago, Michigan, and Kansas. Multiple arrests were made in Virginia, where antidraft organizers were arming themselves. In West Virginia, the Young Men's Anti-Militarism League circulated a flyer urging men to refuse to register for the draft: "Young men, are you going to register for

46 "American Union against Militarism Records," in Bennett and Howlett, *Antiwar Dissent in World War I America*, 57–59; "Anti-Militarists to Register. But League Calls on Them to Claim Exemptions in Legal Way," *New York Times*, May 23, 1917.

47 Emma Goldman, *No Conscription League Manifesto* (New York: Solidarity, 1917), 1. This pamphlet is reprinted in Bennett and Howlett, *Antiwar Dissent in World War I America*, 71–75. For context see "Federal Agents Watch Anti-Draft Efforts Here," *New York Times*, May 31, 1917.

48 "Anti-Draft Plotters Held," *New York Times*, May 23, 1917; Bryant v. United States, 257 F. 378 (5th Cir. 1919).

49 Sidebar to "Anti-Draft Plotters Held," *New York Times*, May 23, 1917.

military service in a foreign country while rich men who brought this war get richer?"[50]

In Michigan, Nathan L. Welch, editor of the *Michigan Socialist*, and nine others were arrested for distributing antidraft leaflets. In Chicago, three men and several women were arrested as part of what officials believed was "a nationwide conspiracy" to encourage draft resistance. Elsewhere in Chicago, antidraft rioters were arrested in Grant Park. In Lynn, Massachusetts, police searched for the people who had posted antidraft leaflets in the city's business district. In Topeka, Kansas, the Justice Department investigated a public meeting of draft objectors. In Kanawha and Boone Counties, West Virginia, federal agents investigated the source of leaflets that had been circulated urging men of conscription age not to register. Hulet Wells, the former president of the Seattle Central Labor Council, was arrested in Seattle and charged with seditious conspiracy for distributing the leaflets of the No-Conscription League.[51]

On May 29, newspapers printed reports of further antidraft activities in Chicago, West Virginia, Seattle, Texas, Detroit, and San Francisco. In addition, twenty-two-year-old Harry Ball had been arrested in Philadelphia for distributing No-Conscription League flyers; in Massachusetts, antidraft flyers were found circulating in Somerville and other cities; and a new lawsuit had been filed in Kansas challenging the constitutionality of the draft.[52] Several people were injured in Toledo, Ohio, in a riot involving two thousand people, as state militiamen fought anticonscription protesters.[53] On May 30, authorities detained and questioned dozens of antidraft organizers in Chicago and Milwaukee.[54]

On May 31, the *New York Times* reported that young American male refugees (that was the word used by the *Times*) were fleeing the country for Canada, Mexico, and Cuba. For days, steamers from the United States had been carrying American men, "many of whom make no concealment of the fact that they left home to escape registration under the conscription act."[55] (Eventually, colonies of these draft evaders formed in both Canada and Mexico.)[56]

In May, students organized the Collegiate Anti-Militarism League, which urged men of draft age not to register on June 5. The group had chapters at

50 "Reveal Plots to Fight Draft; 25 under Arrest," *New York Times*, May 29, 1917; "Wheeling Circulars Urge Young Men Not to Fight," *New-York Tribune*, May 30, 1917.

51 "Reveal Plots to Fight Draft," *New York Times*, May 29, 1917.

52 "Plotters Here against Draft under Watch," *New York Times*, May 30, 1917.

53 "Draft Law Causes Riots," *New York Times*, May 30, 1917.

54 "To Ask Indictment of Chicago Plotters," *New-York Tribune*, May 31, 1917.

55 "Won't Miss Draft on Foreign Soil," *New York Times*, May 31, 1917.

56 "Net for Slackers to Be Nationwide," *New York Times*, September 2, 1918.

Columbia, Oberlin, Yale, the Massachusetts Institute of Technology, Harvard, City University of New York, Union Theological Seminary, Princeton, Amherst, and elsewhere. On May 31, three Columbia students—Owen Cattell, Charles Phillips, and Eleanor Parker—were arrested for printing and distributing the Collegiate Anti-Militarism League's flyer urging men not to register. The flyer said, in part, "For those of us who are opposed to the draft and will refuse to be conscripted there remains only the question of how best to make our resistance felt. If we want to strike with the full strength of our numbers we must refuse to register for conscription—even if it means jail for us.... We urge all conscientious objectors to resist registration—even to the extent of going to prison if necessary." When the three were arraigned in federal court on June 2, the courtroom was packed with supporters of the student antidraft activists.[57]

On the evening on May 31, a capacity crowd of fifteen thousand antiwar and antidraft activists packed Madison Square Garden in New York to oppose conscription—under the watchful eyes of 300 plainclothes detectives inside the Garden and an additional 450 police officers carrying rifles outside the Garden. Five people were arrested and charged with distributing antidraft leaflets, including the anarchists Louis Kramer and Morris Becker.[58] Also on May 31, three people were arrested in Columbus, Ohio, in a raid on a print shop that was turning out antidraft leaflets that were being used across the country. In Kansas City, Kansas, fifty people were charged with a conspiracy to oppose the draft. In Topeka, four people were arrested for antidraft organizing.[59]

On June 1, the front page of the *New York Times* carried four articles about antidraft agitation, under a banner headline that read, "Hit Draft Plots in Many States—Federal Agents Active—Seize Youthful Anti-Conscription Agitators Here on Conspiracy Charge." The lead article explained that "throughout the country, the federal agents were active yesterday to curb the efforts of [antidraft] plotters." The article discussed antidraft actions in Columbus, Kansas City, Boston, and New York. The No-Conscription League was featured prominently: "One of the organizations which is being carefully watched is the so-called No-Conscription League, in the propaganda of which anarchists of the Goldman-Berkman type and certain I.W.W. agitators are prominent." Eight antidraft organizers had been arrested in Kansas City, Kansas, four others in

57 "Hit Draft Plots in Many States; Arrest 3 College Students Here; Police Display Quiets Antis' Meeting," *New York Times*, June 1, 1917; "Anti-Draft Cases before Grand Jury," *Columbia (University) Spectator*, June 4, 1917.

58 "New Police Arms Awe Socialists," *New York Times*, June 1, 1917; "9 More Foes of Draft Jailed Here," *New-York Tribune*, June 3, 1917.

59 "U.S. Begins Round-Up of Draft Foes in Nation; Eight Are Arrested Here," *New-York Tribune*, June 1, 1917.

Topeka, and one each in Lawrence and Olathe.[60] In Detroit, Morris Sugar and Samuel Diamond were arrested for organizing opposition to the draft. Six Socialists there published and circulated a special Anti-Conscription Edition of the *Michigan Socialist* newspaper focused solely on draft resistance. The six men announced that they would refuse to register on June 5 and called on others to do the same.[61]

Another article on the front page of the *Times* on June 1 reported that three men had been arrested in Columbus, Ohio, and charged with capital treason "in connection with anti-registration plotting." One of the three was Ammon Hennacy, then a Socialist Party activist who later was active in the Catholic Worker movement.[62] Hennacy was convicted of the charges and served two years in federal prison. At the end of his sentence, and while still in prison, Hennacy again refused to register and was prosecuted a second time; his second sentence was for nine months.[63]

Antidraft agitation was front-page news in much of the rest of the country on June 1, 1917. A page 1 headline in the *Cincinnati Enquirer* was "Scurrilous Dodgers Found," and the story reported not only the Columbus arrests but also arrests in New York, Chicago, and Kansas City. "The offense," the newspaper said quaintly, "has consisted in the circulation of literature and posters advising young men not to register for the draft."[64] Like the *New York Times*, the *Kansas City* (MO) *Times* had multiple front-page stories about antidraft agitation and arrests. Two stories recounted that in Kansas City, a state court judge had denied a request by two men for an injunction that would have stopped the Missouri governor and Kansas City mayor from assisting in the upcoming registration. While still in the judge's courtroom, the men were arrested for conspiracy to aid and abet draft evasion.[65]

On June 2, antidraft organizers were arrested in New York and New Jersey. In New York, two women were arrested for distributing antidraft literature on Fifth Avenue; a third woman was arrested while handing out antidraft leaflets at a recruiting station in Manhattan. Six people were arrested at an antidraft meeting in Guttenberg, New Jersey; two of the them were clubbed by police.

---

60 "Hit Draft Plots in Many States."

61 Sugar v. United States, 252 F. 79 (6th Cir. 1918).

62 "Make Charge of Treason against Three Prisoners Held in Ohio's Capital for Anti-Draft Plotting," *New York Times*, June 1, 1917; Ammon Hennacy, *The Book of Ammon* (Baltimore: Fortcamp, 1994).

63 R. Alfred Hassler, *Conscripts of Conscience* (New York: Fellowship of Reconciliation, 1942), 30–31.

64 "Scurrilous Dodgers Found," *Cincinnati Enquirer*, June 1, 1917.

65 "Slackers Lose in Court," and "Nip Draft Plots," *Kansas City Times*, June 1, 1917.

Federal agents reported that US mails were being flooded with thousands of antidraft flyers, and the *Times* and *New-York Tribune* carried stories about antidraft protests in Connecticut, Texas, Missouri, and Ohio.[66]

On the evening of Monday, June 4, just before the national registration day, the No-Conscription League held a public rally at the Hunts Point Palace in New York at which both Goldman and Berkman spoke. More than two thousand people crowded the hall, while thousands more milled around outside. Berkman said, "Conscription in a free country means the cemetery of liberty, and if conscription is the cemetery then registration is the undertaker." He predicted that thousands of men would not register the next day and declared, "Neither the soldiers of this country nor the workers have any enemies across the ocean."[67]

On June 5, millions of men aged twenty-one to thirty presented themselves to 4,648 draft boards across the country to register. In his report to the secretary of war, Crowder gave a rosy account of the registration: "That this vast labor should have been accomplished without friction and without the slightest manifestation of antagonism on the part of any disturbing elements is in itself a complete vindication of the loyalty of the American people."[68]

Crowder was willfully averting his gaze from the reality of the situation. On June 5, a total of 9,586,508 men registered, considerably fewer than the 10 million that Crowder had predicted based on US Census and other actuarial data.[69] Scholars using Selective Service records estimate that 10 percent to 15 percent of men who were legally obligated to register during the existence of the World War I draft failed to do so, perhaps totaling as many as 3 million nonregistrants.[70] Even the War Department concluded in 1920 that 337,000 men who were legally obligated to register had failed to do so.[71]

---

66 "More Anti-Draft Agitators in Jail; Anarchists Held," *New York Times*, June 3, 1917; "9 More Draft Foes Jailed."

67 These quotations appear in a transcript prepared by the public shorthand reporter Charles Pickler, who was hired by the No-Conscription League. The transcript was introduced as Government Exhibit 33 in the trial of Emma Goldman and Alexander Berkman, and the quotations appear on pages 489 to 490 of the official transcript. *U.S. Supreme Court Transcript of Record* in Goldman and Berkman v. United States, 245 U.S. 474 (1918) (No. 18-702). Modern editions of transcripts of record are published by Gale for the Supreme Court and are available in many libraries as well as for purchase; some are available online, as are some earlier editions.

68 *PMG 2d*, 2.

69 Ibid., 1.

70 The figures are from Capozzola, *Uncle Sam Wants YOU*, 30, and Chambers, *To Raise an Army*, 211.

71 US Office of the Provost Marshal General, *Final Report of the Provost Marshal General to the Secretary of War on the Operations of the Selective Service System to July 15, 1919*

A second registration was held on June 5, 1918, for men who had turned twenty-one in the year since the first registration. Just more than 900,000 men registered. A third registration was conducted on September 12, 1918, necessitated by the August 31, 1918, amendment to the statute that extended the age of conscription to eighteen to forty-five. About 13 million additional men registered on September 12. In total, 23.9 million men registered in the three registrations.[72]

4 The Draft and Opposition to It

On November 8, 1917, the Selective Service issued a new regulation that created a critically important part of the new draft system, a classification system that encompassed every registrant.[73] The agency created five general classes of men, each designated by its own roman numeral: Class I, Class II, and so forth through Class V. Class I consisted of men immediately eligible to be drafted; Class V consisted of men who were statutorily exempt from conscription. Men who were in Class II could be drafted only after all of Class I had been called; men in Class III could be drafted only after all the men in Class II had been called and so forth. The numbers reflected the importance of the industrial or agricultural work done by the man being classified.[74]

Some classes also had subcategories within a broader class. For example, Class II, married men, had three subclasses: II-A (married men with dependent wives but no children), II-B (married men with dependent wives and children), and II-C (married men who did not support their families financially). These subclasses were meant to be used to determine the order in which men could be drafted within the broader class. In the event, all of the approximately 3 million men drafted during World War I came from Class I.[75] (The Selective Service has always used roman numerals for its classifications, although general circulation periodicals often present them as Arabic numbers: 1A or 1-A instead of I-A, for example.)

It is a measure of how dysfunctional the World War I draft process was that 9.5 million men had registered on June 5, 1917, and induction orders started

(Washington, DC: Government Printing Office, 1919), 12, 53 (hereafter, *PMG 3d*). In their 1978 book, Lawrence M. Baskir and William A. Strauss used the figure 325,000 for World War I nonregistrants. Baskir and Strauss, *Chance and Circumstance: The Draft, the War, and the Vietnam Generation* (New York: Alfred A. Knopf, 1978), 19.

72 *PMG 2d*, 22, 24–26, and 31, table 1.

73 Ibid., 224.

74 Ibid., 2–5, 224.

75 Ibid., 110–12, 2.

going out in September, but the regulation creating the classification system was promulgated on November 8, and the first questionnaire to be used to implement the regulation (that is, actually classify men) was not available until December 15.[76]

In the narrowest sense, this classification system was a means of avoiding the problem that the Confederacy faced when it decided to draft all men aged eighteen to thirty-five and realized, too late, that doing this would deprive the Confederacy of men needed for agriculture, munitions manufacture, and other essential pursuits. In a broader sense, however, the World War I classification system was a means of channeling all men of draft age—not merely the subset who received induction orders—into specific jobs or professions that the government believed to be necessary for the war effort or otherwise useful. The figures from World War I are stark: 24 million men were registered during the war, yet "only" about 3 million received induction orders. With the classification system in place, the government could effectively channel the remaining 21 million men who never received induction orders—more than 87 percent of the total registered—into jobs that the government found desirable for the war effort.[77]

Crowder was clear on this point in his first annual *Report to the Secretary of War*: "The draft itself is an instrument of compelling force in controlling and distributing labor supply throughout the United States." Wilson had said much the same thing himself in a letter dated May 4, 1917: "The idea of the draft is not only the drawing of [some] men into the military service of the Government, but the virtual assigning of [all] men to the necessary labor of the country."[78]

In his second annual report, Crowder elaborated on channeling. The Selective Service regulations, he said, accomplished one of two things: "Either they placed such registrants in the military service, or they forced them, through fear of such service, to engage in a useful industrial or agricultural undertaking."[79]

The value of conscription for channeling the entirety of the male population into areas the government believes to be desirable was not a mere collateral benefit or post hoc rationalization for conscription; the benefits of channeling were much on the minds of members of Congress when they enacted the draft law. At an April 14, 1917, hearing on the draft bill by the House Committee on

76 Ibid., 224, 48.

77 Ibid., 31, table 1.

78 US Office of the Provost Marshal General, *Report of the Provost Marshal General to the Secretary of War on the First Draft Under the Selective-Service Act, 1917, 1918* (Washington DC: Government Printing Office, 1918), 33 (hereafter, *PMG 1st*); Chambers, *To Raise an Army*, 126.

79 *PMG 2d*, 10.

Military Affairs, its members heard from a British officer, Captain Percy George Reginald Benson, of the British Army. England had started the war without conscription and had obtained its first 3 million combat soldiers solely by seeking volunteers. England then switched to conscription, which was preferable to volunteerism, Benson said, in large part because of conscription's ability to channel all men:

> It was evident after a time that universal obligation secured infinitely greater efficiency because the State, if it has every man of military age under an obligation to do his duty by the State, can tell him exactly what to do; and under our volunteer system at the beginning of the war thousands of men from the coal mines, the shipbuilding yards, mechanics and farm laborers, enlisted and went out. Those men were skilled to a very great extent. Afterwards, when urgent necessity for those men came, we had to withdraw a great many men from the front line and bring them back to the workshops where they were urgently necessary. Of course, under a system of universal obligation the State can simply say, "You stay in the workshop, and you go."[80]

This emphasis on using the draft to channel the entire male workforce into areas deemed by the government to help the war effort is the context necessary to understand two additional policies of the Wilson administration.

The first of these policies is reflected in the so-called work or fight order of May 23, 1918. This Selective Service order stated that men who were not working in industries that were essential to the war effort—or who were employed in such industries but were not working hard enough or efficiently enough—would be immediately placed into draft Class I. "Work or fight" meant just that.[81]

To Crowder, the principle was the essence of simplicity: "The man who was industrially useless should lose the preferred classification which he would otherwise have continued to enjoy." And it worked: "The result of the [work or fight] regulations was to destroy idleness and to drive from the nonproductive pursuits all registrants whose deferment hinged upon a change of occupation." Crowder was enthusiastic about the work or fight rule: "The possibilities of this

---

80 *Increase of Military Establishment of 1917*, statement of Captain Percy George Reginald Benson, British Army, 3–4.

81 "Work or Fight, Warning to All on Draft Rolls," *New York Times*, May 24, 1918.

principle are practically limitless.... In time of war, we have only to apply the 'Work or Fight' principle to attain any desired adjustment of the labor supply."[82]

The second policy was the government's ban on volunteers for the army once the draft was up and running. By order of the War Department, voluntary enlistments in the army were banned as of December 15, 1917, and voluntary enlistments into the navy and Marine Corps were banned as of August 8, 1918.[83] It might seem that—in the midst of a war in which the government had huge needs for larger fighting forces—the government should be delighted to accept volunteers for the military. But the government wanted to control the labor of *every* man of draft age. The problem with permitting voluntary enlistment is that sometimes the wrong man might seek to enlist; this is what Crowder called "the pernicious effect of volunteer recruitment." He explained, "Fired with ... patriotism ... large numbers of men whose energies were of infinitely more value in the industrial world than on the battle field, abandoned their productive labors for volunteering recruitment." This was a terrible thing. Allowing men to volunteer simply "took no heed of [the] economic value" of the men volunteering. This was to be remedied, as Crowder quaintly put it, by "eliminating indiscriminate volunteering." If the only purpose of the Selective Service was to get a certain number of soldiers into the military, the government would not have banned voluntary enlistments.[84]

On July 13, 1917, Wilson announced that the first draft call under the new statute would be for 687,000 men.[85] During World War I, a lottery determined the order in which men were drafted within their Selective Service class, with the first lottery conducted on July 20; it encompassed the men who had registered in June. In all, the Selective Service held three lotteries during the war, one for each of the three draft registrations. The second lottery was held on June 27, 1918, for those who had registered on June 5; the third lottery was held September 30, for those who had registered on September 12. A fair amount of publicity and hoopla attended the lotteries, with various notables drawing some of the early numbers. In the first and second lotteries, Secretary of War Baker drew the first number; in the third lottery, Wilson drew the first number. Other well-known participants included senators, representatives, and Crowder.[86]

82 *PMG 2d*, 17, 15.
83 Ibid., 6.
84 Ibid., 13, 6.
85 *PMG 1st*, 5.
86 *PMG 2d*, 41–44.

In late July 1917, after the first lottery and the first draft call, draft board members often found that they were not popular in their home communities. As a result, draft board members started resigning in significant numbers, which undermined the government's attempt to use draft boards to put a friendly local face on conscription policies that emanated from Washington. On August 16, 1917, the War Department prohibited draft board members from resigning without first obtaining their governor's approval.[87]

The first men were inducted into the army under the Selective Service law in September 1917; the last men were drafted in November 1918, the month that the armistice was signed. Table 4.1 shows the numbers drafted each month throughout the war.

Opposition to registration, conscription, and the entire Selective Service apparatus continued after the first registration on June 5, 1917. In his December 1917 year-end *Report to the Secretary of War*, Crowder reported that the government had arrested 5,870 people for not registering. Of these, 2,095 were being prosecuted criminally, a huge number of prosecutions for the Department of Justice at that time. Fourteen thousand two hundred twelve men had failed to appear for induction or for preinduction physical examinations. Three hundred forty-three people were being prosecuted for "indulging in antidraft propaganda throughout the country," and another 180 were arrested for aiding and abetting those who evaded or refused registration or conscription. Of that latter group, 34 had been convicted by year's end, "with heavy sentences usually imposed."[88]

In June 1917, shortly after the first registration, Lawrence Pearson, a Socialist barber in Roanoke, Virginia, was arrested for distributing an antidraft leaflet entitled *A Rich Man's War and a Poor Man's Fight*.[89]

On June 14, Emma Goldman delivered an antiwar and anticonscription speech in New York at a packed Forward Hall that held about a thousand people (the hall was named for the *Forward*, a Yiddish-language socialist newspaper). Goldman started her speech by saying, "I deeply regret that I cannot speak to you in the language I have always spoken from this platform [Yiddish]"; instead, she said that she would speak in English so that the police who were monitoring her talk would understand her. Goldman declared that "conscription is a crime and oppression and an outrage upon reason." On September 18,

87 "Crowder Curbs Resignation of Draft Boards," *New-York Tribune*, August 17, 1917.

88 *PMG 1st*, 25.

89 Jeanette Keith, *Rich Man's War, Poor Man's Fight: Race, Class, and Power in the Rural South during the First World War* (Chapel Hill: University of North Carolina Press, 2004), 85.

TABLE 4.1 Number of men drafted by month in World War I

Month	Number drafted
September 1917	296,678
October 1917	163,493
November 1917	35,721
December 1917	20,320
January 1918	23,288
February 1918	83,779
March 1918	132,484
April 1918	174,377
May 1918	373,063
June 1918	301,941
July 1918	401,147
August 1918	282,898
September 1918	262,984
October 1918	249,185
November 1918	8,938
Total	**2,810,296**

SOURCE: US OFFICE OF THE PROVOST MARSHAL GENERAL, *SELECTIVE SERVICE IN WARTIME: SECOND REPORT OF THE PROVOST MARSHAL GENERAL TO THE SECRETARY OF WAR*, 223, TABLE 79.

Note: The figure for men drafted in October 1918 is the sum of lines 20 and 23 in table 79, and the figure for men drafted in November 1918 is the sum of lines 21 and 24.

1917, US marshals and New York City police broke up another anticonscription meeting of fifteen hundred people at Forward Hall.[90]

The next day, police and US marshals raided the office of the No-Conscription League, arrested Goldman and Berkman, and seized copies of Goldman's *Mother Earth* and Berkman's the *Blast*, magazines that contained anticonscription articles. Goldman and Berkman were indicted on a single count of conspiracy to violate the conscription statute. Their trial commenced on June 27; on July 9, 1917, a "patriotic jury took thirty-nine minutes to return a verdict of

90 Emma Goldman, "Speech against Conscription and War," Emma Goldman Archive, University of California, Berkeley, https://www.lib.berkeley.edu/goldman/pdfs/Speeches-SpeechAgainstConscriptionandWar.pdf, last accessed November 14, 2022; "Marshal Breaks Up Anti-Draft Meeting," *New York Times*, September 10, 1917.

guilty."[91] Goldman and Berkman were given a maximum sentence of two years in prison.[92] After their arrests, the Post Office used Title XII of the Espionage Act to ban *Mother Earth* and the *Blast* from the mails.[93] This effectively killed the publications.[94]

In early August there was antidraft rioting in Texas, and in the mountainous area of western North Carolina, residents of two towns rose up en masse and armed themselves to resist the draft.[95]

Early August 1917 also saw the uprising in rural Oklahoma known as the Green Corn Rebellion, an antidraft uprising organized by the socialist and interracial Working Class Union that may have had loose connections to the Wobblies.[96] The uprising was concentrated in the Oklahoma counties of Seminole, Hughes, and Pontotoc. Starting on Friday, August 3, and continuing for several days, five hundred Black and white farmers—mostly poor and poorly educated sharecroppers—grouped together and armed themselves to oppose the draft. Their plan was to march to Washington and force Congress to repeal the draft law. They meant to eat green corn and barbecue stolen cattle en route, which gave rise to the name Green Corn Rebellion. The armed draft resisters did not get far, but it took dozens of posses comprising one thousand men in all, drawn from seven counties, a full week to put down the rebellion.

---

91 Paul L. Murphy, *World War I and the Origin of Civil Liberties in the United States* (New York: W. W. Norton, 1979), 213–14.

92 Leonard D. Abbott, *The Trial and Conviction of Emma Goldman and Alexander Berkman*, https://www.lib.berkeley.edu/goldman/pdfs/PublishedEssaysandPamphlets_TheTrialandConvictionofEmmaGoldmanandAlexanderBerkman.pdf, last accessed April 26, 2022; "Emma Goldman and A. Berkman behind the Bars," *New York Times*, June 16, 1917.

93 Murphy, *World War I and the Origin of Civil Liberties*, 214.

94 Glassgold, introduction to *Anarchy!*, xxxv–xxxvi.

95 "Draft Resisters Can Be Executed," *New York Times*, August 5, 1917.

96 The Industrial Workers of the World (IWW), whose members were called Wobblies, was a radical syndicalist union. Ironically, as Melvyn Dubovsky makes clear in his majestic *We Shall Be All: A History of the Industrial Workers of the World* (Chicago: Quadrangle Books, 1969), 356–57, although the IWW "had always preached ... anti-militarism and anti-patriotism," the union never took any formal position on the World War I draft and "left conscription a matter of individual conscience." On the IWW, see Patrick Renshaw, *The Wobblies: The Story of the IWW and Syndicalism in the United States* (Chicago: Ivan R. Dee, 1967), and Peter Cole, David Struthers, and Kenyon Zimmer, eds., *Wobblies of the World: A Global History of the I.W.W.* (London: Pluto Press, 2017). For a compilation of materials from IWW newspapers, pamphlets, and other publications, see Joyce L. Kornbluh, *Rebel Voices: An I.W.W. Anthology* (Ann Arbor: University of Michigan Press, 1964). Nigel Sellars argues that the IWW had no formal connection to the Green Corn Rebellion and in fact did not even recognize the Working Class Union as a legitimate union. Sellars, "With Folded Arms? Or with Squirrel Guns? The IWW and the Green Corn Rebellion," *Chronicles of Oklahoma* 77, no. 2 (Summer 1999): 150–69.

Several armed skirmishes were fought during that time; one man was killed and another wounded in the fighting. As soon as fighting subsided in one town, rioting would break out in another. Among other things, the protesters cut telephone and telegraph lines, burned railroad bridges, and dynamited the Frisco trestle bridge over the South Canadian River between Frances and Ada, Oklahoma. In the end, 220 men were indicted. Some leaders were sentenced to six years, others to ten years. The man who had dynamited the bridge got six years. Many rank-and-file participants received sentences of a year or less.[97]

The government was not idle during this time. On August 4, in response to organized, and sometimes armed, draft resistance in Texas, Georgia, North Carolina, and Oklahoma, Crowder announced that draft resisters were deserters in time of war and that they faced the death penalty by firing squad. This was no empty threat; seventeen draft resisters were sentenced to death during World War I (though none was executed).[98]

In Providence, Rhode Island, three men publicly refused to register for the draft. The government registered them anyway; when they received induction orders that autumn, the men refused to be inducted. They were court-martialed for desertion and sentenced to twenty years.[99]

In the autumn of 1917, Joseph Stilson, a Philadelphian, wrote, published, and circulated a pamphlet entitled *Let Us Not Go into the Army*, which urged men to resist conscription. Stilson was also the editor of a left-wing, Lithuanian-language newspaper, *Kova*, which reprinted the pamphlet and other antidraft articles. The federal government indicted Stilson—as well as his printer, Joseph Sukys—for violations of the Selective Service and espionage statutes. Sukys tried to defend himself by saying that he had written nothing and published nothing; he just ran the printing presses. The judge dismissed Sukys's argument by explaining the law of conspiracy: when multiple people partake in a conspiracy, all are equally guilty, regardless of how minor a role they played. Stilson was sentenced to three years for publishing the antidraft pamphlet,

97 Charles C. Bush, "The Green Corn Rebellion" (master's thesis, University of Oklahoma, 1932); Nigel Anthony Sellars, "Treasonous Tenant Farmers and Seditious Sharecroppers: The 1917 Green Corn Rebellion," *Oklahoma City University Law Review* 27, no. 1 (Spring 2002): 1097; "Band of 30 Ready to Battle Posses; Fire on Officers," *Muskogee* (OK) *Times-Democrat*, August 13, 1917; "Draft Resisters Face Charges of Treason—Death," *Hartshorne* (OK) *Sun*, August 9, 1917; "W.C.U. Resist Draft Law," *Latimer County* (OK) *News-Democrat*, August 10, 1917; Fite and Peterson, *Opponents of War*, 40–41; "Posse Takes Nineteen in Rocky Ford Battle," *Tulsa* (OK) *Democrat*, August 5, 1917.

98 "Draft Resisters Can Be Executed."

99 Fite and Peterson, *Opponents of War*, 122.

Sukys to three months for running the presses. The US Supreme Court upheld the convictions and sentences.[100]

In May 1918, an armed band of more than fifty men was hiding out from the draft in Bee Creek Gulf, Tennessee. To raise money, the men sold moonshine whisky; they also had a sideline of robbing local residents. In June, cavalry troops were sent in to clear out the draft resisters; one man was killed in the ensuing armed skirmish. Of this episode, the historian Jeanette Keith writes that the "situation illustrates on a small scale the problems that the national government faced in mobilizing men and popular support for the First World War throughout the rural South."[101]

∴

Black Americans, who suffered virulent racism in the broader society, were made to suffer again from the systemic racism of the Selective Service machinery. In World War I, the US military was racially segregated—"the organization of the Army placed colored soldiers in separate units," as Crowder put it.[102] Nevertheless, many Blacks supported the draft and hoped that, by acquitting themselves well in combat, they might advance the cause of civil rights. W.E.B. Du Bois, editor of the NAACP magazine, *Crisis*, expressed this view with much hope: "Never again will darker people of the world occupy just the place they have before." Indeed, Du Bois believed that Secretary of War Baker had done a great service for the civil rights movement by establishing a segregated, Blacks-only officer training camp in Des Moines, Iowa.[103]

Some Black leaders who were more radical disagreed. A. Philip Randolph and Chandler Owen edited the *Messenger*, a magazine that the Department of Justice decried as "the most dangerous" Black publication in America. The view of Randolph and Owen was that white superpatriots could "volunteer to go to France, if they are so eager to make the world safe for democracy. We would rather fight to make Georgia safe for the Negro."[104]

---

100 United States v. Stilson, 254 F. 120 (E.D. Pa. 1918), *aff'd, sub nom.* Stilson v. United States, 250 U.S. 583 (1919). See also Fite and Peterson, *Opponents of War*, 31.

101 Keith, *Rich Man's War*, 162–64.

102 *PMG 2d*, 191.

103 Arthur E. Barbeau and Florette Henri, *The Unknown Soldiers: Black American Troops in World War I* (Philadelphia: Temple University Press, 1974), 7, 11, 57–58. For a general history of the struggle to desegregate the US military, with some attention paid to the US Selective Service, see Gerald Astor, *The Right to Fight: A History of African Americans in the Military* (Cambridge, MA: Da Capo Press, 1998).

104 Unsigned editorial, *Messenger*, November 1917, 31, reprinted in Barbeau and Henri, *Unknown Soldiers*, 12.

Of the total number of men registered for the World War I draft, 9.63 percent were Black. Of the Black registrants, 51.65 percent were put into Class I (immediately eligible to be drafted), but only 32.53 percent of the white registrants were so classified. Within Class I, 36 percent of Blacks received induction orders, while only 24 percent of whites did.[105]

Baker and Crowder had repeatedly promised African Americans that Blacks would participate fully in combat, albeit in segregated units. Had that promise been kept—which it was not—it might have supported the view of those civil rights leaders who hoped to advance their cause by demonstrating the valor of Black soldiers in combat. But the white military brass considered Blacks to be so inferior that they could not be trusted in combat; although there were two Black combat units (the Ninety-Second and Ninety-Third Divisions), "almost all the black draftees would serve in the military equivalent of chain gangs," Arthur E. Barbeau and Florette Henri write in *The Unknown Soldiers: African-American Troops in World War I*. Blacks were given all the most menial jobs but for the most part were not allowed into combat; the army even had a written policy that explained that "white men released by [Black laborer troops] from day work, would be free to spend their entire time being instructed for combat service."[106]

Southern draft boards routinely drafted rural Blacks who owned their own land but not African Americans who worked white-owned land as sharecroppers. This provided a double economic benefit for white landowners: it both removed unwanted (Black) competition and maintained their labor supply of sharecroppers. On several occasions, the government used segregated Black units as strikebreakers. Although the government had abolished bounties for draftees, it did pay fifty-dollar bounties to those who turned in draftees who failed to report for induction. Southern whites, sometimes colluding with Post Office personnel, withheld induction orders from Black registrants and then collected the bounty for the "deserters" whom they subsequently turned in.[107]

In addition to the as many as 3 million men who failed to register, Crowder estimated that, in all, 337,649 men who were drafted had failed to report for induction.[108] He believed this represented a huge success because 337,649 induction refusers constituted a mere 1.41 percent of the nearly 24 million registrants.

105 *PMG 2d*, 191–93; Barbeau and Henri, *Unknown Soldiers*, 36.
106 Barbeau and Henri, *Unknown Soldiers*, 90.
107 Ibid., 37, 70, 89–90, 92–93.
108 *PMG 3d*, 52–53, table 18.

Crowder did the arithmetic correctly; however, he performed the wrong calculation. The relevant calculation is what percentage the 337,649 induction refusers represent of the 3,004,516 men *who received induction orders* (not of the 23 million men who had registered).[109] That calculation yields a much greater percentage: 11.23 percent of men who received induction orders during World War I refused induction—and that is *in addition to* the millions of men who could not receive induction orders because they had failed to register.

5 The Slacker Raids

One unique feature of conscription in the United States during World War I was the "slacker raids" conducted throughout the country during much of 1918. *Slacker* was a widely used term for those who were, in one way or another, not in compliance with Selective Service law. Often these were men who had quietly refused to register, but the term could include those who had evaded service in another way, such as avoiding induction (or obtaining a deferment by means of pretense or dishonesty). The slacker raids were enormous dragnets, often encompassing an entire city, carried out by hundreds, or even thousands, of constabulary men of one stripe or another—federal agents of the Bureau of Investigation (forerunner of the FBI), local police, active-duty service members, and, not incidentally, members of the semiofficial American Protective League.

The first slacker raids occurred in Minneapolis and Pittsburgh on March 26, 1918. On July 6, a company of New Jersey militia raided a meeting of the American Union Against Militarism in Paterson, New Jersey, and arrested thirty-one alleged slackers at bayonet point.[110] Authorities staged a multiday raid in Chicago from July 11 to July 14, stopping 10,000 men and interrogating an estimated 150,000 people. Agents stopped all men who appeared to be of draft age—on the street, on buses and in train stations; in movie houses, ballparks, stores, and parks; on boat landings, and even in private homes—and asked to see their draft cards; agents detained 20,000 who did not produce their cards. By the end of the four days, courts, jails, and federal buildings in Chicago were crammed to overflowing.[111] Stopping tens of thousands of men without either probable cause or warrants was a clear violation of the Fourth Amendment.

---

109 The figure of 3,004,516 is the sum of two numbers provided by Crowder: the 337,649 men who received induction orders and refused induction *plus* the 2,666,867 who received induction orders and were inducted. The latter figure appears in *PMG 2d*, 223, table 79.

110 Capozzola, *Uncle Sam Wants YOU*, 44.

111 "Round-Up at Ball Park," *New York Times*, July 12, 1918; Capozzola, *Uncle Sam Wants YOU*, 44; Emerson Hough, *The Web* (Chicago: Reilly & Lee, 1919), 143–44.

Slacker raids were held in Galveston, Texas, on July 3 and August 28.[112] On July 21, two hundred US marshals, local New York police, and soldiers conducted a slacker raid at Coney Island, stopping thousands of men and arresting 653.[113] Slacker raids were carried out in Cleveland and Dayton, Ohio; Davenport, Iowa; Detroit, Minneapolis, New York, St. Louis, and St. Paul.[114] On August 2, authorities arrested more than 300 men in Trenton, New Jersey, during simultaneous slacker raids at every theater, hotel, and saloon in town. Two weeks later, 300 Bureau of Investigation agents, local police, and members of the American Protective League (APL) raided the Atlantic City boardwalk, piers, restaurants, and cafés. According to one newspaper report, "The roundup was thorough, and every youth who appeared within draft age was taken into custody unless he carried a [draft] card."[115]

On September 2, an enormous slacker raid commenced in the New York metropolitan area, including suburban New York counties and northern New Jersey. A force of 20,000 to 25,000 men participated: the Bureau of Investigation, US marshals, local police forces, and the APL. They stopped hundreds of thousands of men in all manner of public places. Every major commercial and residential block of the city was covered. By the second day of the raid, 10,000 to 12,000 men had been detained in Manhattan; 8,000 to 10,000 in Brooklyn and Queens; 2,200 in the Bronx; and 12,000 in northern New Jersey—in all, more than 40,000 men in custody, all without warrants, all without probable cause, all in violation of the Fourth Amendment. Yet, despite the enormous dragnet, relatively few actual slackers were caught.[116]

One of the most remarkable aspects of the slacker raids was the central importance of the APL. The APL was not a legal entity; it was not a creature of federal or state statute or regulation of any kind or type. As Emerson Hough candidly stated in *The Web*, the "A.P.L. is not and never was a part of any state or national arm, service, department or bureau"; it was, instead, a "wholly unofficial organization." APL members, in short, were vigilantes. Indeed, Hough's *Authorized History of the American Protective League* proudly refers to its extensive work during the slacker raids as the work of "vigilantes."[117]

---

112 Capozzola, *Uncle Sam Wants YOU*, 45.

113 "Get 653 Slackers at Coney Island," *New York Times*, July 22, 1918.

114 Hough, *Web*, 49.

115 "Arrest 300 in Trenton in Slacker Round-Up," *New York Times*, August 3, 1918; "Slacker Raid at Seashore," *New York Times*, August 16, 1918.

116 "Start Drive Today for Draft Slackers," *New York Times*, September 3, 1918; "Seize 20,000 Here in Slacker Search," *New York Times*, September 4, 1918; "30,000 in N.Y. Slackers Net; 600 Are Held," *New-York Tribune*, September 4, 1918.

117 Hough, *Web*, 12, 144.

Yet the APL was a curious species of vigilante. The US Department of Justice endorsed it, and the APL worked closely with the Bureau of Investigation. Its secret oath of membership described the APL as "organized with the approval and operating under the direction of the United States Department of Justice, Bureau of Investigation." That was true. Its stationery and letterhead contained identical language. In fact, after the war, the Selective Service itself acknowledged that the APL "was set up under the direction of the Department of Justice."[118] Although the APL was not affiliated in any way with the US Secret Service, APL members carried badges that said, "American Protective League—Secret Service."[119] All in all, the APL was a unique hybrid of official and totally unofficial; it worked hand-in-glove with federal authorities, including the Selective Service, but was independent, so the government could disclaim its blatantly illegal conduct.

The APL boldly asserted that it knew full well that "breaking and entering a man's home or office place without warrant is burglary," and it boasted that during its relatively brief existence, it carried out literally thousands of such clandestine, warrantless, illegal searches of homes and offices: "The League examined all of the personal and business correspondence of thousands of men who were never the wiser ... in every city in America."[120]

Today, historians widely criticize the APL.[121] However, at the time it was a welcome tool for the government. It could, and did, do many useful—and wildly illegal—things that the government itself may have been more hesitant to do directly. And the organization was especially useful when it illegally detained a suspected slacker (rather than turning the suspect over to federal or state authorities) because, when the APL held a young man, seeking a writ of habeas corpus would not do the detainee a whit of good.

6 Conscientious Objection

In this chapter, and in the succeeding chapters, two broad taxonomies ought to be kept in mind. The first taxonomy considers what men who are religiously or morally opposed to war are—and are not—willing to do when faced with military conscription. These men fall into three broad categories.

---

118 Earl L. Kirchner, *Enforcement of the Selective Service Law*, Selective Service Special Monograph No. 14 (Washington, DC: Government Printing Office, 1950), 12.

119 Hough, *Web*, 12, 29–61, 120, 143, 166, 201; D. Kennedy, *Over Here*, 82.

120 Hough, *Web*, 163–64.

121 D. Kennedy, *Over Here*, 82.

- Some men willingly enter the military and perform noncombatant service, perhaps as battlefield medics or office workers. Such men sometimes reason that they oppose killing, but helping those who are wounded in war is a good thing. Other men who hold this position apply a Kantian categorical imperative to their decision: the world would surely be a better place if everyone refused to bear arms or hurt their fellow human beings.
- Some men are not willing to enter the military but are willing to do alternative, noncombatant service so long as it is under civilian organization and control. These men often reason that the purpose of every soldier is to aid the broader war effort. Medics are meant to heal wounded soldiers so that they can fight again; administrators push the papers necessary to get weapons and ammunition to the soldiers at the front. These men are not willing to aid the overall war effort, even if they are not the ones pulling the triggers on the rifles. However, since they are not opposed to serving their country, these men are willing to work in some civilian capacity that is in the public interest to aid the commonweal.
- Still other men refuse to cooperate in any way with a conscription apparatus, especially in time of war. During World War I, these men were often called absolutists; during the Vietnam War, men taking this position were sometimes called nonregistrants or draft resisters. Sometimes absolutists take their position because they oppose conscription itself, viewing it as involuntary servitude or slavery. More often absolutists view accepting *any* role in the conscription system—even outside the armed forces and under civilian control—to be unacceptably linked to military conscription. Just this sort of reasoning led Cyrus Pringle (see chapter 2) and some other Quaker war objectors to decline to pay a commutation fee during the Civil War; they viewed the fee as a tax on their religious beliefs and believed that it was wrong to cooperate in any way with a system (conscription) whose sole purpose was to obtain men for warfare.

These categories are not firm or ironclad. Some men may fit somewhere between the first and second categories in that they are willing to perform certain noncombatant tasks in the army but not others. Sometimes war objectors themselves are unsure in which category they fit and will start out in a noncombatant role (in or outside the army), only to find that such service rankles their conscience and that they must become absolutist noncooperators. (*Noncooperator* refers to those who refused to register for the draft, would not apply for conscientious objector status, and generally refused to have any interaction whatever with the conscription apparatus.)[122] The line between doing identical medical work in the same hospital as either a private in uniform or a civilian not in uniform is a fine one indeed and may well present an important ethical line to one man but not to another, or the same man may change his mind about his willingness to cross that line from one time to another. For example, Johannes Klaasen, a Mennonite conscientious objector, arrived at Camp Travis (Texas) in the autumn of 1917 and, with a noncombatant certificate from his local draft board, worked in the camp kitchen and store. In December, however, Klaasen, "after much personal soul searching," decided he could no longer do this noncombatant work for the army because he was, after all, aiding the war effort.[123]

The second broad taxonomy concerns the origin of the beliefs that make a man a CO or draft resister. In the case of all twentieth-century conscription laws in the United States, the government has attempted to draw distinctions between war objectors based upon the origins (and extent) of their beliefs.

Some statutes recognize COs only if their beliefs are religious, they are members of a pacifist religious sect (such as Quakers, Mennonites, or Brethren), and they are opposed to all war (not merely a particular war). The conscription statute of May 18, 1917, was of this type; it defined a CO as a man "who is found to be a member of any well-recognized religious sect or organization … whose existing creed and principles forbid its members to participate in war in any form." This formulation excludes religious Catholics and Methodists who believe that the teachings of Jesus preclude participation in war (and excludes Jews who hold similar religious beliefs stemming from their reading of the Old Testament).[124]

Other laws could recognize a right to conscientious objection where the CO is a religious pacifist (opposed to all war) but not require the CO to belong to any specific sect or religion. The views of the Rhode Island General Assembly

122 As I show in succeeding chapters, during twentieth-century drafts, there were often lively disputes between those who believed that the proper response to conscription was to register, cooperate, and seek to be recognized as a conscientious objector and those who believed that total noncooperation was correct. Often such noncooperation was clandestine, but sometimes it was open and public.

123 Gerlof D. Homan, *American Mennonites and the Great War, 1914–1918* (Waterloo, ON: Herald Press, 1994), 109.

124 For discussions of activities of the historic peace churches—Quakers, Mennonites, and Brethren—during World War I, see Albert N. Keim and Grant M. Stoltzfus, *The Politics of Conscience: The Historic Peace Churches and America at War, 1917–1955* (Eugene: Wipf & Stock, 2000); Rufus M. Jones, *A Service of Love in Time of War: American Friends Relief Work in Europe, 1917–1919* (New York: Macmillan 1920); and Homan, *American Mennonites and the Great War*.

when considering the draft US Constitution in 1789 fell into this category, as the legislature would have provided an exemption for "any Person religiously scrupulous of bearing arms," without the necessity of membership in any specific church or denomination.[125]

Yet other laws could recognize a right to conscientious objection for secular pacifists, those who oppose all war of any type but base their objection on non-religious moral, ethical, or philosophical grounds. The Indiana Constitution of 1851 fell into this category, as it provided that "no person conscientiously opposed to bearing arms shall be compelled to do militia duty," as did early constitutions of Iowa, Illinois, Kentucky, and Pennsylvania. This category could include political objectors.

Finally, other laws could recognize a right for COs who were selective objectors. These men do not oppose all war, and thus are not pacifists, but may oppose a particular war. No United States conscription statute has ever recognized selective objectors as COs.

At this point, the reader might object that these fine-point distinctions are too obscure to carry much meaning, that they raise issues akin to medieval religious arguments about how many angels can dance on the head of a pin. However, bear three points in mind. First, these are exactly the distinctions that successive US conscription statutes drew during the twentieth century. Second, this meant that these were the distinctions that men who were, or wanted to be, COs had to consider. Third, these very distinctions—and the perceived absurdity of them—sometimes led some men to decide that the only proper path for them was to be absolutist noncooperators with the entire draft apparatus.

A different but related question is where in the law a "right" to conscientious objection comes from (if there is one). On this question federal courts have always been clear. There is no constitutional right to conscientious objection; it is a matter of "legislative grace." That is, Congress may *create* a right to be a CO when it passes a draft law if, and only if, it wants to do so, and it can set whatever standards it wishes in deciding who qualifies.

The origin of this rule derives from two interwar cases, *Schwimmer* and *Macintosh*, that arose in the context of naturalization.[126] In both cases,

---

125 *Elliot's Debates*, 335.

126 United States v. Schwimmer, 279 U.S. 644 (1929). The decision in *Schwimmer* was 6–3, with justices Oliver Wendell Holmes, Louis Brandeis, and Edward T. Sanford dissenting. See also *United States v. Macintosh*, 283 U.S. 605 (1931). *Macintosh* was decided 5–4, with Chief Justice Charles Evans Hughes and Holmes, Brandeis, and Harlan F. Stone dissenting. For a discussion of the *Schwimmer* and *Macintosh* cases within the broader context of other conscientious objection and naturalization cases at the time, including that of Martha Jane Graber, a religious Mennonite who was initially denied citizenship because of

foreign-born pacifists seeking US citizenship refused to take an oath stating that they would bear arms to defend the United States and were denied citizenship for that reason alone. In both cases, the applicants for citizenship appealed, saying that there is a constitutional right to be a CO and not bear arms. In both cases, the applicants lost, with the Supreme Court finding that there is no constitutional right to be a CO. That doctrine was imported from the naturalization context to the conscription context. As a result, a long line of cases from the US Supreme Court and from lower courts holds that there is no constitutional right to be a CO from conscription.[127]

Thus, it is almost certain that the courts would have found to be constitutional a conscription statute that made no provision whatever for conscientious objection. As this and the following chapters will demonstrate, there have been repeated efforts to have the federal courts find a constitutional right to conscientious objection, but those efforts have never been successful.[128]

∴

In considering the matter of conscientious objection during the eighteen months of conscription during World War I, two overarching themes emerge. First, provisions for COs were not well planned or carefully thought out. Instead, provisions for COs were not merely ill-conceived but often not thought of at all, and even when planned, they were often poorly executed and sometimes contradictory. Second, and probably as a result of the first problem, treatment of COs was wildly inconsistent and chaotic.

The first error—and one of the most egregious—made by Crowder and the newly created Selective Service System with regard to COs was creating much of the conscription apparatus before Congress had permitted them to do so. The chronology here is crucial. On April 6, Congress declared war. On April 7, the administration submitted a *draft* conscription bill to Congress. In April and May, both houses of Congress debated language and amendments for what would end up as the CO section of the law. On May 18, the conscription bill

---

her pacifism, see Duane Stoltzfus, "Nurse without a Country: When a Mennonite Who Refused to Bear Arms Prevailed in Her Bid for Citizenship," *Mennonite Quarterly Review* 93, no. 3 (July 2019): 335–62.

127 Hamilton v. Regents of the Univ. of Calif., 293 U.S. 245, 264 (1934); United States v. Newman, 44 F. Supp. 817, 822 (D.C. Ill. 1942); Hammond v. Lenfest, 398 F.2d 705, 715 (2d Cir. 1968).

128 For a contrary view, see Kastenberg, *To Raise and Discipline an Army*, 106. Kastenberg asserts that, in order to survive constitutional scrutiny, the World War I conscription statute *had* to contain a provision exempting conscientious objectors. I discuss this issue further in chapter 8.

became law. But both Crowder and the Selective Service started to assemble the apparatus before the law was enacted. In his 1920 book, Crowder said, "The whole plan ... was communicated to the governors of all the states in a confidential letter dated April 23, 1917, nearly a month before the Selective Service Bill became law."[129]

Having the registration forms printed before Congress had provided statutory authorization to do so was illegal. Moreover, the War Department knew at the time that it was illegal. In secret testimony about the conscription bill before the House Military Affairs Committee, Baker had stated, point-blank, that printing the forms without congressional authorization was illegal.[130] The result of this illegal conduct was that millions of registration forms were distributed to draft boards *that provided no space and no opportunity whatever for men to indicate that they were COs*. The thinking on the part of the Selective Service—to the extent that there was any thinking at all—seemed to be that, since the statute contemplated that COs would eventually be required to perform noncombatant military duty, Selective Service officials need not be bothered with identifying men who claimed to be COs.[131]

It was left to a private peace organization, the American Union Against Militarism, to advise men how they might attempt to register as COs, that is, try to alert their draft boards to their CO claim at the time of registration. In the days leading up to the June 5 registration, an AUAM leaflet told men, "When you register, state your protest against participation in war." The problem was that the government had provided no way at all to do that. As the AUAM leaflet conceded, men were forced into the position of "stating that protest in such form as they may think best."[132] On June 5, 1917, the Selective Service registered 9.6 million men using forms that provided no means for a man to indicate whether he wanted to be recognized as a CO.

On June 30, 1917—more than three weeks after the first registration had been completed—the Selective Service issued a regulation describing how local draft boards should process registration forms from men who claimed to be COs or, rather, how local draft boards *should have* handled forms already received and processed. The regulation stated that COs should have been asked to provide two separate sworn affidavits. One affidavit should have come from

---

129 Crowder, *Spirit of Selective Service*, 121; *PMG 1st*, 11.

130 *Increase of Military Establishment of 1917*, statement of US Secretary of War Newton Baker, 68.

131 *PMG 2d*, 56. For general information about conscientious objectors, see Thomas, *Conscientious Objector in America*, 96–100, and Fite and Peterson, *Opponents of War*, 122.

132 American Union Against Militarism, "To Men of Military Age Opposed to War!" flyer reprinted in Bennett and Howlett, *Antiwar Dissent*, 57–59.

the CO himself, attesting to his beliefs; the second affidavit should have come from the clerk or minister of a church or denomination attesting to the man's membership. But, of course, none of this had actually been done.[133]

Many, and perhaps most, draft board members had never heard of religious sects "whose existing creed and principles forbid its members to participate in war in any form"; no one ever explained or made clear how these draft boards were to implement the new regulation. In addition, many draft board members who had heard of pacifist religious sects believed, as a matter of principle, that such religious beliefs ought to be ignored. In his first *Annual Report to the Secretary of War*, Crowder reported that a "very large majority of the district boards recommend that religious objectors should not be discharged. Evidently the feeling is widespread that the religious-creed objection is in many cases not genuine."[134]

After the war was over, Crowder said it would have been "impracticable" to help local draft boards with any guidance, so "each board decided for itself on the facts of each claim." As a result, draft boards in places that had significant Quaker or Mennonite populations granted many CO requests, but most draft boards, whose members had never heard of Quakers or Mennonites, approved no CO requests. If and when a draft board determined that a man satisfied the legal requirements for being recognized as a CO, the draft board was supposed to issue him a noncombatant certificate. In practice, this sometimes happened and sometimes did not.[135]

Given the utter lack of any system or organization for addressing the CO claims of registrants, it is remarkable that during the war 64,693 men made claims to be recognized as COs. Of this number, the Selective Service recognized and approved 56,830 claims.[136] Yet, of that number, local draft boards issued only 20,873 noncombatant certificates—that is, a mere 36.7 percent of the men recognized as COs by their draft boards received the official, written recognition of that fact.[137] The War Department never attempted to explain why so few noncombatant certificates were issued, but the failure is utterly in keeping with the general inconsistency and lack of planning applied to all matters related to COs.

133 "Selective Service Rules and Regulations promulgated June 30, 1917," sec. 20, exhibit 1, reprinted in US Secretary of War, *Statement Concerning the Treatment of Conscientious Objectors in the Army* (Washington, DC: Government Printing Office, 1919), 35.

134 *PMG 1st*, 58.

135 *PMG 2d*, 56.

136 Ibid., 57.

137 US Secretary of War, *Statement Concerning the Treatment of Conscientious Objectors*, 3, 15, 24.

On October 10, 1917, the army issued a confidential order directing that COs at the various military bases were to be segregated from the larger body of conscript soldiers and that "these men [the COs] will be handled with tact and consideration." The order was confidential because the War Department feared that "had [it] been made public [it] might have been the means of creating even more objectors." The War Department need not have been worried about this; given the actual treatment accorded to COs, there was little chance that the War Department was going to incite many new men to join the ranks of the COs.[138]

In September 1917, involuntary inductions of registered men started; 296,678 men were conscripted, inducted, and shipped to military bases. The next month 163,493 more men were conscripted and sent to army bases. The drafting of men continued every month and, by March 1918, fully 755,763 men had been drafted. Yet it was not until March 23, 1918, that Wilson issued an order defining what constituted noncombatant service. That meant that between the first drafting of men in September 1917 and the president's order at the end of March 1918, nobody had any idea what should or would be done with thousands of men at army bases all over the country who claimed to be COs.

After the war, the Department of War tried to explain away its disorganization and lack of planning for COs by claiming falsely that the whole matter of COs was new and therefore unanticipated. "The problem [of COs] being a new one, we had to make adjustments as we went along, which accounted for a number of orders and instructions," said F. P. Keppel, third assistant secretary of war, who throughout the war was in charge of handling matters pertaining to COs. But Keppel's statement was not true. Constitutional provisions and statutes dating to before the ratification of the US Constitution had clearly recognized the issue of conscientious objection. The US government had difficulty dealing with COs during World War I, but the issue was not a new one.[139]

Wilson's March 23 order declared that three specific areas of military service were to be deemed noncombatant: service in the Medical Corps, including in combat units at the front; service in the Quartermaster Corps in the United

138 Confidential Department of War Memorandum, exhibit 6, reprinted in US Secretary of War, *Statement Concerning the Treatment of Conscientious Objectors*, 37. In his book *For Peace and Justice*, Charles Chatfield misquotes the confidential order as "respect and consideration" (71), instead of "tact and consideration." More important, and egregiously, Chatfield cites the memorandum uncritically, saying that it was a good-faith effort aimed "at meeting the concerns of some pacifist churches" (71).

139 US Secretary of War, *Statement Concerning the Treatment of Conscientious Objectors*, 9.

States or in war zones but behind the front; and service in the Engineer Corps, also in the United States or in war zones behind the front.[140]

Wilson's March 23 order contained one additional—and remarkable—provision. The president ordered that CO treatment be accorded to two groups of men. The first group was the religious COs recognized by statute—members of a religious sect or organization defined in section 4 of the law. The second category was those "who object to participation in war because of conscientious scruples" but are not members of a religious organization. That is, although Congress had determined that COs must not merely be religious but must also be members of recognized peace churches, Wilson—by executive order and with no statutory authority—extended CO recognition to secular pacifists. The National Civil Liberties Bureau (which I further discuss later this chapter) recognized the importance of Wilson's executive order. In a pamphlet the bureau issued on June 1, 1918, shortly after Wilson's order, the NCLB said: "Although the provisions of our draft act relating to conscientious scruples apply only to members of organized religious sects opposed to war ... the Government's policy has been much more liberal. The President['s] ... 'Executive Order' ... is a conspicuous contribution to the reconciliation of individual conscience and the State under conscription."[141]

One result of the lack of any clear policies by the War Department between the first draft in September 1917 and Wilson's order in March 1918 was that how COs were treated depended substantially on the bases to which they were assigned. Camps Upton (New York), Sherman (Ohio), and Custer (Michigan) may have had some of the best conditions for COs; among the worst were Camps Grant (Illinois), Dodge (Iowa), and Cody (New Mexico).[142] Overall, conditions for COs were often terrible. There were many cases of COs being forced into uniforms, beaten savagely, starved, exposed to extreme cold or heat without adequate provisions, and hazed by mobs of soldiers who just naturally hated COs.[143] COs were threatened and sometimes attacked with a variety of weapons, including bayonets; one CO, John Kos (Camp Grant), had sixty flesh

140 Ibid., 38–39, exhibit 11; Scott H. Bennett, "Conscientious Objectors: Conscience, Courage, and Resistance," in *A History of American Literature and Culture of the First World War*, ed. Mark Van Wienen and Tim Dayton (Cambridge: Cambridge University Press, 2021), 221.

141 National Civil Liberties Bureau, *The Facts about Conscientious Objectors in the United States (under the Selective Service Act of May 18, 1917)* (New York: National Civil Liberties Bureau, 1918), 3.

142 Thomas, *Conscientious Objector in America*, 129; Homan, *American Mennonites and the Great War*, 101.

143 National Civil Liberties Bureau, *Facts about Conscientious Objectors in the United States*.

wounds from repeated stabbings.[144] COs were hanged (but not killed) by ropes tied around their neck, threatened with immediate execution, immersed in the muck of latrines.[145]

During the war, 504 COs were court-martialed, with most of those courts-martial occurring during this early period.[146] In many circumstances, the way these courts-martial came about was complete happenstance. With no coordinated plan for how to deal with COs, an officer would ask a CO to do something perfectly ordinary: peel some potatoes, don a uniform, sign a form. The CO would refuse and then would be court-martialed for refusal to follow orders. At Fort Riley (Kansas), a group of twelve Mennonites was ordered to cut down one sunflower plant. They refused, were court-martialed, and each sentenced to twenty-five years in prison.[147]

One measure of the wildly inconsistent treatment of COs is the widely different sentences handed to them during the war—all for essentially the same offense, not following an order that the CO deemed would require cooperating with the military. Of the 504 COs court-martialed during World War I, one was acquitted on a technicality and the remaining 503 were convicted (50 of these convictions were later overturned). Seventeen were sentenced to death, and 142 were sentenced to life in prison. One was sentenced to 99 years, and several to 40 or 50 years. Eighty-nine COs were sentenced to 10 years, and 130 were sentenced to either 20 or 25 years. Yet plenty of COs received sentences of one or two years, and some received sentences of less than a year.[148]

On March 16, 1918, Congress passed the Furlough Law, which was all of one sentence long. It allowed the secretary of war, at his discretion, to furlough men enlisted in the army to "civilian occupations and pursuits."[149] Although, as written, the law applied to all enlisted men, the statute was widely understood to apply only to COs.

---

144 Thomas, *Conscientious Objector in America*, 146.

145 Ibid., 144.

146 US Secretary of War, *Statement Concerning the Treatment of Conscientious Objectors*, 8–9. Fully 371 of the 504 courts martial occurred during just the first two months of conscription, September 1917 through November 1917. *PMG 2d*, 61.

147 Thomas, *Conscientious Objector in America*, 104, 168.

148 US Secretary of War, *Statement Concerning the Treatment of Conscientious Objectors*, 8–9; Thomas, *Conscientious Objector in America*, 179.

149 An Act To authorize the Secretary of War to grant furloughs without pay and allowances to enlisted men of the Army of the United States, Pub. L. No. 65-105, 40 Stat. 450 (March 16, 1918).

After the Furlough Law was passed, the War Department did nothing for more than two months. Then, on June 1, 1918, in belated response to the new law, the War Department created a three-person board of inquiry, consisting of one military man from the army's Judge Advocate General's Department, and two civilians. The original Board of Inquiry was composed of Major Richard C. Stoddard; Judge Julian W. Mack, of the Seventh Circuit Court of Appeals; and Harlan Fiske Stone, who then was the dean of the Columbia Law School. Major Walter Guest Kellogg later replaced Stoddard.[150]

The purpose of the Board of Inquiry was to examine each CO who had refused noncombatant army service and determine the man's sincerity. For men deemed to be sincerely opposed to war in any form, the order provided that the men could be "furloughed" to civilian work in agriculture or industry. For many Mennonite COs, agriculture furlough was a felicitous plan, as most of them came from farming backgrounds. The board could also, "in exceptional cases," assign men to work in the Friends Reconstruction Unit in France. Creating the Board of Inquiry was the first coordinated effort by the War Department to consider what should be done with these COs—and it came more than a year after the conscription statute was enacted, more than eight months after COs had started arriving at army camps, and after more than two months of inaction after the Furlough Law was passed, but less than six months before the end of the war.

Between June 1918 and January 1919, the Board of Inquiry met and examined 2,294 COs at military bases all over the country. After the armistice was signed in November 1918, the board expanded the scope of its work to include interviews with absolutist noncooperators who, by that time, had been concentrated at Fort Leavenworth, Kansas (and were among the 2,294). Of the total number interviewed, the Board of Inquiry deemed the overwhelming majority, 1,978 of 2,294, to be sincere. Of this number, 1,500 were given furloughs to farms or factories, 88 were assigned to the Friends Reconstruction Unit, and 390 were willing to accept noncombatant military duty.[151]

Part of the idea behind the Board of Inquiry was to provide consistency of results. As Kellogg boasted, "The Board was consistent in its methods and its decisions. All objectors came before one and the same Board."[152] Even in the

150 US Secretary of War, *Statement Concerning the Treatment of Conscientious Objectors*, 10–11, 41–42, exhibit 16; Walter Guest Kellogg, *The Conscientious Objector* (New York: Boni and Liveright, 1919), 25; Harlan Fiske Stone, "The Conscientious Objector," *Columbia University Quarterly* 21, no. 4 (October 1919): 253, 257–58.

151 US Secretary of War, *Statement Concerning the Treatment of Conscientious Objectors*, 24–25.

152 Kellogg, *Conscientious Objector*, 27.

narrowest sense, not all COs came before the same board. Since the board considered itself to have a quorum of one member, COs actually came before seven different boards—three boards with one member each, three different boards with different combinations of two members each, or the full board with all three men. More important, the board treated different men differently. Some men were interviewed alone, others were interviewed in groups of different sizes. Some men were interviewed for a minute or two before being declared sincere.[153] Other men were questioned at length and thrown highly legalistic trick questions by Stone. Stone liked to ask a CO whether he had ever used a postage stamp since the United States had entered the war. If the CO answered yes, this was taken as a sign of insincerity because the penny for the stamp went to the government.[154] The board often probed the avowed pacifism of COs with hypothetical questions such as what he would do if his mother were being raped; one man was asked what he would do if a "big nigger" raped his mother.[155]

Perhaps the greatest problem with the board was the fundamental attitude that its members had toward conscientious objection as a religious or philosophical position and toward individual COs. Today, Harlan Fiske Stone is a revered figure, widely viewed as a "wise and visionary" chief justice of the US Supreme Court and a hero of Roosevelt's New Deal.[156] One scholarly article, almost reverential in tone, entitled "Harlan Fiske Stone: In Defense of Individual Freedom," focused exclusively on Stone's wartime service on the Board of Inquiry and concluded that Stone "brought sympathy and understanding to the relatively novel and vexing problem of conscientious objectors." Jeremy K. Kessler, of Columbia Law School, writes that Stone "imbued the national bureaucracy with democratic norms of deliberation and dissent" and that he "went out of [his] way to accommodate minority views during World War I."[157]

However, the unmistakable impression one gets from Stone's own discussion of his service on the Board of Inquiry is rather different. Indeed, Stone's plainly stated views are so revealing that they are worth considering in some detail.

After the war, Stone published an article about his service on the board in the *Columbia University Quarterly*. At the outset, Stone identifies the basic

---

153 Thomas, *Conscientious Objector in America*, 110–11.

154 Capozzola, *Uncle Sam Wants YOU*, 60.

155 Homan, *American Mennonites and the Great War*, 137.

156 Michael McConnell, "Symposium: Fidelity in Constitutional Theory," *Fordham Law Review* 65, no. 4 (March 1997): 1269, 1273, refers to Justice Stone as a "revered figure." Barry Cushman, "Rethinking the New Deal Court," *Virginia Law Review* 80, no. 1 (February 1994): 201, refers to Stone as a "wise and visionary Justice."

157 Alpheus Thomas Mason, "Harlan Fiske Stone: In Defense of Individual Freedom, 1918–1920," *Columbia Law Review* 51 (February 1951): 147; Kessler, "Administrative Origins," 1083, 1090.

problem with COs: these were men who held "peculiar beliefs, due to an almost infinite variety of causes, of which religious training, environment, ill-balanced mentality, not to say sheer stupidity, lack of self-discipline, false social and political theories were the most frequent." Stone had little patience for "religious conscientious objectors.... Their average mentality was low ... all in all they presented a depressing example of dense ignorance of what was going on in the world, and stolid indifference to those moral and political questions which were so profoundly stirring the minds and hearts of their fellow countrymen." Stone complained of "the tedium of examination of long lines of stolid, bovine-faced Dunkards, Mennonites, Hutterites, and the like, each one supremely interested in the salvation of his own soul even though the world perish. Many of the religious objectors were plainly fanatics, with abnormal mental experiences, requiring the attention of a physician or psychiatrist."[158]

For Stone, however, it was the "non-religious objectors who brought to the Board its real perplexities.... They were sublime egotists and glib talkers. Their self-confidence was monumental."[159]

After his detailed and lengthy explication of the intellectual shortcomings and moral wrongheadedness of COs, Stone declares, with not the slightest hint of irony or self-awareness, that "the writer has thus far expressed no opinion of the merits of [the CO's] cause."[160]

Walter Guest Kellogg, the military man on the panel, held similar views about COs. To Kellogg, COs were variously morons, mental weaklings, stupid, dull, glib, egotistical, and self-centered.[161]

The personal views held by members of the Board of Inquiry—that most COs were, variously, mentally ill, stupid, bovine-faced, strange, abnormal, stubborn, peculiar, rigid, narrow-minded, obstinate, hypocritical morons—could not but have influenced their meetings with the COs and their dispositions of their cases.

Of the various different types of draft objectors, the absolutists often experienced the worst treatment. The cases of four Hutterite COs from South Dakota, drafted together in July 1918, presents an exemplar of the type of treatment that many COs—especially the absolutists—faced.

158 Stone, "Conscientious Objector," 254, 260, 262.
159 Ibid., 263–65.
160 Ibid., 268.
161 Kellogg, *Conscientious Objector*, 67–94.

Hutterites are a pacifist Anabaptist denomination (as are Mennonites) founded in sixteenth-century Austria. In the nineteenth century, many Hutterites settled in South Dakota and other midwestern states, where they formed tightly knit farming communities that practiced communal ownership of tilled lands.[162]

On the national registration day, June 5, 1917, four Hutterite men registered at the draft board in Beulah, South Dakota. Three of the men were brothers: David, Michael, and Joseph Hofer; the fourth man, Jacob Wipf, was a brother-in-law. All four were married and had children. Under Selective Service law, this should have made all of them exempt from the draft—except that the Hutterite practice of communal holding of farmland led the four men to conclude that they could not truthfully claim that their families were *solely* dependent on them for support. Each man told the draft board, "My creed forbids any military service." But that did them not a whit of good. On May 25, 1918, the four men were drafted and put on a military train to Camp Lewis (Washington). En route to camp, other recruits attacked the four Hutterites; the men were generally roughed up and had their beards forcibly shaved off. For Hutterites, having a beard is a matter of religious obligation. While the government did not organize these attacks, it also did nothing to protect the draftees under its control.[163]

Once at Camp Lewis, the Hofers and Wipf refused to sign a paper entitled "Declaration of Soldier," because they did not consider themselves to be soldiers, and they refused to line up in military formation. As a result, they were put into the guardhouse. Typical treatment for COs in the guardhouse of a military camp was a diet of bread and water and bare cells without beds or other furniture.[164]

On June 10, 1918, the four Hutterites were tried together by court-martial; they were charged with disobeying orders for refusing to sign the "Declaration of Soldier." All were quickly convicted and sentenced to twenty years.[165]

In July 1918, the four Hutterites were transferred to a military prison, Pacific Branch, Disciplinary Barracks (on Alcatraz Island in San Francisco Bay), where they were placed in solitary confinement in the Hole. The Hole at Alcatraz consisted of cold, dark, damp, underground dungeons with no natural or artificial light, no bed, no furniture of any kind, and no sanitary facilities. Some cells had a metal pail for excrement. The men were fed bread and water. The cells were

162 Duane C. Stoltzfus, *Pacifists in Chains: The Persecution of Hutterites during the Great War* (Baltimore: Johns Hopkins University Press, 2013), 2–5, 14–18.

163 Ibid., 7–10, 18–20.

164 Ibid., 23–29.

165 Ibid., 77–103.

infested with vermin, including bedbugs, and rats danced on the men's sleeping bodies at night. The men refused to don uniforms, so they were clad only in underwear. For the first four and a half days in the Hole, they received no food. The men were sometimes beaten; on one occasion, Michael Hofer was beaten into unconsciousness. But perhaps the worst aspect of their treatment was that for eight or nine hours a day, the men were chained to the bars by their hands and arms, with the chains drawn up so that the men's feet barely touched the floor.[166]

In November 1918, immediately after the armistice was signed, the four Hutterites were transferred to the army's Disciplinary Barracks at Fort Leavenworth (Kansas), where the War Department was consolidating many of the absolutists. By this time, the four men were in seriously weakened physical condition from their treatment at Alcatraz. They arrived at Leavenworth late at night on November 19 and were forced to march in the cold to the prison. Again, the men were shackled to the bars of their cells for nine hours each day, strung up so that their toes barely touched the floor. On November 29, Joseph Hofer died of his mistreatment, leaving a widow and two children. As a final insult, the army dressed Joseph Hofer's corpse in the military uniform Hofer had refused to don in life before it shipped the body home to his family. On December 2, Michael Hofer died of his treatment. As if to underline the wholly inconsistent and haphazard treatment accorded to COs, the surviving brother, David Hofer, was released without any explanation on December 2 to return home. Jacob Wipf was not released until April 13, 1919, as part of a broader release of COs.[167]

On November 21, 1918, the National Civil Liberties Bureau, which advocated on behalf of COs in prison, published a pamphlet entitled *Political Prisoners in Federal Military Prisons*. The NCLB provided the names of specific prisoners and detailed the abuses they were suffering. The pamphlet concluded with this statement: "The use of torture in military prisons … is as unnecessary as it is barbarous."[168]

Asserting that the US government was using torture on nonviolent COs in military prisons was then and is now a shocking allegation, yet the treatment to which COs were deliberately subjected for long periods comports with the ordinary, everyday meaning of the word *torture*, with the dictionary definition of the word, and with the definition of torture under international law. The ordinary, everyday meaning of *torture* is the deliberate infliction of extreme pain. The dictionary definition is "the inflicting of severe pain

166 Ibid., 104–37.

167 Ibid., 156–83, 184–210.

168 National Civil Liberties Bureau, *Political Prisoners in Federal Military Prisons* (New York: National Civil Liberties Union, November 21, 1918).

(as from burning, crushing, or wounding) to punish, coerce, or afford sadistic pleasure."[169] And part 1, article 1 of the international treaty on torture, the Convention Against Torture and Other Cruel, Inhuman or Degrading Treatment or Punishment, says that "the term torture means any act by which severe pain or suffering, whether physical or mental, is inflicted on a person."

Howard Moore, a political CO imprisoned at the Fort Leavenworth Disciplinary Barracks and elsewhere, later published a memoir about his World War I experience. Moore detailed the treatment he and other COs received at Fort Upton on Long Island. When Moore first arrived at the army base in late April 1918, a sergeant took offense at his noncooperation and threw him out of a second-floor window. Each night other soldiers terrorized the COs by shooting randomly into the barracks where the COs were housed together. One CO who made the mistake of asking for seconds in the mess hall was beaten unconscious.[170]

In his postwar memoir of his experience on the Board of Inquiry, Kellogg acknowledged that COs were sometimes mistreated. This is what he said: "It is not surprising that, in a certain few cases, the patience of the officer was so exhausted by the maliciously annoying attitude of various objectors in his charge that he lost his temper and maltreated them."[171]

∴

Once the armistice was signed in November 1918, the NCLB and several peace churches mounted a pressure campaign for the Wilson administration to grant an amnesty to COs in military and civilian prisons. The Board of Inquiry was assigned to interview the absolutists who had been concentrated in Leavenworth; in January 1919, at the recommendation of the board, the War Department released 113 COs from Leavenworth.[172] There was significant national protest against the release of these COs. Through an administrative error, they were released with back pay. Naturally, payment to these men also stirred

169 Merriam-Webster's Collegiate Dictionary, 11th ed., https://www.merriam-webster.com/dictionary/torture, last accessed November 11, 2022.

170 Howard W. Moore, *Plowing My Own Furrow* (Syracuse, NY: Syracuse University Press, 1993), 99, 103.

171 Kellogg, *Conscientious Objector*, 84.

172 The date these 113 men were freed from Leavenworth is a matter of some debate. The US Department of the Army states that these men were released on January 5, 1919, as described in US Secretary of War, *Statement Concerning the Treatment of Conscientious Objectors*, 11. However, Norman Thomas says in *Conscientious Objector in America* (20) that the date of their release was January 23, 1919. The latter date is confirmed by "War Objectors Freed from Leavenworth," *New York Times*, January 28, 1919.

public outcry, although the COs—believing it was wrong and hypocritical to accept money from the War Department—either returned the money to the War Department or donated it to the Red Cross.[173]

The War Department granted no general amnesty, but through the rest of 1919 and into 1920, it reviewed the CO cases individually. The department reduced many of the longer sentences (some more than once) and commuted some sentences to time served. All the death sentences were commuted, and no CO was executed. Additional groups of COs were released, with the last COs freed on November 23, 1920, two days before Thanksgiving. Howard Moore was in the last group to be released.[174]

∴

In the immediate aftermath of the enactment of conscription, several new organizations were created to address the new situation. Two were meant specifically to assist men of draft age. One was the NCLB, which started out as a committee of the American Union Against Militarism. The pacifist Roger Nash Baldwin was the NCLB's initial staffer; he was a public draft resister who was imprisoned the following year. By the autumn of 1917, the organization had become independent of its parent organization. The other newly created organization was the Bureau of Legal Advice (BLA), founded by Martha Gruening, a lawyer, and the pacifist socialist feminists Frances Witherspoon and Tracy Mygatt. Witherspoon served as the executive secretary of the BLA; Mygatt and a local CO, Bruno Grunzig, were volunteer staffers; the chief counsel was Charles Recht. The NCLB and the BLA had offices (on different floors) of the Ginn Building at 70 Fifth Avenue in New York; their respective boards shared some (but not all) members, and they cooperated in their efforts. When the BLA began in May 1917, the NCLB provided the BLA's part-time secretarial help.[175]

Significantly, these two organizations invented draft counseling. Draft counselors met, one on one, with men of draft age to advise them of their options and rights under the Selective Service law. Many men of draft age were immigrants with varying knowledge of English; negotiating arcane regulations written in stilted English was naturally difficult for them. Draft counselors

173 US Secretary of War, *Statement Concerning the Treatment of Conscientious Objectors*, 11–12; Thomas, *Conscientious Objector in America*, 221–24.

174 US Secretary of War, *Statement Concerning the Treatment of Conscientious Objectors*, 11–12; Thomas, *Conscientious Objector in America*, 221–24, 248–49; Moore, *Plowing My Own Furrow*, 158.

175 Frances H. Early, *A World without War: How U.S. Feminists and Pacifists Resisted World War I* (Syracuse, NY: Syracuse University Press, 1997), 90–94.

helped them secure deferments or exemptions to which they were legally entitled. With regard to COs, draft counselors could advise men in advance what steps they might take to increase the chances of having their CO claims recognized by their local draft board.[176]

The BLA and NCLB also provided lawyers to represent antidraft activists, COs, and nonregistrants. The BLA staff also often kept in touch by mail with COs after they had been inducted in the army. This had two distinct but related salutary effects. It provided much-needed emotional support for COs, who often endured horrible conditions in the army. It also provided outside supporters with accurate firsthand information about the mistreatment that COs confronted, which allowed the supporters to publicize those conditions and advocate for reforms.

The two organizations coordinated their work and created a division of labor. The BLA handled draft counseling in New York while the NCLB established satellite efforts in other parts of the country. This aspect of the NCLB's work was especially effective: "Within a few months, the Bureau claimed 120 cooperating attorneys throughout the United States."[177] The NCLB also published pamphlets for national distribution, including *Conscription and the "Conscientious Objector" to War* (May 1917) and *The Facts about Conscientious Objectors in the United States* (June 1918), although some of the information for NCLB publications came from the correspondence that BLA staff maintained with imprisoned COs.[178]

Today, the National Civil Liberties Bureau is far better known than the short-lived Bureau of Legal Advice. There are, no doubt, multiple reasons for this, including that in 1920 the NCLB became the American Civil Liberties Union (ACLU), which today has grown into the nation's most prominent civil liberties advocate and litigator. In addition, the historian Frances Early argues that sexism is an important reason why the male-led NCLB is better known today than the women-led BLA.[179]

One other organization that came into existence in 1917 in response to the draft was the American Friends Service Committee. AFSC had the dual purposes of providing direct humanitarian assistance to civilian war victims in devastated Europe and of providing a means of alternative service—not under

176 Ibid., 27, 64–72, 93.

177 Laura Weinrib, "Freedom of Conscience in War Time: World War I and the Limits of Civil Liberties," *Emory Law Journal* 65, no. 4 (2016): 1051, 1092.

178 National Civil Liberties Bureau, *Facts about Conscientious Objectors*; Early, *A World without War*, 92–93, 116–18.

179 Early, *A World without War*, 51–54; Ernest Freeberg, *Democracy's Prisoner: Eugene V. Debs, The Great War, and the Right to Dissent* (Cambridge, MA: Harvard University Press, 2010).

military control and auspices—for Quaker and other COs during the war. The first organizational meeting for AFSC was held in Philadelphia on April 30, 1917, several weeks before conscription was enacted. Vincent Nicholson was named the executive secretary and an influential Quaker, Rufus Jones, was made chairman.[180]

During World War I (and, indeed, for the rest of the twentieth century) one of the criticisms leveled against COs was that they were slackers who only sought personal safety by evading military duty. In this context, it is worth noting that many AFSC volunteers went to Europe before they were drafted because they were so eager to serve their country and humanity but nonviolently in time of war. Other men who had failed their draft physical examinations nevertheless volunteered for AFSC work in Europe. As Jones put it, "A man with an imperfect eye which made his aim inaccurate for shooting could perfectly well build houses for French refugees." AFSC volunteers sometimes found themselves under fire at or near the front lines.[181]

Eventually 600 men served in Europe with the AFSC during and after World War I. Eighty-nine were COs directed to the AFSC by the Board of Inquiry. Local draft boards sent 100 more to the AFSC. Four hundred additional men volunteered without being obliged to do so by law. AFSC volunteers constructed and equipped hospitals; worked to feed, clothe, and house refugees; rebuilt schools and provided teachers; and provided services to large numbers of newly created orphans. They worked in the Somme and Marne regions in the aftermath of the famous battles at those sites.

The AFSC continues to exist today. It continued its work of providing relief to civilian war victims during World War II, and in 1947, the AFSC was the corecipient (with the [British] Friends Service Council) of the Nobel Peace Prize, in recognition of its relief work in Europe during and after World War II.

∴

Once the armistice was signed on November 11, 1917, the conscription apparatus came to a halt almost immediately. That same day, the Selective Service canceled all draft calls for the army. On November 16, all physical examinations for draftees stopped, and on December 10, 1917, local draft boards closed.[182]

---

180 This account of the origins of the American Friends Service Committee is based on Jones, *A Service of Love in Time of War,* and William J. Frost, "Our Deeds Carry Our Message: The Early History of the American Friends Service Committee," *Quaker History* 81, no. 1 (Spring 1992): 3–51.

181 Jones, *A Service of Love in Time of War*, 57.

182 *PMG 3d*, 8, 7.

7 Draft Prosecutions during World War I

In August 1917, three months after the draft law was enacted, Charles T. Schenck was the general secretary and Dr. Elizabeth Baer was secretary of the Socialist Party in Philadelphia, when the party published and distributed an antidraft leaflet.[183] The headline on the front of the flyer declared, in capital letters, "LONG LIVE THE CONSTITUTION OF THE UNITED STATES," followed by "Wake Up, America! Your Liberties Are in Danger!" (see Figure 4.2). Directly under this headline, set off from the rest of the text, was the text of the Thirteenth Amendment. The front of the flyer went on to say: "The Thirteenth Amendment to the Constitution of the United States, quoted above, embodies this sacred idea [that slavery is wrong]. The Socialist Party says that this idea is violated by the Conscription Act. When you conscript a man and compel him to go abroad to fight against his will, you violate the most sacred right of personal liberty, and substitute for it what Daniel Webster called "despotism in its worst form."

The reverse side of the flyer said, "If you are conscientiously opposed to war, if you believe in the commandment 'thou shalt not kill,' then that is your religion, and you shall not be prohibited from the free exercise thereof."

On September 15, Schenck and Baer were indicted under the Espionage Act on three counts for producing and mailing the flyer.[184] In December, they were tried, convicted, and sentenced. They appealed to the US Supreme Court and made an argument substantially based upon the First Amendment. In their brief to the Supreme Court, their lawyers said that "free and general discussion of public matters [is] essential to prepare the public for an intelligent exercise of their rights as citizens and to subject those in power to scrutiny." In contrast, the government's brief to the Supreme Court dismissed the First Amendment argument, saying, the "First Amendment of the Constitution [is] not involved in this case" and "Defendant's [*sic*] claim of unconstitutionality is frivolous."[185]

The Supreme Court agreed with the government and sent Schenck and Baer to prison for publishing and distributing their leaflet. In a unanimous decision written by Justice Oliver Wendell Holmes, the court ruled that the First Amendment did not protect Schenck and Baer. Nothing in the opinion indicated that any of the Supreme Court justices, including Holmes, thought

---

183 Schenck v. United States, 249 U.S. 47 (1919), affirming United States v. Schenck, 253 F. 212 (E.D. Pa. 1918).

184 *U.S. Supreme Court Transcript of Record* in Schenck v. United States, 249 U.S. 47 (1919) (No. 18-347) (hereafter Schenck transcript); indictment, Schenck transcript, 5–9, and judge's charge to the jury, 64–65.

185 Brief for the United States, Schenck transcript, 12.

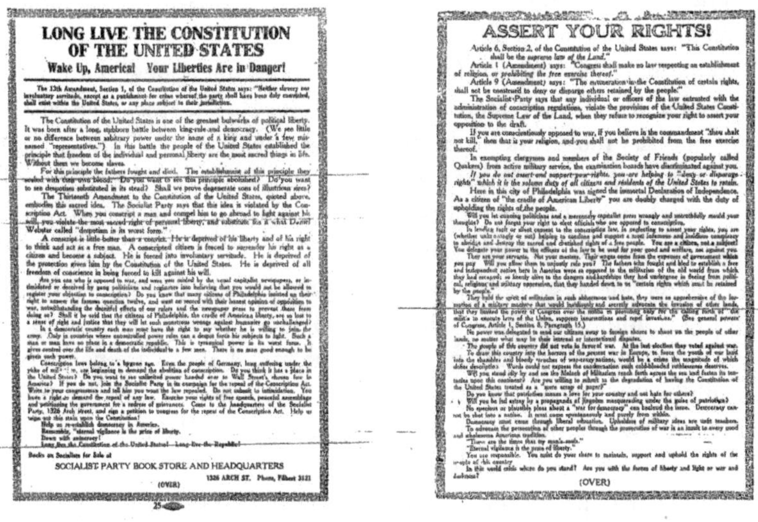
LONG LIVE THE CONSTITUTION OF THE UNITED STATES

Wake Up, America! Your Liberties Are in Danger!

The 13th Amendment, Section 1, of the Constitution of the United States says: "Neither slavery nor involuntary servitude, except as a punishment for crime whereof the party shall have been duly convicted, shall exist within the United States, or any place subject to their jurisdiction.

The Constitution of the United States is one of the greatest bulwarks of political liberty. It was born after a long, stubborn battle between king-rule and democracy. (We see little or no difference between arbitrary power under the name of a king and under a few misnamed "representatives.") In this battle the people of the United States established the principle that freedom of the individual and personal liberty are the most sacred things in life. Without them we become slaves.

For this principle the fathers fought and died. The establishment of this principle they sealed with their own blood. Do you want to see this principle abolished? Do you want to see despotism substituted in its stead? Shall we prove degenerate sons of illustrious sires?

The Thirteenth Amendment to the Constitution of the United States, quoted above, embodies this sacred idea. The Socialist Party says that this idea is violated by the Conscription Act. When you conscript a man and compel him to go abroad to fight against his will, you violate the most sacred right of personal liberty, and substitute for it what Daniel Webster called "despotism in its worst form."

A conscript is little better than a convict. He is deprived of his liberty and of his right to think and act as a free man. A conscripted citizen is forced to surrender his right as a citizen and become a subject. He is forced into involuntary servitude. He is deprived of the protection given him by the Constitution of the United States. He is deprived of all freedom of conscience in being forced to kill against his will.

SOCIALIST PARTY BOOK STORE AND HEADQUARTERS

(OVER)

ASSERT YOUR RIGHTS!

Article 6, Section 2, of the Constitution of the United States says: "This Constitution shall be the *supreme law of the Land*."

Article 1 (Amendment) says: "Congress shall make no law respecting an establishment of religion, or *prohibiting the free exercise thereof*."

Article 9 (Amendment) says: "The enumeration in the Constitution of certain rights, shall not be construed to deny or disparage others retained by the people."

The Socialist Party says that any individual or officers of the law entrusted with the administration of conscription regulations, violate the provisions of the United States Constitution, the Supreme Law of the Land, when they refuse to recognize your right to assert your opposition to the draft.

If you are conscientiously opposed to war, if you believe in the commandment "thou shalt not kill," then that is your religion, and you shall not be prohibited from the free exercise thereof.

In exempting clergymen and members of the Society of Friends (popularly called Quakers) from active military service, the examination boards have discriminated against you.

If you do not assert and support your rights, you are helping to "deny or disparage rights" which it is the solemn duty of all citizens and residents of the United States to retain.

Here in this city of Philadelphia was signed the immortal Declaration of Independence. As a citizen of "the cradle of American Liberty" you are doubly charged with the duty of upholding the rights of the people.

(OVER)

FIGURE 4.2 After the Socialist Party chapter in Philadelphia published this antidraft flyer, the party officers Charles T. Schenck and Elizabeth Baer were prosecuted and sent to prison for the crime of speaking or writing against conscription. The flyer appears in the case record of *Schenck v. United States*, 249 U.S. 47 (1919).

that the defendants' First Amendment argument was at all serious. The entire opinion was only six paragraphs, and Holmes rejected the First Amendment argument in exactly one paragraph: "The most stringent protection of free speech would not protect a man in falsely shouting fire in a theatre and causing a panic."[186] To a unanimous Supreme Court, printing a flyer that said, "Long live the Constitution of the United States" and that quoted (accurately) various constitutional provisions was exactly like shouting "Fire!" in a crowded theater.

• •
•

Jacob Frohwerk was accused in a thirteen-count indictment of writing twelve articles in a Missouri German-language newspaper, the *Staats Zeitung*, that were critical of the war and were likely to cause people to resist the draft. One article, the Supreme Court tells us, "begins by declaring it a monumental and inexcusable mistake to send our soldiers to France." Another, published shortly

186 Schenck, 249 U.S. at 52.

after the Green Corn Rebellion in Oklahoma, deplores "the draft riots in Oklahoma" and uses "innuendo" to criticize conscription. This article also described "the sufferings of a drafted man."[187]

The Supreme Court acknowledged that "it does not appear that there was any special effort to reach men who were subject to the draft" but neglected to point out that most Americans of draft age could not read the newspaper, because it was written in German. Frohwerk was sentenced to twelve terms of ten years each, with the sentences to run concurrently. Once again, Holmes wrote the unanimous Supreme Court opinion.

On June 16, 1918, the Socialist leader Eugene V. Debs delivered a two-hour speech in Canton, Ohio. Although long, the speech was extremely bland. It scarcely mentioned either the war or the draft; rather, it was a general pæan to the theoretical virtues of socialism. Even the Supreme Court recognized this: "The main theme of the speech was socialism, its growth, and a prophecy of its ultimate success." However, the nation was at war and Debs did begin by saying that he had just visited two Socialists who were in jail for draft opposition. Debs was indicted on thirteen counts of violating the Espionage Act, with one count alleging that Debs's speech would encourage opposition to the draft. Debs was convicted and sentenced to twenty years. Holmes wrote the unanimous Supreme Court opinion affirming both the conviction and sentence.[188]

The Schenck, Frohwerk, and Debs cases were not outliers. During World War I, there were more than 1,900 criminal prosecutions involving antiwar or anti-draft speeches, newspaper or magazine articles, leaflets and flyers, pamphlets, and books.[189] In the two-year period between June 30, 1917, and June 30, 1919, 877 people were convicted. Although some of those 1,900 cases did not directly involve the draft, many did.[190]

Orville Anderson of South Dakota said to George Rundal, who was of draft age, "Wilson is a murderer in the first degree. He is murdering, not only the Germans, but his own people, the American people, as well, and he is violating the Constitution of the United States by drafting men and sending them to fight in Europe." Anderson was convicted of violating the Espionage Act and was sentenced to four years in prison and a fine of $1,000.[191]

---

187 Frohwerk v. United States, 249 U.S. 204, 207 (1919).

188 Debs v. United States, 249 U.S. 211, 212 (1919). Eugene V. Debs (1855–1926) was an early supporter of the IWW and was the Socialist Party's candidate for president five times. For an excellent biography of Debs, see Nick Salvatore, *Eugene V. Debs: Citizen and Socialist*, (Urbana: University of Illinois Press, 1982).

189 Zechariah Chafee, *Freedom of Speech* (New York: Harcourt, Brace & Howe, 1920), 1.

190 Walter Nelles, *Seeing Red: Civil Liberty and Law in the Period following the War* (New York: American Civil Liberties Union, 1920), 4.

191 Anderson v. United States, 264 F. 75, 76 (8th Cir. 1920).

John T. Dunn, Theodore Hiller, and Adolph Fred Yanyar were the three draft refusers in Rhode Island. By late 1917, they were serving twenty-year sentences in the federal penitentiary in Atlanta. At a public meeting called the People's Forum in Providence on January 13, 1918, Joseph Coldwell said, "Dunn, Yanyar, and Hiller ... are victims of a damnable system of government, for which those who support it are directly responsible.... Dunn, Yanyar, and Hiller are guilty of no crimes. They merely refused to become uniformed murderers." For this statement, Coldwell was prosecuted, convicted, and sentenced to three years in prison.[192]

On July 17, 1917, Kate Richards O'Hare, a "professional Socialist lecturer" (according to the court), delivered an address at a public meeting in Bowman, North Dakota. Before an audience of one hundred people she said that "any person who enlisted in the army of the United States for service in France would be used for fertilizer, and that is all he was good for; and that the women of the United States were nothing more nor less than brood sows, to raise children to get into the army and be made into fertilizer."[193] O'Hare was sentenced to five years. At her sentencing, Judge Martin Wade said, "The American people are not going to stand idly by and see these boys that are marching away to the front, shot in the back by cowards and traitors."[194]

In April 1918, Scott Nearing and the American Socialist Society were indicted for their roles (Nearing as author, the society as publisher and distributor) in producing and distributing a pamphlet entitled *The Great Madness*. The pamphlet was primarily opposed to the war, but chapter 8 addressed conscription. The four-count indictment against Nearing and the society under the Espionage Act was the familiar one of obstructing enlistment and recruiting into the army and conspiring to do so. The trial occurred in February 1919, after the war had ended.[195]

The trial judge was Julius B. Mayer, who had by then handled many draft cases, including those of Emma Goldman, Alexander Berkman, and Roger

192 Coldwell v. United States, 256 F. 805, 807 (1st Cir. 1919).

193 O'Hare v. United States, 253 F. 538 (9th Cir. 1918). Kate Richards O'Hare (1876–1948) was a Socialist Party activist and lecturer. For a contemporaneous account of her case, see National Civil Liberties Bureau, *The Conviction of Mrs. Kate Richards O'Hare and North Dakota Politics* (New York: National Civil Liberties Bureau, 1918). See Kathleen Kennedy, *Disloyal Mothers and Scurrilous Citizens: Women and Subversion during World War I* (Bloomington: Indiana University Press 1999), for a contemporary view of the "implications her case held for the meaning of motherhood and its relationship to women's participation in public culture" (21).

194 Walter Nelles, *Espionage Act Cases: New Law in the Making as to Criminal Utterances* (New York: National Civil Liberties Bureau, July 1918), 47.

195 *The Trial of Scott Nearing and the American Socialist Society*, trial transcript in American Socialist Society et al. v. United States, 266 F. 212 (2d Cir. 1920) (No. 19-186) (hereafter, Nearing trial transcript).

Nash Baldwin. Mayer made little pretense of hiding his bias against the defendants. At one point he coached the prosecutor on what objections he ought to make, and on another occasion Mayer took over cross-examination from the prosecutor when Mayer believed that the prosecutor was not doing a good enough job. Nevertheless, two of the four counts against the defendants had to be dismissed, and the jury acquitted Nearing on both of the remaining counts. The society was convicted on one count and fined.[196]

A modern reader may well be puzzled as to how any of these cases can be squared with the First Amendment's guarantee of free speech. How could the Supreme Court send Charles Schenck to prison for publishing a flyer that says, "Long Live the Constitution of the United States"? How could Joseph Coldwell be sentenced to three years for publicly saying that three Rhode Island draft resisters had "merely refused to become uniformed murderers"? The answer is quite simple. At the time, the uniform understanding of the free speech provision of the First Amendment was that it prevented only prior restraint—that is, the government's prohibiting someone from saying or publishing something in advance of publication. According to this traditional understanding, the First Amendment did not preclude punishing someone after the fact for saying or publishing something wrong or bad.

This traditional understanding had roots in both civil and criminal law. On the civil side, an allegation of slander was (and is) a civil action against someone who supposedly said or published something both false and harmful about another person; the First Amendment was understood as doing nothing to bar slander lawsuits. (The same understanding obtains today.) On the criminal side, Congress had made sedition and speaking publicly against the draft criminal offenses. The First Amendment did not protect a speaker from criminal charges brought by the government if the speaker said things that the government had outlawed—because those criminal charges were brought only after the speaker had uttered the offending words. Because the government had not prevented the speaker from speaking at all, there was no prior restraint and therefore no First Amendment violation.

This traditional understanding of the First Amendment derived from Lord Blackstone, the eighteenth-century British judge. Freedom of the press, said Blackstone, is "laying no previous restraint upon publications, and not in

196 Nearing trial transcript, 173, 163.

freedom from censure for criminal matter when published. Every free man has an undoubted right to lay what sentiments he pleases before the public; but if he publishes what is improper, mischievous, or illegal, he must take the consequences." Justice Holmes had said much the same thing a decade before US entry into World War I in the Supreme Court case *Patterson v. Colorado*. The purpose of the First Amendment, said Holmes, is merely "to prevent ... previous restraints upon publications," but it does "not prevent the subsequent punishment" of illegal speech.[197]

In the foregoing cases no one was prevented in advance from speaking. They were merely sentenced to prison after the fact for saying things that were illegal.

8 Legal Cases Pertaining to the Conscription Statute

The leading US Supreme Court case on the draft during World War I—indeed, the most important conscription case ever decided by the US Supreme Court—is *Selective Draft Law Cases* (*Arver*). In this case, the Supreme Court consolidated and rendered one opinion in six substantially identical cases from different men all challenging the constitutionality of the draft.[198]

It was sensible for the Supreme Court to consolidate these cases. Each of the cases arose in the same manner: a man of draft age publicly refused to register for the draft on June 5, 1917. Each man was promptly arrested and charged with one count of not registering. At trial, each acknowledged that he was of draft age and had not registered, but his lawyers asserted that the conscription statute was unconstitutional. Each was convicted and given a maximum sentence of a year in prison. The appeals went straight to the Supreme Court. Several defendants were represented by the same set of lawyers, and in the Supreme Court these lawyers submitted identical briefs in several cases, asserting the same constitutional arguments. One of the men was represented at both the trial court level and on appeal to the Supreme Court, by Harry Weinberger, an NCLB lawyer. Another NCLB lawyer, Walter Nelles, filed a (single) amicus curiæ brief in the consolidated cases in which he highlighted one specific constitutional argument that the bureau was keen to emphasize. (Amicus curiæ is Latin for "friend of the court." So-called amicus briefs can be filed by people or organizations that are not parties to a proceeding, if they believe they have something valuable to contribute to the issue[s] under discussion in a case.)

---

197 William Blackstone, *Commentaries on the Laws of England* (Oxford: Clarendon Press, 1769) 4:151; Patterson v.Colorado, 205 U.S. 454, 462 (1907).

198 Selective Draft Law Cases (Arver), 245 U.S. 366 (1918).

In the trial courts and on appeal to the Supreme Court, the *Arver* defendants made four principal constitutional arguments. First, they argued the constitutional distinction between the article 1 congressional power to call up state militias and the separate power to raise an army but by volunteers only (see chapter 1).[199] Second, the defendants argued that the draft was a form of involuntary servitude barred by the Thirteenth Amendment. They explained that the amendment, by its plain language, specifically bars both "slavery"—the amendment clearly refers to chattel slavery of Blacks in the former Confederacy—and "involuntary servitude." There would have been no reason to include "involuntary servitude" if the amendment were intended to reach only the former institution of slavery. The lawyers quoted Justice Joseph Story's famous nineteenth-century treatise, *Constitutional Law*: "The 13th Amendment forbids not merely the slavery known to our laws, but all kinds of involuntary servitude not imposed in punishment of a public offense."[200]

The defendants' lawyers also made two separate First Amendment arguments pertaining to religion. A word of explanation is required here. The First Amendment says, "Congress shall make no law respecting an establishment of religion, or prohibiting the free exercise thereof." That is, the First Amendment has two religion clauses, and they provide two different protections. The establishment clause says that, in the United States, we do not have, and cannot have, an established government religion the way the Church of England is the official religion in England. A law that stated, "The Mayor of New York must be a Presbyterian" would violate the establishment clause. The free exercise clause says that everybody in the United States is allowed to practice ("exercise") any religion they wish. Michael is allowed to be Jewish, and Peter is allowed to be a Jain. In no case is the government allowed to interfere with your right to practice your religion.

In the *Arver* case, the defendants argued that the conscription statute violated both provisions. First, the statute exempts as COs *only* those—like Quakers, Mennonites, and Brethren—who are members of a "well-recognized religious sect" that forbids "its members to participate in war in any form." This favors adherents of some religions (Quakers and Mennonites) over those of other religions (Baptists and Catholics) and thus violates the establishment clause. Second, for religious pacifists who are not Quakers, the conscription statute interferes with the free exercise clause, because the law makes it

199 See pages 7 to 8 in the brief to the US Supreme Court filed on behalf of Louis Kramer and Alfred Grahl in *Arver*, contained in *United States Supreme Court Transcript of Record, Arver v. United States* [245 U.S. 366 (1918)] (hereafter, Arver transcript).

200 Kramer and Grahl brief, Arver transcript, 25–40.

impossible for them to practice their pacifist religious beliefs; the law does this by forcing these people into the military.[201]

These two religion clause arguments were the sole focus of the separate amicus brief filed in *Arver* by Walter Nelles for the NCLB. Nelles started by quoting an article, "War's Heretics," by Norman Thomas in *Survey* magazine (August 11, 1917) in which Thomas explained that "conscientious objectors include Christians, Jews, agnostics and atheists; economic conservatists [*sic*] and radicals; philosophic anarchists and orthodox socialists." Nelles went on: "The Conscription Act, by [requiring] violation of conscience, prohibits the free exercise of religion to all conscientious objectors, whether their objection rests upon their duty towards God or their duty towards Man.... Since a man's religion is thus in effect synonymous with the beliefs he holds sacred, an *exercise of religion* occurs whenever he does or refrains from doing anything whatever by reason of belief and under penalty of self-disgrace.... Conscientious refusal to take part in the war is ... an exercise of religion."[202] At the time, this was a novel argument. However, other conscription opponents later in the twentieth century often used similar arguments (see chapters 5 to 7).

The Supreme Court ruling came on January 14, 1918; Chief Justice Edward White, who had been a Confederate soldier during the Civil War, delivered the opinion of the unanimous court. As discussed in chapter 1, the court found the power to conscript in the constitutional power of Congress "to raise and support armies" and "to make all laws which shall be necessary and proper for carrying into execution the foregoing powers." In a passage often cited since, the court said, "As the mind cannot conceive an army without the men to compose it, on the face of the Constitution the objection that it does not give power to provide for such men would seem to be too frivolous for further notice."[203] This was at least a plausible conclusion.

However, the rest of White's opinion was a hodge-podge of irrelevancies, disingenuous citations, mistaken citations, and out-and-out evasions. The court observed that some other countries had conscription and provided an alphabetical list of them, from Argentina to Turkey. At the time of the ruling, the United States was at war with three countries that the Supreme Court listed: Austria-Hungary, Bulgaria, and Germany. The court was at pains to cite articles 57 and 59 of the April 16, 1871, Constitution of the German Empire, put in place by Count Otto von Bismarck, but the court did not explain how this could be

201 Ibid., 6–7.

202 Amicus brief filed by Walter Nelles, attorney, for the National Civil Liberties Bureau (NCLB), Arver transcript, 3, 7–8 (emphasis in the original).

203 Arver, 245 U.S. at 377.

relevant to interpreting the US Constitution at a time when the United States was at war against the German Empire and "Prussianism."

Next, the court cited colonial era constitutions of Vermont, New York, Massachusetts, Delaware, Maryland, and Virginia pertaining to conscription of state militias but did not bother to address the argument of the defendants—that the constitutional power of Congress to call forth the state militias was different from the methods for creating a standing army.

Next the court stated that other courts, in earlier decisions, had found conscription to be constitutional. The court cited *Kneedler* but disingenuously failed to acknowledge that there were two *Kneedler* opinions and that only one had found the draft to be constitutional.

In saying that earlier decisions had upheld the constitutionality of conscription, the *Arver* court also cited nine Civil War decisions from state courts in four Confederate states: Georgia, Texas, Alabama, and North Carolina. Four of these cases had nothing to do with the issue before the Supreme Court in *Arver*, the constitutionality of conscription. Both *Fitzgerald v. Harris* (Georgia) and *Gaitlin v. Walton* (North Carolina) addressed whether the Confederate Congress's post hoc revocation of allowing substitutes presented a contracts-clause problem (see chapter 3).[204] *In re Emerson* (Alabama) considered whether a man in the Confederate army could be drafted into the Alabama state militia under Alabama law.[205] *Ex parte Hill* was a real embarrassment. The only issue in the case was whether a local Alabama probate judge had the authority to issue a writ of habeas corpus on behalf of a conscript. The Alabama Supreme Court ruled that, while Alabama state judges do have the authority to issue the writ, local probate judges do not.[206] *Hill* did not address, consider, or decide anything about the constitutionality of conscription. The remaining five Confederate cases cited by the US Supreme Court in 1917 in *Arver* did, in fact, address the constitutionality of conscription in the Confederacy. Still, these Confederate states had committed treason against the United States by seceding. It is distinctly odd to read an opinion of the US Supreme Court that cites the views of treasonous mutineers as a way of explaining the correct and proper meaning of the US Constitution.

In the last two sentences of the *Arver* opinion, the court tells the reader there are some matters it will *not* address. The court declines to address either of the First Amendment religion arguments "because we think its unsoundness is too apparent to require us" to discuss them. The court also declines

204 Fitzgerald v. Harris, 33 Ga. Supp. 38 (1864); Gaitlin v. Walton, 60 N.C. 333 (1864).

205 Reported as *in re* State *ex rel.* Graham, 39 Ala. 437 (1864).

206 *Ex parte* Hill, 38 Ala. 429 (1863).

to address the Thirteenth Amendment (involuntary servitude) argument because military service is the "supreme and noble duty" of every citizen, and the suggestion that the Thirteenth Amendment would bar forced conscription "is refuted by its mere statement." Thus, the Supreme Court did not even consider—much less address—three of the core constitutional arguments the defendants presented.

Nevertheless, despite its manifest shortcomings, *Arver* remains today the most important US Supreme Court case concerning the constitutionality of conscription. For the remainder of the twentieth century, federal courts, including the Supreme Court, would repeatedly invoke the statement that "the mind cannot conceive an army without the men to compose it."

∴

As I have noted, the World War I conscription statute made refusing to register a civilian misdemeanor punishable by a year in prison. However, a man who refused induction was deemed to be in the army (even if he did not believe that he was a soldier) and was therefore susceptible to court-martial by the army. The courts consistently upheld this understanding, most notably in *Franke v. Murray*.[207]

∴

The 1917 conscription statute stated that Selective Service decisions "shall be final" (except for rare appeals to the president pursuant to Selective Service regulations).[208] That did not prevent draftees aggrieved by local draft board decisions from seeking writs of habeas corpus. During the Civil War, courts in both the Union and Confederacy issued inconsistent rulings on the identical issue. This pattern continued into World War I. In some cases, federal courts ruled that, because of the plain language of the Selective Service statute, courts had no jurisdiction to intervene or grant writs of habeas corpus for draftees who disagreed with draft board decisions. In *United States ex rel. Bertalini v. Mitchell*, the court in the Eastern District of New York said, "A determination by the local or district draft board is final." In *United States ex rel. Pascher v. Kinkead*, the Third Circuit Court of Appeals ruled the same way. However, when presented with the identical issue, the court in the Eastern

207 Franke v. Murray, 248 F. 865 (1918).

208 An Act To authorize the President to increase temporarily the Military Establishment of the United States, May 18, 1917, Pub. L. No. 65-12, 40 Stat. 76 (1917).

District of Tennessee ruled the opposite way in *Ex parte Fuston* and granted a writ of habeas corpus; in *Ex parte Cohen*, the court in the Eastern District of Virginia did the same thing.[209]

Different courts also inconsistently decided a related but not identical jurisdictional issue. The issue was whether a civilian federal court has jurisdiction over a claim brought by a soldier in the army. In *Ex parte Beck*, the district court in Montana ruled that a civilian court does not have jurisdiction, but a few weeks later, in *United States ex rel. Koopowitz v. Finley*, a court in the Eastern District of New York reached the opposite conclusion on the same question.[210]

9 Conclusion

The Selective Service System during World War I was extremely efficient. During a mere eighteen months, starting with no preexisting legal foundation or personnel infrastructure, the newly created system conscripted 2.8 million men. Had the war continued, the Selective Service could have drafted additional millions of men in the following months and years. (On the date of the armistice, 1.4 million additional men in Class I had passed their physical examination but had not yet been inducted; another 1.2 million men in Class I were awaiting their physical examinations; and more than 6 million more were not yet classified but would likely have ended up in Class I).[211] But the efficiency came at a steep price.

Part of that price was legal. As Provost Marshal Enoch H. Crowder bragged after the war, part of the efficiency of the system derived from the War Department's setting up the system and printing the registration forms before Congress had acted. This was not merely illegal, but, as Secretary of War Baker testified to Congress, the War Department knew it to be illegal while doing it. That knowledge, however, did not stop or even slow down the process.

Part of the price was logistical. The unseemly and illegal haste to create a conscription apparatus led to a particular problem for the tens of thousands of men who claimed to be COs, because the registration forms, which were printed before Congress had decided on the parameters of the CO provision,

209 Bertalini v. Mitchell, 248 F. 997, 1000 (E.D.N.Y. 1918); United States *ex rel.* Pascher v. Kinkead, 250 F. 692 (3d Cir. 1918); United States *ex rel.* Pascher v. Kinkead, 250 F. 692 (3d Cir. 1918); *Ex parte* Furston, 253 F. 90 (E.D. Tenn. 1918); *Ex parte* Cohen, 254 F. 711 (E.D. Va. 1918).

210 *Ex parte* Beck, 245 F. 967 (D. Mont. 1917); United States *ex rel.* Koopowitz v. Finley, 245 F. 871 (E.D.N.Y. 1917).

211 *PMG 3d*, 14.

provided no way for a man to indicate that he was a CO. On June 5, 1917, 9.5 million men registered, using forms that took no account of COs. It fell to the AUAM to improvise imperfect ways—ways not contemplated by the War Department and not accounted for on the registration forms—for men to claim that they were COs. And fully 750,000 men had been drafted and sent to army camps before Wilson and the War Department belatedly decided, on March 23, 1918—eight months before the end of the war—what to do with the thousands of COs in those camps. Additional months would pass before any of the theoretical plans for treatment of COs would become realities. In the meantime, more than 500 COs were court-martialed. Seventeen were sentenced to death and another 142 received life sentences—while many, for exactly the same offense, were given sentences of a year or less.

Yet these specific problems that COs faced were merely a reflection of the broader problems faced by all men dealing with the hastily contrived Selective Service System. The Wilson administration and its War Department were adamant that the Selective Service System was designed to be an efficient, modern, supposedly scientific means of channeling all the nation's manpower into areas that served the war-making capabilities of the nation—if not in the army, then in the manufacture of munitions or the production of food crops on the home front. That was the purpose of the elaborate classification system and the administration's "work or fight" order. Yet the June 5, 1917, registration date had long come and gone before the classification system was created on November 8, 1917—and another month went by before the questionnaires needed to implement the classification system were printed on December 15, 1917. These delays made it much more difficult for the government to achieve what the War Department itself touted as one of its main goals: the channeling of manpower.

Over the years, scholars have come to view the US experience with conscription during World War I as highly successful. Writing in 1940, Edward Fitzpatrick wrote that the World War I draft was an "amazing success."[212] In 1987, John Whiteclay Chambers II said, "Unlike the Civil War experience, the World War I draft proved quite successful in its major goal of raising the wartime army."[213] It is true that World War I conscription avoided the issues presented by allowing substitution and paying bounties. There are, however, at least three problems with the conclusion that the effort was highly successful.

212 Fitzpatrick, *Conscription and America*, 34.

213 Chambers, *To Raise an Army*, 179.

- At the most fundamental level, a stunning 11.23 percent of men who were drafted refused to go. This does not betoken an amazing success. The level of illegal resistance to the World War I draft is underappreciated today.
- Merely raising a wartime army was not the only goal of the wartime Selective Service System. An equally important goal for the Wilson administration and its War Department was the effective channeling of the entire male workforce into areas deemed to serve the military necessity. In that important function, Selective Service was far less successful, in large part because of the delayed roll-out of the mechanisms to accomplish it.
- There may be a distinction between *success* and *efficiency* that is worth observing and preserving. As I have noted, the World War I Selective Service System was surely efficient, but the efficiency came at a price. Thus, whether that efficiency translates into a successful effort is not clear. As many as three million men who were legally obligated to register may not have done so, and hundreds of thousands of men received induction orders but evaded, refused, or escaped induction. The Justice Department was saddled with thousands of draft-related prosecutions, and men of draft age fled the country and established colonies in Canada and Mexico.

An important point is that, in order to make the conscription system work efficiently, the federal government was forced to rely on an unofficial army of a 250,000 lawless vigilantes who engaged in widespread, overtly criminal behavior in trying to track down the millions of men evading the draft at the various stages of registration, classification, and induction. The APL conducted thousands of illegal warrantless searches of homes and offices in its effort to track down men it called slackers, and it raised much of its money through protection rackets that today are associated with the Mafia.

A government system that incites disobedience by millions of its people, one that deliberately creates and carefully nurtures an army of criminal vigilantes to achieve its efficiency may not be able to also claim that it was successful.

Several legacies of World War I conscription were enduring. The Selective Service classification system that originated then continued for the rest of the twentieth century whenever the United States had conscription, as did the closely linked concept of manpower channeling. The American Friends Service Committee and the Women's International League for Peace and Freedom, the peace organizations founded during the war in direct response to the draft, still exist today. The ACLU, which started as the NCLB to address the draft, also is still active. The Espionage Act, passed to deter opposition to the World War I draft, remains on the books and is still used, in part, to suppress dissent.

Methods of draft resistance that were pioneered during World War I—including publicly refusing to register and publicly refusing induction—were models for draft resisters for the rest of the century. Draft counseling, which involved thousands of draft counselors and hundreds of thousands of young men of draft age during the Vietnam War, was invented by Witherspoon, Mygatt, and the Bureau of Legal Advice during World War I.

Some contemporary scholars, viewing the government's response to opposition to conscription during World War I, mistakenly believe that that response was more liberal and progressive than it in fact was. For example, Jeremy Kessler writes that wartime officials came "to recognize and respect the rights of individual conscience, rights held by all citizens, whether sectarian or nonsectarian, religious or nonreligious." He argues that administrators came "to manage resistance to the draft in a manner that preserved, and even enhanced, the democratic character of the national state." He says that progressive administrators "went out of their way to accommodate minority views during the war." The genesis of Kessler's mistake is that he takes at face value the vacuous statements the administrators made, rather than considering whether the statements were true. In particular, Kessler makes far too much of such documents as the War Department's October 10, 1917, confidential memorandum announcing that COs ought to be treated with "tact and consideration."[214]

COs were, by and large, not treated with "tact and consideration." They were treated badly, often tortured for prolonged periods, and sometimes they died as a consequence of their mistreatment. COs were beaten, sometimes into unconsciousness, shot at, bayoneted, starved, held for extended periods in vermin- and rat-infested cells without sanitary facilities or beds, and manacled in stress positions for eight to nine hours a day. These circumstances were not isolated events; such treatment occurred in hundreds of cases at facilities all across the country *after* the date of the "tact and consideration" memo. The grieving young widows of Joseph and Michael Hofer would not have been much comforted if they had been told that buried somewhere in a War Department file was a confidential memorandum that said that their husbands should have been treated with tact and consideration.

214 Kessler, "Administrative Origins," 1083, 1087, 1088, 1090.

CHAPTER 5

World War II

World War II started on September 1, 1939, when Germany invaded of Poland. At that time, the size of the US Army was 242,648 people, and the entire US military (including the navy and both the army and navy air units) consisted of 404,550 people.[1] The army was 90.1 percent larger than when the United States had entered World War I (when it has only 127,588 people), although the population of the country had grown by 41 percent during the intervening three decades to 131.67 million in the 1940 census (versus 93.4 million in the 1910 census).[2] Nevertheless, at 242,642 in 1939, the size of the army was comparatively small in relation to the number of men drafted through the end of the war (9.92 million), as well as to the size of the army (8.27 million) and the size of the entire US military at war's end (12.3 million).[3]

On May 31, 1940, on the eve of the evacuation of the British Expeditionary Force from Dunkirk, President Franklin D. Roosevelt asked Congress, before adjourning, to pass legislation authorizing him to call the National Guard into active service in the army.[4] On August 27, Congress gave Roosevelt the authority he had requested, and on August 31, Roosevelt issued an executive order to bring the National Guard into the active-duty army.[5] By September 30—as a result of new voluntary enlistments and the call-up of the National

1 US Selective Service System, *Selective Service in Wartime: Second Report of the Director of Selective Service, 1941–41* (Washington, DC: Government Printing Office, 1943) (hereafter, SS *2d*), 63, table 30.

2 "About the 1940 Census," National Archives and Records Administration, https://1940census.archives.gov/about/, last accessed April 6, 2022; US Census Bureau, *Sixteenth Census of the United States. Vol. 1: Population, General Report and Analysis* (Washington, DC: Government Printing Office, 1913), 5.

3 US Selective Service System, *Selective Service and Victory: The Fourth Report of the Director of Selective Service, 1944–1945, With a Supplement for 1946–1947* (Washington, DC: Government Printing Office, 1948) (hereafter, *SS 4th*), app. 136, 590, and app. 157, 617.

4 *SS 2d*, 63; "The President's Message," *New York Times*, June 1, 1940.

5 Joint Resolution to strengthen the common defense and to authorize the President to order members and units of reserve components and retired personnel of the Regular Army into active military service, Pub. Res. 76–96, 54 Stat. 858 (1940), § 2 (this resolution authorizes the call-up of the National Guard); Exec. Order No. 8530, 54 Stat. 858 (1940); "60,000 Guardsmen Summoned for Active Duty September 16; 6,000 from New York State," *New York Times*, September 2, 1940.

 | DOI:10.1163/9789004546684_007

Guard—the army had tripled in size to 745,601 people, and the US military, including the navy, had more than doubled, to 1.02 million people.[6]

1 The Statute

The World War II conscription statute was called the Burke-Wadsworth bill after its two main sponsors, Senator Edward R. Burke (D-NE) and Representative James W. Wadsworth, Jr. (R-NY).[7] The Senate bill (S-4124) was introduced on June 20, 1940, and the House bill (H-10132) was introduced a day later.[8] The bill had been drafted primarily by a private individual, the lawyer Grenville Clark of the Military Training Camps Association, a private organization that supported military preparedness and conscription and believed that the Roosevelt administration was not doing enough to prepare for war. Clark closely coordinated the drafting of the language of the bill with a young army major, Lewis B. Hershey, who in 1940 headed up the Joint Army Navy Selective Service Committee. After World War I, the committee had developed contingency plans for a conscription apparatus should Congress reauthorize a draft.[9]

Once introduced, the bill moved swiftly through Congress. The Senate Committee on Military Affairs held hearings in July, and the full Senate first passed the bill on August 28, 1940. The House Committee on Military Affairs held hearings in July and August, and the bill passed the full House on August 29.[10] On September 14, following a Senate-House conference, both houses approved

6 SS *2d*, 64.

7 On the World War II draft generally, see George Q. Flynn, *The Draft 1940–1973* (Lawrence: University Press of Kansas, 1993), and Paul Dickson, *The Rise of the G.I. Army, 1940–1941: The Forgotten Story of How America Forged a Powerful Army before Pearl Harbor* (New York: Atlantic Monthly Press, 2020).

8 US Selective Service System, *Selective Service in Peacetime: First Report of the Director of Selective Service 1940–41* (Washington, DC: Government Printing Office, 1942) (hereafter, SS *1st*), 23.

9 *Compulsory Military Training and Service* [*Act*] *of 1940: Hearings Before the Committee on Military Affairs of the United States Senate on S. 4164*, 76th Cong., 3d sess. (1940), 6–9 (hereafter, *Senate Hearings*); J. Garry Clifford and Samuel R. Spencer, Jr., *The First Peacetime Draft* (Lawrence: University Press of Kansas, 1986), 5, 71–73. Lewis B. Hershey (1893–1977) was director of the Selective Service System from 1941 to 1970. A hagiographic book on Hershey is George Q. Flynn, *Lewis B. Hershey, Mr. Selective Service* (Chapel Hill: University of North Carolina Press, 1985), and a very sympathetic one is Nicholas A. Krehbel, *General Lewis B. Hershey and Conscientious Objection during World War II* (Columbia: University of Missouri Press, 2011).

10 *Hearings Before the Committee on Military Affairs of the United States House of Representatives on H.R. 10132*, 76th Cong., 3d sess. (1940) (hereafter, *House Hearings*).

FIGURE 5.1 Franklin D. Roosevelt signing the World War II conscription bill into law
CREDIT: GETTY IMAGES / BETTMANN / CONTRIBUTOR

a somewhat revised Burke-Wadsworth bill. On September 16, Roosevelt signed the bill into law (see Figure 5.1).[11]

The United States did not formally enter the war until after the Japanese attack on Pearl Harbor in December 1941, and whether enactment of the conscription statute fifteen months earlier did (or did not) constitute the creation of a peacetime draft is an interesting question. On the one hand, the book by J. Garry Clifford and Samuel R. Spencer, Jr., on the enactment of the statute—and, especially, of the crucial role played by Grenville Clark and the Military Training Camps Association—is entitled *The First Peacetime Draft*.[12] On the other hand, Freda Kirchwey, writing in the *Nation* of September 7, 1940, said,

11 *SS 1st*, 25; Selective Service and Training Act of 1940, Pub. L. No. 76-783, 54 Stat. 885 (1940).

12 Clifford and Spencer, *First Peacetime Draft*. Arthur A. Ekirch, Jr., also refers to this as the first peacetime draft in *The Civilian and the Military: A History of the American Antimilitarist Tradition* (Colorado Springs, CO: Ralph Myles, 1972), 261, as does Charles DeBenedetti in *The Peace Reform in American History* (Bloomington: Indiana University Press, 1980), 139.

"It is foolish to call the Selective Training and Service Act a measure for 'peacetime conscription.' The United States is not waging military war, but it is only nominally at peace. Openly and officially we have identified ourselves with the nations fighting against Hitler." The reality of the situation may have been captured most accurately in the first paragraph of the first annual report provided by the Selective Service to Roosevelt on the new draft. The Selective Service said, "It was the first peacetime conscription law in our history, though it was passed in the shadow of war."[13]

The ambiguity of the situation allowed those testifying against the bill in Congress to denounce it as an unprecedented peacetime draft, while those testifying in support of the bill emphasized the emergency posed by the European conflict.[14] Neither side spoke at the committee hearings with perfect foresight. Peace activists, including Dorothy Day of the Catholic Worker movement and John Nevin Sayre of the Fellowship of Reconciliation (FOR), confidently asserted that the United States itself could not be attacked.[15] Meanwhile, supporters of conscription argued volubly that it was impossible that conscripted American boys could ever be sent to Europe.[16]

Hershey played a significant role in both the Senate and House committee hearings. Although he did not reveal this fact to Congress during his extended and detailed testimony, the government in 1940—just as it had done in 1917—had already illegally printed Selective Service registration forms before the draft was enacted, indeed even before the Burke-Wadsworth bill had been introduced in Congress. The printing was supposed to be a government secret, but word leaked out, and Sevellen Brown, a reporter for the *Providence (RI) Journal*, asked the War Department about the forms. The War Department lied

---

13 Freda Kirchwey, "The Draft Bill," *Nation*, September 7, 1940, 184; *SS 1st*, 3.

14 *House Hearings*, statement of Emily L. Eaton, representing the New York State branch of the Women's International League for Peace and Freedom, 211; statement of Norman Thomas, 240; statement of Rev. Charles F. Boss, Jr., of the Commission on World Peace of the United Methodist Church, 421; statements of Dr. James B. Conant, president of Harvard University, 22, and Grenville E. Clark, 13.

15 *Senate Hearings*, statements of Dorothy Day, 153, and John Nevin Sayre, 246 . For an interesting biographical sketch of Sayre, see Charles F. Howlett, "John Nevin Sayre and the Fellowship of Reconciliation," *Pennsylvania Magazine of History and Biography* 114, no. 3 (July 1990): 399–421. For background and general context on the US antidraft movement during the World War II era, see Scott H. Bennett, *Radical Pacifism: The War Resisters League and Gandhian Nonviolence in America, 1915–1963* (Syracuse, NY: Syracuse University Press, 2003), chaps. 4–5; Lawrence S. Wittner, *Rebels against War: The American Peace Movement, 1933–1983* (Philadelphia: Temple University Press, 1984), chaps. 2–5; and R. Alfred Hassler, *Conscripts of Conscience* (New York: Fellowship of Reconciliation, 1942).

16 *Senate Hearings*, 166.

and denied that it had printed them. But Brown had a copy of the forms, and the newspaper printed part of a form with its front-page story, along with the War Department's obviously dishonest denial.[17]

In their congressional testimony on the conscription bill, Hershey and other War Department officials made clear that, in supporting conscription, they were thinking both about getting some men for the army and channeling other men into war-related fields of national importance. Hershey said, "We believe … with orderly, scientific process of selective service we can first pick men in sufficient numbers at the time we want them from those who should go [into the army], and by the same method we can leave the men that should stay at their tasks, whether in essential industry, in vital agriculture, or in some other employment which contributes materially to the national interest."[18]

Hershey was clear that some men would want to volunteer but would not be allowed to do so—that was their tough luck—because the Selective Service was about channeling the entire male population to the job or function that the government, not the individual, thought best: "The idea is that a man serves where he can best serve the country and that is not his choice. He may want to go, and he may want to get away from that shop, but that is unfortunate."[19]

Secretary of War Henry L. Stimson's testimony to Congress about the Burke-Wadsworth bill sounded much like General Crowder's annual reports to Wilson during World War I on the subject of channeling:

> The men who are the leaders in all branches of normal life are the men who are the most patriotic in time of danger and … those especially valuable citizens, under such a voluntary system, make a rush to enlist and fight, when they might be more useful … if they stayed home and kept running the machinery of business. And it is the business of selective service to compel them to stay home, even against their will…. The compulsory system … [takes] into account not only the battle needs of the country but also the needs for supplies and armament.[20]

General William E. Shedd III, of the Army General Staff, emphasized to Congress that conscription was necessary to prevent what had happened in England during World War I when the country was crippled by having the

17 Sevellen Brown III, "Blank for Draft in U.S. Prepared," *Providence (RI) Journal*, June 6, 1940, 1, 6; Dickson, *Rise of the G.I. Army*, 101.

18 *House Hearings*, statement of Lewis B. Hershey, 117.

19 Ibid., 121.

20 Ibid., statement of Henry L. Stimson, 384.

"wrong" men volunteer: "England experienced difficulty during the first World War, when she was recruiting her army by voluntary enlistment and found she had to return from the army to industry and agriculture hundreds of thousands of men because her industry had been crippled when these men volunteered." Conscription, Shedd explained, solves this problem by providing a mechanism, through deferments for certain registrants, to more nearly control the work of all men, not merely those who are drafted into the military.[21]

In his congressional testimony, Hershey also emphasized the key role that local draft boards would play. "We place full responsibility upon the local board, the neighbors who decide who shall go and who shall stay," Hershey said, using language that would return to haunt the Selective Service during the Vietnam era (see chapter 7).[22]

1.1 *The Statute's Conscientious Objector Provision*

To their credit, even before the Burke-Wadsworth bill was written or introduced, the historic peace churches—Quakers, Mennonites, and Brethren—had recognized the movement toward war and conscription, both in public sentiment and the likelihood of congressional action. On January 10, 1940, a delegation of Quakers, Mennonites, and Brethren met with Roosevelt to urge that, in the event of a return to conscription, significant changes be made from the World War I policies regarding the treatment of COs. The church delegation left a letter with the president that said in part, "In the last war the United States Government finally authorized such nonmilitary humanitarian service to be substituted for military service, and furloughed conscientious objectors to this relief work or farm labor. But this arrangement was provided only after months of confusion and distress."[23]

The historic peace churches were wise and foresighted to approach the government well before an actual proposal for conscription was introduced into Congress. Although they were correct that the World War I experience had produced "confusion and distress," they rather understated the point. In fact, during World War I chaos had ensued because nobody in government had given the slightest thought to appropriate handling and treatment of COs until hundreds of them were already in military prisons and were being

---

21 Ibid., statement of William E. Shedd III, 536.

22 *House Hearings*, Hershey statement, 116.

23 E. Raymond Wilson, "Evolution of the C.O. Provisions in the 1940 Conscription Bill," *Quaker History* 64, no. 1 (Spring 1975): 3; Mulford Q. Sibley and Philip E. Jacob, *Conscription of Conscience: The American State and the Conscientious Objector* (Ithaca, NY: Cornell University Press, 1952), 45. For the complete text of the letter to Roosevelt, including the signers, see Sibley and Jacob, *Conscription of Conscience*, app. A, 485–86.

beaten, starved, abused, and tortured. By seeking to address in advance what could obviously become a huge problem if not properly considered, the peace churches were performing a substantial public service.

However, the forethought of the peace churches was not rewarded. Although the Joint Army Navy Selective Service Committee had been working on a conscription bill since 1926, when the Burke-Wadsworth bill was introduced in 1940, the section pertaining to COs was a verbatim copy of the (wholly unsatisfactory) wording in the World War I conscription statute. The 1940 bill had all the bad features of the CO provision in the World War I draft, including no alternative civilian service and no CO status even for religious objectors who were not members of historic peace churches (let alone nonreligious moral, philosophical, or political objectors).[24]

What the peace churches, and many others, wanted instead was a set of provisions that more nearly mirrored the conscription law, the National Service (Armed Forces) Act, that had gone into effect in Britain on September 3, 1939, when Britain had entered the war. That statute, which was far more generous to COs than any US conscription statute has ever been (including to the present day), contained several salient provisions.[25]

One was that the British statute contained no limiting definition of what a "conscientious objector" was. That meant that COs did not have to belong to one of the traditional peace churches and did not have to be religious; moral, philosophical, and political reasons were satisfactory. In fact, under the British law, a CO did not even have to be an absolute pacifist; objection to a particular war was sufficient. If a man believed he was a CO, he was. To be sure, (civilian) tribunals were established to examine and rule on the good faith of CO claims; this was done to separate men with deeply held principles from casual draft evaders.

In addition, under the British law, men who had successfully established their sincerity would be placed into one of three (not two) different categories. Again, the CO himself decided in which category he would be placed. In the first category, a CO could enter the army as a noncombatant. In the second category, a CO could perform alternative civilian service under civilian, not military, control and supervision. In the third category, COs who were opposed

24 *Senate Hearings*, 3, and testimony of Hershey, July 12, 1940, 374; Clifford and Spencer, *First Peacetime Draft*, 41–43; *House Hearings*, 3.

25 National Service (Armed Forces) Act, 1939, 2 and 3 Geo. VI, c. 81 (1939). (This is a citation to British law.)

to any involuntary servitude or forced service were permitted to perform no service at all, even civilian service under civilian control.[26]

Naturally, this last provision provided a great deal of difficulty for the American peace activists trying to influence Congress during the summer of 1940—the provision that allowed a man to do no service at all. And the sloppy, and sometimes downright inaccurate, statements of the peace activists lobbying Congress at the time did not help the situation. E. Raymond Wilson, one of the Quakers leading negotiations about the bill's provisions for COs with members of the Senate and House Military Affairs Committees and with the War Department, claimed the British law that the churches and others were touting would "grant complete exemption to the 'absolutists'—those who were unwilling to register or give any cooperation." That was wrong. The British law required all men to both register and prove the sincerity of their beliefs. But the testimony of the peace activists reflected great confusion about what constituted an absolutist and how to treat them. Howard K. Beale, representing the ACLU, stated that the five hundred COs imprisoned in US Army stockades during World War I had been "conscientiously opposed, completely and absolutely opposed, to any kind of service during the war." Paul Comly French, representing the Quakers, said the same thing.[27] This, too, was wrong. Many, and probably even most, World War I COs would have accepted alternative civilian service. In fact, many World War I COs did accept alternative service when it was belatedly offered to them.

The result of the sloppy talk by CO advocates was that some members of Congress may have understandably, but incorrectly, gotten the impression either that the CO advocates were promoting a wholly unreasonable and untenable position (COs could choose to ignore the law and refuse to register at will and with impunity) or that the numbers of men who would refuse even alternative civilian service were much larger than they actually were—and that they would insist on performing no service at all. Of course, it is also true that, even if the CO advocates had spoken more clearly and had been properly understood, the idea that men could simply choose not to serve in any way at all—just because they did not want to—was anathema to the foundational

26 Ibid. Regarding British law in general, see Julien Cornell, "The British Draft: A Comparison," chap. 17, in *The Conscientious Objector and the Law* (New York: John Day, 1943). Cornell was a civil liberties lawyer who represented many draft violators during World War II. See also Julien Cornell, *Conscience and the State* (New York: John Day, 1944). For a remembrance of Cornell, see Charles F. Howlett, "Julien Davies Cornell: Gentle Quaker, Determined Litigator," *Friends Journal* 53, no. 5 (May 1, 2007): 6–8.

27 Wilson, "Evolution of the C.O. Provisions," 4; *Senate Hearings*, statement of Howard K. Beale, 186, and statement of Paul French, 202.

philosophy of the Selective Service as enunciated by Hershey and the War Department: that, while the *manner* of service could vary, the *fact* of service was to be universal.

Shedd, of the Army General Staff, suggested that the Quakers meet with Colonel Victor O. Kelliher of the War College to discuss revisions of the language about COs in the Burke-Wadsworth bill and, on July 24, 1940, in the midst of the hearings on the House side, French and Wilson met with Kelliher at the War College to do that. On July 25, one day after their meeting, French and Wilson presented compromise text, developed jointly by the Quakers and the War Department, to the House committee.[28] In their meeting, the Quakers had gotten much, but not all, of what they were seeking. The compromise proposal would delete the original CO provision in the bill and substitute the following language:

> Nothing contained in this act shall be construed to require any person to be subject to combat training or service in the land or naval forces of the United States who, by reason of religious training and belief, is conscientiously opposed to participation in war in any form, and is found to be a bona fide conscientious objector as hereinafter provided. All persons claiming such conscientious objections shall be listed on a register of conscientious objectors at the time of their classification by a local board, and persons so registered shall be at once referred to the Department of Justice for inquiry and hearing.... The Department shall, after hearing, if the objection be sustained, recommend that the objector shall (1) be assigned to noncombatant service in such capacity as the President may designate as noncombatant, or (2) if found to be a conscientious objector opposed to war in any form, that he may be assigned to work of national importance under civilian direction, or (3) if said objection is not sustained, the local board shall be immediately notified thereof, and the name of such objector shall then be removed from the register of conscientious objectors, and such objector thereafter be subject to classification.[29]

Note that the Quakers got a big win with the deletion of the language that COs must be members of a "well recognized religious sect whose creed or principles

28 US Selective Service System, *Conscientious Objection:* Special Monograph No. 11 (Washington, DC: Government Printing Office, 1950), 75–76; Sibley and Jacob, *Conscription of Conscience*, 48–49.

29 *House Hearings*, 211; Wilson, "Evolution of the C.O. Provision," 8.

forbid its members to participate in war." However, the Quakers had also urged that CO status be extended to nonreligious objectors, those whose beliefs were moral, political, or philosophical. The Quakers did not prevail in that regard.[30]

Also note that COs still had to be absolute pacifists, opposed to all wars; selective objection to a particular war did not qualify a man to be a CO.

The Quakers did succeed in having alternative civilian service included as an option for COs. This was hugely important and represented a major improvement over the World War I statute. However, the Quakers had also pushed for an option that would have allowed men opposed to any service at all to be completely excused, as the British draft law allowed. Not surprisingly, the War Department did not acquiesce to this.

The compromise language also simply failed to address some issues. One was exactly how alternative civilian service might be performed. Would men have to find their own individual job situations, subject to government approval? Or would some sort of civilian camps be established for COs that paralleled military camps? If so, who would establish the camps? And who would run them? The government? Some nongovernmental agency (or agencies)? Another issue not addressed was whether—and, if so, how—COs performing alternative civilian service might be compensated or paid. According to Wilson, there had been "almost no discussion" about these questions between the Quakers and the military, and no discussion at all of these questions during the congressional hearings.[31]

One additional important change to the CO provision was made before the bill was signed into law. In the floor debate on the House side on September 6, Representative Francis Walter (D-PA) offered a floor amendment that removed the registries for COs; it also removed the role of the Justice Department in investigating claims of conscientious objection and gave that role to the local draft boards (that is, the Selective Service System). The Walter amendment was adopted and found its way into the final legislation. At the time, the Quakers were bitterly disappointed to lose the civilian registries, which they regarded as crucially important, but they may have overestimated the importance of the eleventh-hour change. The civilian draft boards created by the statute were, after all, civilian too. The Quakers correctly viewed draft boards as closely aligned with the military; the entire Selective Service System was to be headed by an active-duty military officer, and the Selective Service worked closely with the War Department. Still, in time of war—and a broadly popular war, at that—the real-world difference between results obtained from civilian draft

30 Wilson, "Evolution of the C.O. Provision," 7.
31 Ibid., 13.

board members and civilian Justice Department functionaries may have been slight.[32]

The relevant portion of the CO provision in the final statute said:

> Nothing contained in this Act shall be construed to require any person to be subject to combatant training and service in the land or naval forces of the United States who, by reason of religious training and belief, is conscientiously opposed to participation in war in any form. Any such person claiming an exemption from combatant training and service because of such conscientious objection whose claim is sustained by the local [draft] board shall, if he is inducted into the land or naval forces under this Act, be assigned to noncombatant service as defined by the President, or shall, if he is found to be conscientiously opposed to participation in such non-combatant service, in lieu of such induction, be assigned to work of national importance under civilian direction.[33]

The section went on to outline an appeal provision for COs whose claims had been denied, with an investigation (and nonbinding recommendation) to be made by the Justice Department but with the appeal to be decided by a Selective Service appeals board.

Thus the World War II CO statute differed from the World War I statute in that it no longer required membership in a historic peace church, but it retained a religion requirement. The law did not recognize secular ethical or political COs. It retained the requirement that a CO be an absolutist pacifist. Peace advocates achieved a major victory in that, for the first time in US history, a draft law provided for alternative civilian service. Had this been provided for by statute during World War I, much grief might have been avoided. Left entirely unaddressed was the matter of how (and by whom) alternative civilian service might be run and whether (and how) those COs might be compensated.

1.2 *Other Provisions of the Conscription Statute*

The remaining provisions of the World War II conscription statute, enacted in 1940, were fairly straightforward.

It provided that all male citizens and all male aliens aged twenty-one to thirty-six had to register for the draft at a time designated by the president for

---

32 US Selective Service System, *Conscientious Objection*, 84; Wilson, "Evolution of the C.O. Provision," 10; Clifford and Spencer, *First Peacetime Draft*, 222.

33 Selective Service and Training Act of 1940, § 5(g).

that purpose. It also provided that all male citizens and all male alien declarants in that age range were liable to be conscripted into the army involuntarily. (A declarant was an alien who had declared his intention to seek US citizenship.) That is, the law specified that all male aliens had to register, but only male alien declarants were liable to conscription.[34]

The law limited the number of draftees to 900,000.[35] It provided that the term of active service for draftees would be twelve months, to be followed by ten years in the reserves—but the president could extend the term of active service at will if it was "in the interests of national defense."[36] This language obviously was a blank check for the executive branch. The law also had an interesting provision that showed the importance Congress placed on the Monroe Doctrine. It provided that draftees "shall not be employed beyond the limits of the Western Hemisphere except in the Territories and possessions of the United States, including the Philippine Islands."[37] That is, draftees would not be sent to Europe to fight against Germany but may be used freely in Central and South America.

The law also provided an absolute exemption for clergy and divinity students.[38]

The all-important channeling provision provided deferments for men working in war-related industries and agriculture necessary to sustain a war effort. As the Selective Service candidly stated in its first *Annual Report*, the "primary responsibility" of the agency was "to determine who is to serve the Nation by training for the armed forces, and who is to serve in industry, agriculture and government. All must serve." In furtherance of the channeling purpose of the Selective Service, the law stated that any exemption or deferment existed only as long as the reason for it existed: if a man changed jobs, he could lose a deferment that had been based on his previous job.[39]

The statute also allowed deferments for men with dependents and for men "found to be physically, mentally, or morally deficient or defective." That last provision meant that you would not be drafted if you were missing a leg, blind in both eyes, or gay.

---

34 Ibid., §§ 2, 3.

35 Selective Training and Service Act of 1940, Pub. L. No. 76-783, 54 Stat. 885 (September 16, 1940), § 3(a).

36 Ibid., §§ 3(b) (twelve-month service, extendable by the president) and 3(c) (ten years in the reserves).

37 Ibid., § 3(e).

38 Ibid., § 4(d).

39 Ibid., § 4(e), (f).

The law prohibited bounties, substitutes, or payment of commutation fees. These provisions may appear quaint in a 1940 statute, but it is easy to understand the reason for including the section.

The law specified that the employers of drafted men would have to take them back after a draftee's term of service was over. Given the number of men drafted during the war, this became an important provision for many people after the war ended.[40]

The measure created the Selective Service System to carry out all the functions of the draft, including registering, classifying, and drafting men. The basic operating unit of the agency would be the local draft board. Every county would have at least one draft board, consisting of three or more civilians, who had to live in the geographical area served by the board.

The same section also provided that Selective Service could issue all regulations necessary to carry out the statute, including deciding the order in which to draft men, once again making the Selective Service an administrative agency (see chapter 4). In the event, there were 6,443 local draft boards, one for each of the 3,070 counties in the United States, plus additional boards for metropolitan areas of more than thirty thousand people. In its first month of operation, Selective Service issued six volumes of regulations, covering everything from how men would be registered, how they would be classified, what forms the system would use, and how the system would be organized internally to the auditing standards the Selective Service would use.[41]

The law stated that the local draft boards would determine "all questions or claims" relating to registration, classification, and drafting and that "the decisions of such local boards shall be final," except for the appeal to the Selective Service appeal boards.[42] The meaning and constitutionality of these words would be heavily litigated during and after the war.

Under the law, draft evasion or refusal; violating Selective Service regulations; counseling, aiding, or abetting others to do so; or interfering with the Selective Service "by force or violence" were felonies punishable by as long as five years in prison, a fine of as much as $10,000, or both. All these violations were to be triable in civilian federal courts, not by military courts-martial. This was an effort to avoid one of the problems that had existed with the draft during World War I. The language containing this provision was quite clear; it said: "No person shall be tried by any military or naval court-martial in any case arising under this Act unless such person has actually been inducted."[43]

40 Ibid., §§ 4(e), 7, and 8.
41 Ibid., § 10; US Selective Service System, *Conscientious Objection*, 91.
42 Selective Training and Service Act of 1940, § 10.
43 Ibid., § 11.

Nevertheless, as I will show, it took a Supreme Court decision relatively late in the war to force the government to observe this provision.

1.3 *Amendments to the Conscription Statute*

Congress made several amendments to the conscription statute, mostly after the United States had entered the war.[44] On August 18, 1941, as the United States moved closer to war, Congress passed an amendment that lifted the limit of 900,000 men to be drafted under the original conscription statute. The same amendment also cut down on the length of time for which the president could extend draftees' term of service. The original law granted permission for unlimited extensions ; now Congress limited such extensions to a total of eighteen months.[45]

Within days of entering the war, Congress passed two amendments to the conscription statute, one each on December 13 and December 20. What is notable about these amendments is that, although they worked a thorough-going change to conscription in the United States, the texts of the amendments were quite short. This was possible because the September 1940 law had established the entire detailed structure for the Selective Service System.

The December 13 amendment did two things. The original statute had limited the use of conscripted troops to the western hemisphere; this amendment lifted that restriction, permitting conscripted forces to fight anywhere in the world. It also retroactively (and prospectively) lifted the previous limit on the length of service of conscripts (and volunteers) in all branches of the armed forces. Now draftees (both those going into the army and COs going into alternative service) would be held to service until the end of the war, plus an additional six months.[46]

The December 20 amendment did three additional things. It changed the ages of men who were required to register with the Selective Service to all men aged 18 to 65 (formerly 21 to 36) , and it expanded the age of men susceptible to conscription to 20 to 45 (also formerly 21 to 36).[47]

---

44 In addition to the amendments discussed in the text, several other amendments made minor or technical changes to the underlying legislation. See, for example, An Act Providing for certain deferments and exemptions under the Selective Training and Service Act of 1940, Pub. L. No. 77-206, 55 Stat. 621 (1941).

45 An Act To extend the periods of service of persons in the military service, Pub. L. No. 77-213, 55 Stat. 626 (1941).

46 An Act To provide for the extension of enlistments in the Navy in time of war, Pub. L. No. 77-338, 55 Stat. 799 (1941).

47 An Act to amend the Selective Service Act of 1940, Pub. L. No. 97-360, 55 Stat. 844 (1941). The upper age limit had been reduced to twenty-eight in a different, minor congressional amendment on August 16, 1941. An Act Providing for certain deferments and exemptions

The December 20 amendment also allowed the Selective Service to make *any* "rules and regulations" pertaining to conscription and deferment that are "in the national interest." Combined with the provision that all men aged 20 to 45 were liable to conscription (and all men 18 to 65 had to register, in contemplation of a wider draft), this amendment gave the government, and specifically the Selective Service, enormous power to channel nearly the entire male workforce of the nation into areas that contributed to the war effort.

Thus, within a few days after Pearl Harbor, all men aged 18 to 65 had to register, and any man aged 20 to 45 could be conscripted. The conscription law now placed no limits on the numbers of men that could be drafted and no limit on their term of service. And there were no limits on the classification system that would be used to channel men into certain jobs or away from other jobs.

A year later, on November 13, 1942, Congress passed the next substantial amendment to the draft law. This amendment lowered from twenty to eighteen the age of men who could be drafted, removed any bar to the drafting of people convicted of minor criminal offenses, and provided a deferment to all men found to be "necessary to and regularly engaged in an agricultural occupation or endeavor essential to the war effort." This last section of the law was called the Tydings Amendment, after its primary advocate, Senator Millard E. Tydings (D-MD).[48]

On December 5, 1943, Congress enacted what at the time was called the Fathers' Draft Bill but might more properly have been called the Fathers' Deferment Bill.[49] This provided that men who had married before December 8, 1941, and had a child born before September 15, 1942 (that is, conceived before Pearl Harbor), would not be drafted until all men who were not fathers had been drafted. The bill had been much discussed, and the date limitations were put in so as not to reward men who married or fathered children in order to avoid the draft.

under the Selective Training and Service Act of 1940, for publicity [*sic*] with respect to classifications, and for other purposes, Pub. L. No. 77-206, 55 Stat. 621 (1941).

48 An Act to Amend the Selective Service and Training Act of 1940, Pub. L. No. 77-772, 56 Stat. 1018 (1942). Millard E. Tydings should not be confused with his son, Joseph Tydings, who represented Maryland in the US Senate after his father.

49 US Selective Service System, *Selective Service as the Tide of War Turns: The Third Report of the Director of Selective Service* (Washington, DC: Government Printing Office, 1945), 53 (hereafter, *SS 3d*); An Act amending the Selective Service and Training Act of 1940, Pub. L. No. 78-197, 57 Stat. 596 (1943).

2 The Draft in Operation

By any reasonable measure, US conscription during World War II was an enormous undertaking. A total of 46.8 million men were registered in six groups, as reflected in Table 5.1. The first five groups consisted of all men of certain ages who had to register on specific dates; the sixth, and last, group consisted of men newly turning eighteen. The total male population of the United States was 66 million in the 1940 Census, so 70.9 percent of all men registered.

A few words of explanation will be helpful in understanding Table 5.1. The first large registration was held on October 16, 1940, just a month after Congress passed the conscription statute; the government was wasting no time. This first registration was for men aged twenty-one to thirty-five, in accordance with the newly enacted law.

The second registration (on July 1, 1941) was much smaller, only for men who had turned twenty-one since the first registration.

The third, fourth, and fifth registrations were for men aged eighteen to sixty-five after the statutory amendment of December 20, 1941, changed the age range of men required to register. In the third registration (February 16, 1942), men aged 20 to 21 and 37 to 44 were obliged to register. In the fourth registration (April 27, 1942), which was quite large, men aged 45 to 64 were obliged to register. Note, however, that men in this age group, while obligated to register, were never subject to conscription. And in the fifth registration, on June 30, 1942, all men then aged 18 to 20 were registered.

The sixth, and final, registration was not a single-day affair—it required younger men to register as they reached their eighteenth birthday.

The number of men drafted during the World War II era also was impressive, almost ten million men; Table 5.2 shows the number of men drafted annually.

TABLE 5.1 Total registrations in World War II

First registration (Oct. 16, 1940)	16,878,723
Second registration (July 1, 1941)	849,915
Third registration (Feb. 16, 1942)	8,635,236
Fourth registration (Apr. 27, 1942)	14,369,325
Fifth registration (June 30, 1942)	2,871,121
Sixth registration (thru Dec. 31, 1945)	3,232,832
Total	**46,837,152**

SOURCE: US SELECTIVE SERVICE SYSTEM, *SELECTIVE SERVICE AND VICTORY*, 465, FIG. 49.

TABLE 5.2 Number of men drafted in World War II

10/1/40–12/31/41	950,848
1942	3,041,872
1943	3,344,740
1944	1,612,492
1945	970,484
Total	**9,920,346**

SOURCE: US SELECTIVE SERVICE SYSTEM, *SELECTIVE SERVICE AND VICTORY*, 590, FIG. 136.

In addition to the 9.9 million men who were drafted, nearly 5 million men entered the military as so-called voluntary enlistees (during the periods when enlistments were allowed). In fact, a substantial percentage of enlistments were not voluntary at all but were made under duress of the draft.[50]

Unpaid civilian members of the 6,443 local draft boards did most of the work of conscription. All local draft board members were male, selected by their governor, subject to approval by the Selective Service.[51]

The conscription statute allowed the Selective Service to create an internal mechanism for appeals by men who disagreed with local draft board determinations. The Selective Service created a two-tiered system. It established 274 regional appeal boards; these affirmed the decisions of the local draft board in more than half the cases they heard. In theory, men whose appeal was turned down could appeal to the president. In fact, Roosevelt delegated this authority to an army officer, Hershey, who in turn delegated the authority to a group of army officers. Thus, the so-called presidential appeals were actually heard exclusively by men in military uniform.[52] This was a clear violation of the conscription statute, which required that civilians hear all appeals. This fact was recognized at the time. The ACLU and others vigorously protested the military's participation but to no avail. Multiple court challenges to this blatant violation of the Selective Service statute were unavailing.[53]

On February 13, 1943, a three-person delegation of peace movement leaders met with Paul McNutt, head of Roosevelt's War Manpower Commission, to

50 *SS 4th*, 588, fig. 135.

51 *SS 1st*, 61.

52 *SS 4th*, 39–42.

53 American Civil Liberties Union, *Conscience and the War* (New York: American Civil Liberties Union, 1943), 16. See, for example, *United States ex rel. Brandon v. Downer*, 139 F.2d 761 (2d Cir. 1944), and *Giese v. United States*, 143 F.2d 633, 635–36 (D.C. Cir. 1944), *aff'd per curiam* by an equally divided court in *Giese v. United States*, 323 U.S. 682 (1945).

discuss the composition of this Presidential Appeal Board. Because the Selective Service was then, briefly, under the commission, McNutt was technically Hershey's superior in the government hierarchy. The delegation was composed of Dorothy Detzer of the Women's International League for Peace and Freedom, Norman Thomas of the Socialist Party, and Roger Nash Baldwin of the ACLU. The activists presented McNutt with a written memorandum that said in part, "Contrary to the provisions of the Selective Service Act, the Presidential Appeal Board dealing with conscientious objectors consists of five military officers. According to the law, this should be a wholly civilian function." A footnote in the memorandum quoted the relevant statutory provision. The activists were not successful. The discussion "became quite heated and McNutt expressed the opinion [that] the conscientious objectors were fortunate in not being lined up against a wall and shot."[54]

McNutt's private comment on this occasion may be shocking, but it was not out of character. On another occasion, McNutt told a public audience that he supported "the extermination of the Japanese people in toto." When asked whether he meant just the Japanese military or the entire civilian population (genocide is universally recognized as a crime against humanity), McNutt was careful to clarify that he meant the entire population.[55]

On October 14, 1940, Roosevelt nominated Clarence Dykstra, president of the University of Wisconsin, to be the director of the Selective Service, and he was confirmed by the Senate the next day. On April 1, 1941, Dykstra resigned, and on July 31, 1941, Hershey became director. Hershey remained in that job until he retired on February 16, 1970.[56]

---

54 National Committee on Conscientious Objectors of the American Civil Liberties Union, memorandum for Mr. McNutt, February 13, 1943, box IX-12-1, folder 3; Roger N. Baldwin to Mr. McNutt, February 16, 1943, box IX-12-1, folder 3; and French to NSBRO [memorandum] no. 120, February 18, 1943, box IX-12-1, folder 3, all in Mennonite Central Committee Archives, Akron, PA (hereafter, Mennonite archives). French served as the executive of the National Service Board for Religious Objectors (NSBRO) for the duration of the war. In that capacity he sent numerous detailed memoranda to his governing board in order to keep its members apprised of NSBRO's work. The memos were dated and numbered sequentially. Because French sometimes sent more than one memo per day, they are identified here by both date and number.

55 John W. Dower, *War without Mercy: Race and Power in the Pacific War* (New York: Pantheon Books, 1986), 55.

56 *SS 1st*, 65; "Hershey Aide, a Retired Colonel, Named Acting Director of Draft," *New York Times*, February 17, 1970.

TABLE 5.3 Selective service classification system during World War II

I-A	Available for military service
I-A-O	Conscientious objector available for noncombatant military service
II-A	Deferred in support of the war effort
II-B	Deferred due to employment in war production
II-C	Deferred due to employment in agriculture
III-A	Deferred for dependency reasons
III-D	Deferred for reason of extreme hardship
IV-A	Deferred, 45 years of age or older
IV-B	Government official deferred by statute
IV-C	Deferred Alien
IV-D	Clergy or clerical student
IV-E	Conscientious objector available for civilian alternative service
IV-F	Rejected for physical, mental or moral reasons

SOURCE: US SELECTIVE SERVICE SYSTEM, *SELECTIVE SERVICE AND VICTORY*, 62–64.

Note: Classifications III-B and III-C do not appear in this table because the War Department collapsed them into other classifications during World War II.

On December 5, 1942, Roosevelt issued Executive Order 9279, prohibiting all voluntary enlistments in both the army and the navy.[57] This confirmed that a central role of the Selective Service was to channel all of the nation's manpower into areas that the government considered to be war related, not merely to obtain that subset of men who were conscripted into the army. After the executive order, the army would no longer be plagued by having "the wrong men" enlist. The executive order remained in effect for the duration of the war.[58]

To further implement the Selective Service's key role in channeling all the country's manpower into war-related pursuits vital to the national interest, the agency created a system of classification of all (registered) men. Table 5.3 shows the basic Selective Service classifications during the war.[59]

57 Exec. Order No. 9279, 7 Fed. Reg. 10177 (1942); see SS *3d*, 205, sec. 4, which discusses the executive order. The full text of Executive Order 9279 appears in app. 27, 332–34.

58 SS *4th*, 78. President Harry Truman revoked Executive Order 9279 on August 29, 1945, in his Executive Order 9605.

59 During the war, small changes were occasionally made to the classification system to reflect statutory changes made by Congress or other changes in circumstances.

The Selective Service further strengthened the channeling function of the draft with a regulation that provided that men who had a draft deferment because they were employed in a war-related industry would be reclassified I-A (immediately eligible to be drafted) if they changed jobs without first obtaining the approval of their draft board (even if the new job would have qualified a man for a deferment had he received draft board approval before switching jobs).[60]

The general policy of the Selective Service was to assign a man the highest classification (that is, the furthest away from I-A) for which he qualified. Thus, if a fifty-year-old man was involved in a war industry, he would be given the higher classification for which he was eligible, IV-A (older than forty-five) instead of the lower classification for which he was also eligible, II-B (involved in war production). The classification system established by the Selective Service fairly tracked the underlying statute. For example, government officials who were draft exempt under the law were classified IV-B; clergy and divinity students, also exempted by law, were classified IV-D. Men in these categories could not be drafted so long as they held those classifications. The two classifications for COs similarly tracked the provisions of the statute: the I-A-O classification was for men who were willing to perform noncombatant roles in the army; the IV-E classification was for men who would perform only alternative civilian service.

During World War II, as during World War I, the Selective Service used a lottery system to determine the order for calling men who were classified I-A. The numbers drawn in the lottery used the sequential numbers that men had been assigned when they registered. That is, every draft board in the country had a man who was number 1 for that draft board; he was the man who had registered first at that board. Boards then had men with the numbers 2, 3, and 4 for that board.[61]

When conscription started in 1940, lottery numbers were important because draft calls were still relatively small and men with high numbers were not immediately drafted. However, once the United States entered the war, draft calls shot up dramatically, and lottery numbers lost importance as everyone who was classified I-A was drafted. In fact, on June 1, 1942, the need for conscripts was so acute that the War Department and Selective Service abolished the previous requirement that army recruits or draftees must have finished the fourth grade. At the end of the war, more than 35 percent of draftees had finished all or part of elementary school but had never attended a day of

60 SS *4th*, 97.
61 SS *1st*, 89–99.

high school, and nearly 29 percent more were high school dropouts. A mere 2.7 percent of draftees were college graduates.[62]

On December 10, 1943, the Selective Service issued a regulation that said that men who were classified I-A (available for regular army service) or I-A-O (COs available for noncombatant army service) and who were in any way "delinquent"—that is, they had not scrupulously followed all Selective Service requirements—would be moved to the front of the list of men to be drafted, regardless of lottery number. This is significant for two reasons. It gives the lie to the Selective Service's repeated assertions during the war that the agency only ran the draft system but left to the Justice Department all enforcement actions against men who violated the law or draft regulations.[63] Also, the World War II regulation, promulgated when Hershey was the director of the Selective Service, foreshadowed a highly controversial regulation issued by the agency in October 1967 during the Vietnam War, when Hershey was still director (see chapter 7).

∴

The situation of US citizens of Japanese ancestry created multiple questions for the government during the war that included, but were not limited to, matters related to conscription. On January 5, 1942, less than a month after Pearl Harbor, the Selective Service placed all US citizens of Japanese descent into classification IV-C. This was absurd, because, according to Selective Service regulations, classification IV-C was for "aliens" and, by definition, these US citizens were not aliens. But issues with the draft were only one part of a broader set of problems faced by Americans of Japanese descent.[64]

On March 24, 1942, the US government imposed an 8:00 p.m. to 6:00 a.m. curfew on people of Japanese descent on the West Coast. The curfew applied to citizens and noncitizens alike, with no particularized suspicion about wrong-doing by any specific individual, and the curfew was imposed by ukase, without any pretense of legislative or other process.[65]

Five days later, on several days' notice, the roundup of Japanese immigrants and US citizens of Japanese descent began on the West Coast; eventually,

---

62 SS *4th*, 120, app. 151, 607.

63 SS *3d*, 169. SS *3d* states: "The enforcement of the law is not a function of the Selective Service System itself, but of the Department of Justice" (211). The report paraphrases the same assertion on page 221, for example.

64 Eric L. Muller, *Free to Die for Their Country* (Chicago: University of Chicago Press, 2001), 41.

65 Ibid., 21.

110,000 people were herded into ten concentration camps operated by the War Relocation Authority.[66]

By late 1943, the War Department's need for men was growing ever more acute and, on January 20, 1944, it announced that the US citizens of draft age among the interned Japanese Americans would be reclassified from IV-C (aliens) to I-A (immediately eligible to be drafted). In February 1944, en masse drafting of the internees began. There was something truly absurd about the War Department's actions. The government had put the internees into concentration camps because it deemed these people far too dangerous to be allowed to walk free in San Francisco; now the government was yanking these supposedly dangerous aliens out of concentration camps, arming them with machine guns and hand grenades, and deploying them with the US Army—albeit it in racially segregated units. This created some controversy, as I discuss later in this chapter.[67]

This book divides draft objectors into two broad categories as a convenient way to distinguish between draft opponents who held two broadly different sets of beliefs, COs and resisters. The principal distinction between the two groups is that the COs cooperated with the Selective Service System and a great many were drafted, either into noncombatant roles in the army or into alternative civilian service. In contrast, resisters did not cooperate with the system; they refused to register and/or refused induction. These men went to prison.

Of course the bifurcation was not absolute. Some men who refused induction had first tried (but failed) to be recognized as COs. Moreover, the distinction in nomenclature (CO versus resister) did not become widely used until the Vietnam era draft. It was not used by conscription opponents during World War II. For example, the *Conscientious Objector* was a World War II newsletter, originally published jointly by the War Resisters League, Fellowship of Reconciliation, and Peace Committee of New York Friends Meeting (and later by an independent editorial board). In its issue for December 1940–January 1941, the *Conscientious Objector* ran a short piece about the sentences that early

66 Ibid., 39, 50. In this context the term *concentration camps* is appropriate as these were "places where persons charged with no crimes were made prisoners, not because of what they had done, but because of what they were: men, women, and children who had been born in Japan, or, more often the children or … grandchildren of persons who had been born in Japan." Roger Daniels, "Introduction: Heart Mountain after Sixty Years," in Mike Mackey, ed., *A Matter of Conscience: Essays on the World War II Heart Mountain Draft Resistance Movement* (Powell, WY: Western History Publications, 2002), 1.

67 Muller, *Free to Die for Their Country*, 64.

nonregistrant resisters received; the article started this way: "The severity of the sentences meted out to those conscientious objectors who refused to register under the Selective Service Act should be a matter of prime concern to all pacifists."[68] Jim Peck was a World War II resister; he never claimed to be a CO or sought CO status from the government, and he served a three-year sentence in federal prison for refusing induction. Yet at the beginning of his World War II prison memoir, Peck says, "I was ... in jail as a result of my stand as a conscientious objector."[69]

3 Conscientious Objectors

According to the Selective Service, more than 70,000 registrants claimed to be COs during the World War II draft (commencing September 1940).[70] Other sources put the number at 95,000 to 105,000. A fair understanding of the aggregate figure requires identification of the several different categories into which these men fell.

First were the COs who were inducted into the armed forces during World War II for noncombatant service. These were the men classified I-A-O by their draft boards. After the war, the Selective Service put this number at 25,000 men; however, two separate contemporaneous published estimates put the figure at 50,000 or more and 60,000, respectively.[71]

Another 12,000 COs, who objected even to noncombatant military service, performed alternative civilian service in Civilian Public Service (CPS) camps. These were the men classified IV-E by their draft boards.

An additional 13,780 men actually were classified as COs, either as I-A-O or as IV-E at one point, but were subsequently reclassified as IV-F (medically unfit

68 "Undue Severity," *Conscientious Objector* 2, no. 6 (December 1940–January 1941): 4.

69 James Peck, *We Who Would Not Kill* (New York: Lyle Stuart, 1958), 12, 58. Peck (1914–1993) was a lifetime peace and civil rights activist long associated with the Congress of Racial Equality and War Resisters League. Peck also wrote an autobiography, *Upperdogs vs. Underdogs* (New York: AMP&R, 1980), and *Freedom Ride* (New York: Simon & Schuster, 1962).

70 US Selective Service System, *Conscientious Objection*, 315.

71 The figure of 25,000 appears in US Selective Service System, *Conscientious Objection*, 315; the figure of 50,000 for those classified as I-A-O is from Winslow Ames, "CO's in Uniform—Non-Combatant Duty," *Reporter* 3, no. 2 (September 16, 1944): 3. Ames, who was a conscientious objector during World War II, was hired by the National Service Board for Religious Objectors (NSBRO) to do an extensive survey of I-A-Os. Another CO, William Fifield, who served in Civilian Public Service Camp 46 in Big Flats, New York, put the number of I-A-Os at 60,000 in an article in the January 1945 issue of *Harper's*. William Fifield, "Report to Harper's," *Pacifica Views* 2, no. 36 (February 9, 1945): 2.

for service) or placed into another category. This comported with the Selective Service policy of assigning men the highest classification (furthest away from I-A) for which they were eligible. The COs in this category were never drafted, either into noncombatant military service or into alternative civilian service, because of that higher classification.

Finally, approximately 20,000 more men claimed, upon registration, to be COs, but the Selective Service never considered their claims because they were immediately placed into another classification upon registration.

Using the official Selective Service figures published five years after the war ended, the foregoing four categories add up to approximately 70,780 registrants claiming to be conscientious objectors during World War II. If one uses one of the higher figures of men classified I-A-O, cited in other published accounts of the time, there were 95,000 to 105,000 COs during World War II.[72]

The educational levels of World War II COs and draftees in general provide an interesting comparison. Of the COs who worked in Civilian Public Service (CPS) camps run by the American Friends Service Committee, 20.3 percent were college graduates, another 27.9 percent had had some college, and 20.2 percent had postgraduate education. That is, the percentage of these COs who had attended college was higher than the percentage of army draftees who had finished high school.

4 Civilian Public Service

The provision in the World War II conscription statute that allowed for alternative civilian service was a major victory for the peace churches, other religious denominations, and secular peace organizations that had pushed so hard for such a provision. But when the Burke-Wadsworth bill was signed into law on September 16, 1940, the victory remained very much an incomplete one, for, despite planning for renewed conscription since 1926, no one in either the military or the rest of the US government had given even a moment of thought to what alternative civilian service might look like or how it could be run. What ultimately emerged as the way alternative service would be operated—that is, CPS camps—was the result of an elaborate set of negotiations between the historic peace churches and the government during the autumn of 1940.[73]

72 US Selective Service System, *Conscientious Objection*, 314–15.

73 Melvin Gingerich, *Service for Peace: A History of Mennonite Civilian Public Service* (Akron, PA: Mennonite Central Committee, 1949), 56; "Memorandum of Conversation with Colonel Hershey, October 15, 1940," in Paul Comly French, *The Origins of Civilian Public Service* (Washington, DC: National Service Board for Religious Objectors, 1940): app. 2, 11; Albert

On September 4, 1940, twelve days before the conscription statute was signed into law, sixteen people from the peace churches (and Walter Van Kirk of the Federal Council of Churches, forerunner of the National Council of Churches) met at the Commodore Hotel in Washington, DC, to consider what alternative civilian service might look like and what their churches could do to help create it. How such a program might be financed was one of the matters the group considered. Paul Comly French, a Quaker, was asked to continue working in Washington on CO-related issues for another sixty days, and the group planned for a larger, follow-up meeting of their churches in Chicago early the next month.[74]

On September 16, 1940, Roosevelt signed the World War II conscription bill into law.

On October 1, 1940, Hershey, newly promoted to colonel, became (a very temporary) acting director of the Selective Service, and, on October 2, Hershey asked French whether Quakers would be prepared to administer alternative civilian service for COs. French took Hershey's query back to the Friends War Problems Committee, and that committee agreed that it would be willing to take on the task.[75]

On October 4 and 5, the larger meeting of the peace churches took place in Chicago; sixty-five Quakers, Mennonites, and Brethren attended. French, following Hershey's suggestion, proposed that Quakers take on the role of running some sort of alternative civilian service program in coordination with the government. The Mennonites and Brethren, however, did not want such a program to be entirely administered by the Quakers and instead proposed "setting up a new organization [that] will be composed of Friends, Mennonites and Brethren" to coordinate the operation of an alternative civilian service program.[76]

On October 11, the newly created agency, the National Council for Religious Objectors, came into being. The chairman was M. R. Zigler (Church of the Brethren), the vice chairman was Orie O. Miller (Mennonite), and the executive secretary (a staff person) was French. A seven-member executive committee, representing the three peace churches, was named. Barely six weeks later, on November 26, the newly created entity was renamed the National Service

N. Keim, *The CPS Story: An Illustrated History of Civilian Public Service* (Intercourse, PA: Good Books, 1947), 25–26.

74 Keim, *CPS Story*, 24–25.

75 *SS 1st*, 65.

76 Report of Meeting of Representatives of HPCs [historic peace churches], October 4 and 5, 1940, Mennonite Home Mission, box IX-7-10, folder 3, Mennonite archives; Keim, *CPS Story*, 26–28; Gingerich, *Service for Peace*, 54–55; and French, *Origins of Civilian Public Service*, 5.

Board for Religious Objectors (NSBRO). (For simplicity, this chapter refers to the entity as NSBRO, even for the brief period before the name change.)[77]

On October 15, 1940, Dr. Clarence Dykstra became the first (nonacting) director of the newly re-created Selective Service System. Hershey continued to serve at the agency.

When NSBRO came into existence on October 11, it faced some urgency in developing a plan for alternative civilian service. The newly enacted conscription statute allowed for alternative service, but the government had made clear both that it had given no thought to the matter and that it would welcome a plan or proposal from the churches.[78]

On October 15, NSBRO representatives met with Hershey. Two important things occurred at that meeting. Hershey told the pacifists that they "had better get a definite plan [for CPS] worked out by the beginning of next week." Hershey then assured the pacifists that he knew that "the Government would have to provide funds for this work." So sure was Hershey on this second point that he pressed the pacifists to do actual calculations of how much money it would cost to pay for staffing CPS and provide the materials for its work so he "would be in a position to intelligently go about securing the necessary funds to do the work." Hershey's assurances regarding governmental funding become significant in light of later events.[79]

In response to Hershey's urging, NSBRO got to work and, on October 24, less than two weeks after coming into existence, gave its written proposal for alternative service to the government. The proposal contemplated offering COs two different kinds of work programs, with different funding streams proposed for each type. In one program, government agencies would both provide civilian work and pay the COs; in the other program, private agencies (that is, the historic peace churches) would provide both work and pay.[80]

On October 29, the government made a counterproposal that would add a third category of work program. This would be a hybrid category in which the churches would run work camps in conjunction with government agencies, with the COs paid by the government. It is important to note here that the government was proposing to NSBRO that, of the three types of CPS camps to

---

77 Keim, *CPS Story*, 27–28; Gingerich, *Service for Peace*, 55–56.

78 Gingerich, *Service for Peace*, 56–57; Keim, *CPS Story*, 28; French, *Origins of Civilian Public Service*, 6.

79 French, *Origins of Civilian Public Service*, app. 2, 11.

80 Gingerich, *Service for Peace*, 56–57; Keim, *CPS Story*, 28; French, *Origins of Civilian Public Service*, 6, 19–20.

be established, two would involve government pay for men doing alternative service.[81]

On November 25, NSBRO accepted the government's three-part counter-proposal. Yet this plan never came into existence. On November 29, Dykstra took the plan to the White House, and "the President expressed instant and aggressive opposition to the plan." Roosevelt thought that the work camps proposed were too cushy an alternative for COs who were avoiding military service and that government funding for the scheme was a politically risky idea. Dykstra told NSBRO that "in view of the President's attitude he was sure that the possibility of providing wages for the [CPS] men was completely out of the question."[82]

On December 5, Dykstra and Hershey, representing the Selective Service, and Lowell Mellett, representing the White House, met with NSBRO representatives in Washington to discuss the situation. Dykstra "asked whether the church agencies would be willing to administer all projects for conscientious objectors and pay the entire cost." Under this arrangement, the government would not provide a salary or stipend for the COs in CPS and would pay no overhead administrative costs of the program. Dykstra and Hershey said that they were reluctant to ask Congress for funds, especially without support from the administration, fearing that to do so might open the door to congressionally imposed limitations or restrictions on the program. If federal funds were used, Dykstra "anticipated that such action would lead to complete government administration, with no opportunity for COs to serve under religious agencies."[83]

NSBRO later reported that "the churches were confronted with a problem as serious as it was unexpected. Up to this point, they had assumed that there would be a variety of alternatives and that the government would carry the main financial responsibility, including maintenance and pay for all COs except a modest number who might want to work completely under religious auspices and would be unwilling to accept government support." The NSBRO's assumption had been reasonable; it conformed to the arrangement that the Selective Service had not only agreed to but that it had actually proposed.[84]

---

81 Paul Comly French to National Council for Religious Conscientious Objectors, November 14, 1940, box IX-7-10, folder 3, Mennonite archives; Gingerich, *Service for Peace*, 56–57; Keim, *CPS Story*, 28; French, *Origins of Civilian Public Service*, 3, 6–7.

82 Gingerich, *Service for Peace*, 56–57; Keim, *CPS Story*, 28; French, *Origins of Civilian Public Service*, 6–7.

83 Gingerich, *Service for Peace*, 57; Keim, *CPS Story*, 28–30. French quotes Hershey directly in "Memorandum of Conversation with Colonel Hershey, October 15, 1940," *Origins of Civilian Public Service*, app. 2, 11.

84 French, *Origins of Civilian Public Service*, 7.

On December 10, the NSBRO board met to consider the new situation. After some discussion, members decided to "accept the responsibility within the limit of our financial ability and make every attempt to carry the program through successfully." The NSBRO meeting adjourned, and the pacifists went directly to Dykstra's office to tell him, and Hershey, their decision.[85] The board's decision may have been a bit premature. The Church of the Brethren did not formally accept the proposal and agree to provide funding for the CPS camps until December 19, and the Mennonite Central Committee did not formally approve the arrangement until December 21.[86]

On December 19, Roosevelt approved the arrangement. The next day, the deal was formalized by an exchange of memoranda between the Selective Service (Dykstra) and NSBRO (French). At the time, NSBRO stated that this would be an experiment, subject to periodic reexamination. NSBRO and the Selective Service formally extended the arrangement in writing in September 1941 and again in January 1943.[87]

With that, an extraordinary situation came into existence. The Quakers, Mennonites, and Brethren—pacifist religious bodies—would not only be operating a part of the nation's military conscription system, they would also *be paying for it*. As the Mennonite historian Albert Keim later wrote, "It was a rash decision for such normally cautious people."[88] And as later events would demonstrate, it was not merely rash but deeply controversial for many COs, especially some of the secular ones. Yet the minutes of the next meeting of the NSBRO Board of Directors, held at the Philadelphia offices of the American Friends Service Committee on December 22, 1940, showed not the slightest appreciation by board members of just how controversial that decision would become.[89]

Several factors help to explain the thinking of the peace churches at the time. In the autumn of 1940, when the agreement was made, Pearl Harbor had

---

85 Minutes of the National Service Board for Religious Objectors (marked "CONFIDENTIAL NOT FOR PUBLICATION"), December 10, 1940, box IX-7-10, folder 3, Mennonite archives; French, *Origins of Civilian Public Service*, 8. The minutes of the NSBRO board meeting for December 10, 1940, at the Hotel Harrington can be found in War Resisters League Papers, RG 040, ser. A, subser. VI, box 23, Swarthmore College Peace Collection (SCPC), Swarthmore College, Swarthmore, PA, and are reprinted in French, *Origins of Civilian Public Service*, 23–24.

86 "Brethren Advisory Committee for COs, December 18, 1940"; "Minutes of the Standing Committee in Special Session, December 18 and 19, 1940"; "MCC Annual Meeting," December 21, 1940, all in French, *Origins of Civilian Public Service*, 22–23.

87 NSBRO press release, December 20, 1940, unlabeled folder adjacent to folder 3 in box IX-7-10, Mennonite archives; Keim, *CPS Story*, 28–29; Gingerich, *Service for Peace*, 58–59.

88 Keim, *CPS Story*, 29.

89 Minutes of the NSBRO meeting, December 22, 1940, box IX-7-10, folder 3, Mennonite archives.

not yet been attacked, the United States had not yet entered the war, and military service—including the term to be served by COs—was still only one year, not the duration of the war plus six months that it would become a year later by amendment to the statute.

Also, even the original NSBRO proposal to the Selective Service had some camps that the churches would run and finance, whereas the government would operate and fund other camps. Nevertheless, the idea of church-run (and church-financed) camps was not wholly contrary to the thinking of the historic peace churches.

In addition, in late 1940, the bitter, harrowing experience of COs in World War I was barely more than twenty years old. The church elders who comprised the NSBRO board believed that having COs in civilian clothes, doing civilian work in civilian camps operated by the peace churches, was a great deal better than having COs placed in the army and abused and tortured—even if the churches had to pay to run the civilian camps.[90] Even looking back on those events today, one is hard-pressed to disagree with this view.[91]

Indeed, in some ways, the peace churches *liked* the idea that they would get to run the CPS camps. The coming of war was a world crisis, one that could not be evaded or avoided merely by being a pacifist. Civilian Public Service reflected "the desire of the [peace] churches for channels through which an increasing sense of community responsibility might be expressed."[92] The peace churches viewed the chance to operate the camps as an opportunity to demonstrate a better way to approach conflict than war. As Mennonites said after the war, "It was necessary during the war that unmistakable witness be borne to the fact that war and the war method is sinful, utterly contrary to the teachings of Jesus, and that therefore, as followers of Christ, we can have no part in it. But the church needed to do more than give an unmistakable 'NO' to the military method. She must, with even greater insistence, point to the better way of Christ, testifying to it by word and work. Civilian Public Service was intended to be such a witness."[93] The Brethren made a similar point: "The design [of CPS] was to portray a new way of citizenship. Brethren hoped to exemplify a constructive way of peace, in contrast to the way of war endorsed by the government."[94]

---

90 Gingerich, *Service for Peace*, vi–vii.

91 Arthur A. Ekirch, Jr., "A Political Prisoner in Wartime," *Peace and Change* 12, no. 1 (April 1987): 77.

92 Leslie Eisan, *Pathways of Peace* (Elgin, IL: Brethren, 1948), 9.

93 Gingerich, *Service for Peace*, ix.

94 Eisan, *Pathways of Peace*, 366.

Another factor in the churches' acquiescence to funding the camps was that they believed that this would allow them to handle the religious education of their members in the camps: "The most important consideration ... in causing the Mennonites to agree that it was wise for them to participate in the management [and financing] of CPS was the fact that the plan gave them the opportunity to direct the religious and educational development of [their] men."[95]

The CPS system thus invented by the Selective Service and the peace churches in the autumn of 1940 saw service by almost exactly 12,000 men during its existence from 1941 through 1947.[96] Not surprisingly, members of the historic peace churches comprised the largest share of the CPS population: 4,665 Mennonite CPSers (38.8 percent), 1,353 Brethren (11.3 percent), and 951 Quakers (7.9 percent), together comprising a majority of men in CPS.[97] Other mainstream Christian denominations also had significant representation: 845 Methodists (7.0 percent), 243 Baptists, 235 Presbyterians, 162 Catholics, and 124 Lutherans.[98] Sixty Jews served in CPS.[99]

How draft boards treated COs varied enormously from board to board. Many draft boards took a dim view of COs and granted CO status to few, if any, registrants. Other boards took the position that COs must belong to one of the historic peace churches, the express language of the statute to the contrary notwithstanding. For example, as the Selective Service itself acknowledged, some draft boards believed (contrary to law) "that a member of the Catholic church could not possibly have a basis for conscientious objection." Indeed, the problem of identically situated men being classified differently from one local board to another was a problem that ran through the entire Selective Service System and affected men in all classifications, not just COs.[100]

---

95 Gingerich, *Service for Peace*, 396, 403.

96 The *Directory of Civilian Public Service*, published by NSBRO after the war, puts the number of men in Civilian Public Service at 11,996. National Service Board for Religious Objectors, *Directory of Civilian Public Service, May, 1941 to March, 1947* (Washington DC: National Service Board for Religious Objectors, 1947), xix. However, I have identified several men who served in CPS during the war whose names did not appear in the NSBRO directory. The US Selective Service System's *Conscientious Objection*, 315, which was published after the war, puts the number at 11,950.

97 National Service Board, *Directory of Civilian Public Service*, xviii–xix.

98 US Selective Service System, *Conscientious Objection*, 318.

99 National Service Board, *Directory of Civilian Public Service*, xix.

100 *SS 2d*, 258, 171.

The almost universally cited figure for the number of CPS camps during World War II is 151.[101] However, this accounts for only the main camps and fails to take into account some of the vagaries of CPS organization. For example, men served alternative service by working on dairy farms in thirty-four counties in fifteen states, but all of these men were designated as attached to a single camp, Camp 97. The dairy herd–testing program had units in fourteen states, but all were considered to be parts of Camp 100. Many camps that the US Forest Service and the National Park Service supervised had multiple spur camps that were not considered separate entities. Thus, the number of *locations* where CPS men worked far exceeded the official count of 151 main camps.[102]

The Mennonite Central Committee ran the largest number of camps of any group, 61 main camps; Mennonite camps also had the largest census overall. The American Friends Service Committee (AFSC) operated 38 main camps, and the Brethren Service Committee (BSC) operated 39 main camps.[103] The Association of Catholic Conscientious Objectors, American Baptist Home Missionary Society, the Methodist World Peace Commission, and other church agencies ran other camps.[104] The populations of all the camps, however, were heterogeneous; a man did not have to be a Mennonite to request service at a Mennonite camp, for example, and a majority of the men serving at BSC camps in fact were not Brethren.[105] At first, there were no government-run CPS camps at all, but eventually the government ran four main camps.[106]

The alternative service done by the men in CPS comprised a range of projects administered and directed by what the Selective Service referred to as "technical agencies." The so-called technical agencies that handed out work to CPS conscripts were civilian government agencies such as the Forest Service, Soil Conservation Service, National Park Service, Bureau of Land Reclamation, and the Fish and Wildlife Service.[107] The Selective Service unilaterally wrote

---

101 Keim, *CPS Story*, 33; Civilian Public Service Camp Newsletter Collection, 1941–44, Robert S. Cox Special Collections, University of Massachusetts Libraries, Amherst. The listings in the National Service Board's *Directory of Civilian Public Service*, x–xvii, give each camp's location, operating agency, and opening and closing dates.

102 US Selective Service System, *Conscientious Objection*, 173–74.

103 National Service Board, *Directory of Civilian Public Service*, x–xvii.

104 US Selective Service System, *Conscientious Objection*, 176; National Service Board, *Directory of Civilian Public Service*, x–xvii.

105 Eisan, *Pathways of Peace*, 49–50.

106 US Selective Service System, *Conscientious Objection*, 176.

107 Ibid., 190.

the rules and regulations governing life in the CPS camps, and the agencies operating the camps had to implement them.[108]

One of the largest areas of work for CPS was in soil conservation. Several of the first CPS camps established had formerly been camps of the New Deal era Civilian Conservation Corps (CCC), and in December 1942, Lieutenant Colonel Franklin McLean of the Selective Service could say of CPS, "The work being done is mainly a continuation of projects originated by the CCC."[109] In 1945, more than eleven hundred CPS men were engaged in soil conservation work.[110] They constructed hundreds of large and small dams, created reservoirs, built irrigation ditches, leveled land with heavy equipment to allow tillage, reseeded pastures, built fences to prevent soil erosion, built ridges and channels on sloping fields, and planted trees.[111]

CPS men also did a lot of work for the US Forest Service, which in 1945 employed more than eighteen hundred CPS men. They planted seeds for tens of millions of trees, nurtured the seedlings, and transplanted young trees to replace areas that had been overlogged or destroyed by fire. CPS crews built sprinkler systems, surveyed public lands, developed public campgrounds, fortified the banks of streams and lakes, and built bridges and repaired roads in national forests. One highly publicized aspect of CPS work for the Forest Service involved about 240 CPS men at base camps in Montana, Idaho, and Oregon who worked as smoke jumpers. These men parachuted into areas where there were forest fires, often started by lightning strikes, and would remain there for days or weeks fighting the fires. Other CPS men worked on fire prevention by building fire roads, creating firebreaks, and clearing underbrush.[112]

More than 2,200 CPS men worked in psychiatric hospitals. At the beginning of the war US psychiatric hospitals housed more than 800,000 patients, yet these facilities were woefully understaffed. Because hospital staff were not involved in manufacturing tanks or battleships, they were ineligible for occupational draft deferments. The result was a true crisis, with some facilities

108 Clarence Dykstra to Roosevelt and French, December 20, 1940, reprinted in Gingerich, *Service for Peace*, app. 9, 442–44; US Selective Service System, *Conscientious Objection*, 176–77; Eisan, *Pathways of Peace*, 365–71; Keim, *CPS Story*, 32–38.

109 ACLU, *Conscience and the War*, 25.

110 US Selective Service System, *Conscientious Objection*, 190, table 9.

111 Keim, *CPS Story*, 49–50; US Selective Service System, *Conscientious Objection*, 168; Gingerich, *Service for Peace*, 108–24; Eisan, *Pathways of Peace*, 88–92.

112 Keim, *CPS Story*, 43–49; Gingerich, *Service for Peace*, 125–48; Eisan, *Pathways of Peace*, 80–85; US Selective Service System, *Conscientious Objection*, 190; Robert C. Cottrell, *Smoke Jumpers of the Civilian Public Service in World War II: Conscientious Objectors as Firefighters for the National Forest Service* (Jefferson, NC: McFarland, 2006).

losing as much as 80 percent of their prewar staff to the army. Eventually, COs provided staff to 41 adult hospitals in 20 states, and another 17 youth facilities in 12 states. Patients were sometimes psychotic and/or violent, almost always malnourished and ill-treated, sometimes incontinent and unwashed, with frequent bedsores and other undiagnosed (and untreated) nonpsychiatric medical issues. COs worked as ward attendants, social workers, office workers, cooks, and on building maintenance crews at these hospitals, routinely putting in workweeks of seventy-two to one hundred hours per week.[113]

Appalled by the conditions they found in psychiatric hospitals, COs agitated within institutions for improved conditions for patients and sometimes took their observations to the press and public. Public agitation by COs at Eastern State Hospital in Williamsburg, Virginia, led to intervention by the governor and some reforms there. Agitation by COs working at the Cleveland State Hospital led to significant press coverage and ultimately a thorough overhaul of the state's psychiatric hospital system by the state legislature. Work by Quaker COs at the Byberry Hospital in Philadelphia led to the creation of the Mental Health Hygiene Program, which later turned into the National Mental Health Foundation. After the war, work done by men serving in the CPS was often credited with widespread improvements in the lot of psychiatric patients in the United States.[114]

More than five hundred COs volunteered to participate in so-called guinea pig medical experiments in which they were the research subjects.[115] One such set of experiments involved what was then called infectious hepatitis and is now referred to as hepatitis A. Infectious hepatitis was "the most widespread epidemic disease of the Second World War," and at the time scientists had little knowledge of the ætiology, transmission, or treatment of the disease. In a series of experiments, CO volunteers were deliberately infected with infectious hepatitis and then extensively studied. One result of these experiments was identification of the virus responsible for the disease as well as the vector, contaminated drinking water.[116]

113 Sibley and Jacob, *Conscription of Conscience*, 134–40; Eisan, *Pathways of Peace*, 205–38; Gingerich, *Service for Peace*, 213–51.

114 Sibley and Jacob, *Conscription of Conscience*, 160–65; Keim, *CPS Story*, 72.

115 For general discussions of the guinea pig experiments, see "Louse Control" and "Nutrition Experiment," *NSBRO Reporter* 1, no. 2 (August 1942): 4; "Seawater Project," *NSBRO Reporter* 1, no. 3 (September 1, 1942): 3; and "Surgeon General," *NSBRO Reporter* 2, no. 19 (April 1, 1944): 3–4.

116 Sibley and Jacob, *Conscription of Conscience*, 144; Alison Bateman-House, "Compelled to Volunteer: American Conscientious Objectors to World War II as Subjects of Medical

Another infectious disease studied by research on COs was malaria. In these experiments, COs were deliberately infected with malaria so that new potential treatments could be tried on them.[117]

5 Controversies Regarding Civilian Public Service

Several controversies pertaining to the nature and operation of CPS were of ongoing concern to the agencies that ran the program and to the COs who populated the camps. At the most fundamental level was the question of whether CPS was part of the military conscription system or a truly civilian form of alternative service.

The starting point of this discussion must be the language of the law. The conscription statute provided that "if the objector is found to be conscientiously opposed to ... noncombatant service, he shall in lieu of such induction be assigned to work of national importance *under civilian direction*" (emphasis added).

An ACLU pamphlet, published in October 1943, protested that the government was not following the law: "From a legal point of view all conscientious objectors in CPS camps are under the direction of Selective Service at all times. Their position is analogous to that of men in the armed forces." In another pamphlet, the ACLU went further: "The control of the camps ... by military officers in Selective Service, [is] apparently in violation of law."[118]

But the ACLU was not correct. The requirement of the statute was that the work of CPS be "under civilian direction." It was. Civilian agencies such as the US Forest Service and the Fish and Wildlife Service determined the overall work that COs did and directed them day to day. A great many COs took over work that the Civilian Conservation Corps had done for years and even lived in former CCC camps. COs built civilian sanitary privies for civilians in the hookworm belt of Florida and served as civilian orderlies answering to civilian physicians in civilian psychiatric hospitals. They planted trees in national forests. Civilian entities like the AFSC and the Mennonite Central Committee

Research" (PhD diss., Columbia University, 2014), 247; "Civilian Public Service Guinea Pigs," box IX-12-1, folder 1, Mennonite archives.

117 "Civilian Public Service Guinea Pigs"; Bateman-House, "Compelled to Volunteer," 248; Sibley and Jacob, *Conscription of Conscience*, 145.

118 ACLU, *Conscience and the War*, 22; American Civil Liberties Union, *Freedom in Wartime: A Report of the American Civil Liberties Union in the Second Year of War* (New York: American Civil Liberties Union, 1943), 37.

ran the CPS camps. By any fair reading, this constituted work "under civilian direction."

Yet in another sense, the young COs who—after arriving in CPS camps eventually decided that they were unable in good conscience to remain there—had a valid point. Civilian Public Service did involve significant cooperation with (or, depending on your viewpoint, subservience to) the Selective Service. While the Selective Service was technically a civilian agency, military men in military garb ran it. In addition, pursuant to the arrangement negotiated between the Selective Service and NSBRO, the Selective Service set all the rules governing life in CPS camps. Army officers held key posts in the administration of CPS and were paid out of the military budget appropriated by Congress. Military men in military uniforms made regular inspections of CPS camps to ensure that the military's rules were being carried out with military precision.[119]

In this context, it is understandable that some CPSers found the situation untenable. When Francis Bates walked out of a CPS camp, he said, "Conscription is fundamentally wrong and an evil I can no longer cooperate with or submit to." When Philip Isely walked out, he said, "By remaining in CPS, I … give acquiesce (and thereby give tacit support) to conscription and continued war." When Paton Price, Philip Kelsey, and Julian Jaynes walked out of Civilian Public Service, they issued a joint statement explaining that they "could no longer acquiesce in conscription" and that CPS is "a portion of the military program."[120]

Three things, then, were simultaneously true about whether CPS really was alternative *civilian* service. The Selective Service was correct that CPS comported with the strict requirement of the conscription statute that CPS work be "under civilian direction." COs were doing nonmilitary work in nonmilitary settings. Also, Selective Service nonregistrants and the CPS walkouts were clearly correct that accepting alternative service in CPS entailed extensive cooperation with the broader conscription system and obedience to Selective

119 Wittner says that Congress attempted to resolve this issue by "amending the law to the effect that military officers directing Selective Service were to be regarded as working in a civilian capacity." Wittner, *Rebels against War*, 75. If that had been true, it would have been an egregious example of exalting form over substance. However, it is far from clear that the December 5, 1943, amendment to the Selective Service statute, to which Wittner apparently referred, actually did what he claims. All the amendment did was provide that a military officer detailed to work in the (ostensibly civilian) Selective Service System did so "without loss of or prejudice to his status" as an army officer. An Act Amending the Selective Training and Service Act of 1940 as amended, Pub. L. No. 78-197, 57 Stat. 596 (1943), § 2(b).

120 "CO Tells Why He Left Camp," *Conscientious Objector* 4, no. 6 (June 1942): 8; Philip Isely to All, February 1943, in War Resisters League (WRL) Records, RG 040, ser. A, 1923–1949, subser. VII, box 33, folder "Special Cases," SCPC; "Five Quit Camp in CPS Protest," *Conscientious Objector* 5, no. 1 (January 1943): 1, 6; Bennett, *Radical Pacifism*, 101.

Service rules and regulations. In this context, noncooperation with CPS was the only ethical route for a man who believed that any cooperation with the conscription system was wrong. Finally, the peace churches were correct that CPS was substantially better than the system that had prevailed during World War I—better for the COs, better for the historic peace churches to which many COs belonged, and better for the nation at large.

At one of the first NSBRO board meetings, held on December 30, 1940, only a few days after the initial NSBRO–Selective Service agreement had been reached, Thomas E. Jones, who would soon be in charge of the AFSC-sponsored CPS camps, offered the following warning: "The program contain[s] the danger of becoming merely a division of the Selective Service System." Jones's warning was prescient. On April 19, 1943, the US Treasury Department ruled that CPS camps were not liable for federal telephone and telegraph taxes because the camps "are considered agencies of the government and are entitled to exemption from taxes imposed by … the Internal Revenue Code."[121]

∴

A closely related controversy addressed the question of whether the peace churches and NSBRO had made an ethical bargain when they had agreed to operate the CPS camps on behalf of the Selective Service. Here some distinctions need to be observed. Many Mennonites in Mennonite-run camps and Brethren in Brethren-run camps happily cooperated with the system and saw no ethical problems in the arrangement. In contrast, many of the men in AFSC- and government-run camps did see a problem and objected vigorously to cooperation with the government. For example, Alex Stach walked out of CPS Camp 14, which was run by AFSC in Merom, Indiana, on February 16, 1942. In his public statement explaining his action, Stach criticized the peace churches for operating CPS and their close association with the Selective Service. It is not "moral for fellow pacifists to act as agents enforcing involuntary servitude for the Selective Service Act and United States Government," Stach wrote. Paton Price, a CPS walkout in 1943, made a similar point: "The Peace Churches have entered a field in which they obviously have no business: that of administering a portion of the military war program."[122] At a conference

121 Minutes of the NSBRO meeting, December 30, 1940, box IX-7-10, folder 3, Mennonite archives; Treasury Department to Collector of Internal Revenue, April 19, 1943, and Internal Revenue Service to the Office of the Collector, May 4, 1943, both in French to NSBRO no. 179, May 24, 1943, box IX-12-1, folder 3, Mennonite archives.

122 "CO Walks Out on CPS Camp," *Conscientious Objector* 4, no. 8 (August 1942): 1, 6; Paton Price, "Abolition of C.P.S.," WRL Records, ser. A, 1923–1949, subser. VII, box 33, folder

of the Fellowship of Reconciliation in Berkeley, California, in the autumn of 1942, Bayard Rustin, then FOR's youth secretary, criticized "pacifist leadership" for "cooperating with the government in the execution and administration of conscription."[123]

Part of the answer to the question of whether NSBRO had made an ethical decision hinged on how closely it and the peace churches were seen to be working with—and carrying out the orders of—the military men running the Selective Service. Contemporaneous memoranda that French, executive secretary of NSBRO, sent to the NSBRO Board of Directors document that he worked extremely closely with the Selective Service when coordinating NSBRO's work in administering the CPS system. French discussed matters frequently—sometimes daily, sometimes several times a day—by telephone and in person with Selective Service Director Hershey; Lewis Kosch, assistant director (in charge of COs); and their military staff. French and the Selective Service staff coordinated on major items like how many camps there would be, where they would be located, and what work the CPS men would do. They also coordinated on seemingly small details like what punishments should be meted out to individual CPS men in specific cases of infractions of rules.[124]

And French went a good deal further than merely coordinating NSBRO's work with the military men who ran the Selective Service. In an August 31, 1942, confidential memorandum to the NSBRO board, French suggested that he propose to Hershey that NSBRO staff should completely take over the work of several Selective Service sections dealing with CPS. "Such a plan would, I think, give us an opportunity to work closer to Colonel Kosch," French wrote. On this occasion, nothing appears to have come of this proposal. In May 1944, however, French again offered NSBRO staff to the Selective Service; this time he made the offer directly to Selective Service, and the Selective Service accepted the offer. Perhaps most shockingly, in December 1945, French again had NSBRO staff do Selective Service secretarial work at Selective Service offices, but this time he did this on his own initiative, without either consulting the NSBRO board or getting its approval. His post hoc memorandum to NSBRO board members informing them of his unilateral action showed that French was well aware that what he had done might not sit well with the board. His memo

---

"Special Cases," SCPC. For a thoughtful postwar analysis of the shortcomings of the relationship between the historic peace churches and the Selective Service, see Henry Geiger, "Judgment of Alternative Service," *Pacifica Views* 3, no. 30 (December 28, 1945): 1.

123 "FOR Speakers Hit Draft," *Conscientious Objector* 4, no. 11 (November 1942): 2.

124 French to NSBRO, no. 256, October 20, 1943; no. 227, August 27, 1943; no. 252, October 7, 1943, all in box IX-12-1, folder 3, Mennonite archives. French states in his October 20 memo: "I spent about two hours with General Hershey on Monday, and about the same length of time with Colonel Kosch yesterday going over the points."

started, "I hesitate to say the following because I am afraid it may be misunderstood by some of you."[125]

How Selective Service officials viewed the relationship with NSBRO and its constituent churches also evolved over time, with a concomitant evolution of how the CPS men in camp (and some of the administrative staff) viewed the situation. As French explained it, in the early days of CPS, the churches enjoyed wide latitude to determine projects and regulate in-camp discipline. Early on, French believed NSBRO's relationship with the Selective Service "was friendly and even cordial." However, after the United States entered the war, the Selective Service grew enormously, and the cordial relationship became strained. The Selective Service became more unilateral, authoritarian, and, in NSBRO's view, arbitrary. By the end of the war, cooperation between the Selective Service and NSBRO had broken down, and NSBRO's staff were feeling entirely excluded from areas where they had previously had considerable authority. This evolution was bound to cause friction.[126]

Once the relationship with the Selective Service had changed, NSBRO and the service committees that ran the camps were frequently placed in the untenable position of carrying out Selective Service mandates that they had had little or no say in creating and in some cases disagreed with. For example, when Americo Chiarito and a handful of other COs at the AFSC-run CPS camp in Elkton, Oregon, went on a work strike to protest "enforced involuntary servitude," AFSC became the enforcer of punitive measures that the Selective Service required. As a result, "the AFSC was cordially detested by many [CPSers]. The project staff had become definitely hostile toward the men, and the feeling was thoroughly reciprocated. Neither credited the other with good faith."[127]

In April 1943, forty CPS men from thirteen camps defied a Selective Service order and held a Conference on Social Action in Chicago where they discussed various ideas for improving the CPS experience. The service committees that ran the attendees' camps followed the existing Selective Service rule that required them to revoke three days of furlough time for each day that each man had been absent without leave to attend the conference. Shockingly, French was so angered by the effrontery of the recalcitrant CPSers that

125 Confidential French to NSBRO (unnumbered), August 31, 1942, box IX-12-1, folder 4, Mennonite archives; French to NSBRO no. 350, May 30, 1944, box IX-1-1, folder 3; French to NSBRO no. 593, December 18, 1945, box IX-12-1, folder 2, all in Mennonite archives.

126 French to NSBRO no. 231, September 2, 1943, folder 3; French to NSBRO no. 381, September 8, 1944, folder 2; French to NSBRO no. 474, April 3, 1945, folder 2; French to NSBRO no. 380, September 8, 1944, folder 2, all in box IX-12-1 in the Mennonite archives. See generally Gingerich, *Service for Peace*, 364.

127 Sibley and Jacob, *Conscription of Conscience*, 273–74.

he urged the AFSC to impose additional penalties on the men from the AFSC camps. The AFSC wisely rejected this petty proposal. The Selective Service, however, through NSBRO, ordered that camp directors revoke all furlough time of all the CPSers who attended the Chicago conference. This was too much for NSBRO. NSBRO refused to convey the Selective Service diktat to camp directors, and fortunately the Selective Service let the matter lie.[128]

The aftermath of the April 1943 Chicago conference was not the only occasion on which French showed an authoritarian streak. After the 1943 meeting between peace activists and Paul McNutt—at which McNutt shared his opinion that all COs ought to be shot—French sent a confidential memo to the NSBRO board about the meeting to which he had not been invited. French had not a word of criticism for McNutt (or for the illegally constituted Presidential Appeal Board that the peace activists had been protesting); instead, French thought that the peace movement's delegation "had done the whole c.o. cause considerable harm," and opined that this "is the kind of thing that I feared would develop" if peace activists not answerable to him meddled in matters pertaining to COs.[129]

∴

One of the ongoing issues for CPS men was the desire of many COs to be given more meaningful work during a time of national and international crisis; the AFSC-run ambulance service in Europe during and after World War I was the exemplar. On January 21, 1943, the army started limiting COs who were classified I-A-O (available for noncombatant duty in the military) to service in the Medical Corps; previously some I-A-Os had served in the Quartermaster Corps and other noncombatant units.[130] As a result of this change, 567 COs "who believed that they had an obligation to share in the suffering of those who had been forced into" the armed services voluntarily gave up their IV-E

128 French to NSBRO no. 160, April 5, 1943, box IX-12-1, folder 3, Mennonite archives; Chicago Conference on Social Action, Juanita Morrow Nelson and Wallace F. Nelson Papers, RG-262, box 6, SCPC. For a discussion of this controversy, see Sibley and Jacob, *Conscription of Conscience*, 264–68.

129 French to NSBRO no. 120, February 18, 1943, box IX-12-1, folder 3, Mennonite archives.

130 Winslow Ames, "CO's in Uniform—Non-Combatant Duty," *Reporter* 3, no. 2 (September 15, 1944): 3, available at the Center on Conscience and War Records, SCPC; Sibley and Jacob, *Conscription of Conscience*, 89.

classifications (alternative civilian service) in order to be reclassified I-A-O so that they could go to the front and serve in the Medical Corps.[131]

Always controversial among CPS men was the bargain that NSBRO and the peace churches had struck with the government in December 1940 that, although drafted by the government into service that the government believed to be of national importance, the CPS men would receive no government pay, dependent allowance, health insurance, or death benefits. More than a third of the CPSers had at least one dependent, hundreds were injured on the job (some lost fingers or legs), and thirty died. Denton Darrow, for example, died on January 2, 1943, while searching for seven navy airmen whose plane crashed into the Pacific Ocean off the coast of Oregon. Warren Duggan was a CPS guinea pig volunteer at Yale University Hospital who died of the infantile paralysis that he contracted as part of an experiment.[132]

As early as December 1941, the NSBRO board knew that "there is much bitter feeling against the National Service Board" among CPS men who thought "that they should not be under a program which forces them into a camp and to put up their own money or accept charity."

From the beginning of Civilian Public Service, the War Resisters League had opposed the deal that the peace churches had struck with the Selective Service. Most WRL members were secular and had reservations about serving in church-run camps. In addition, the league "felt that the government had an obligation to provide for its sons who were conscientious objectors on the same basis as its sons who were in the land and naval services."[133]

---

131 Sibley and Jacob, *Conscription of Conscience*, 89, 523n8. They put the number of such COs at "about 600." The precise figure was 567 between the inception of CPS in December 1940 and July 31, 1945, as depicted in *SS 4th*, 179, table 27.

132 *Hearing before the Senate Committee on Military Affairs on S-675*, 78th Cong., 1st sess. (1943), 22–23; National Service Board, *Directory of Civilian Public Service*, 34; Sibley and Jacob, *Conscription of Conscience*, 222–23; "Pacifist Drowns Attempting to Rescue 7 Missing Navy Men," *Conscientious Objector* 6, no. 2 (February 1943): 2; and "Drowned," *Reporter* 1, no. 9 (January 15, 1943): 1. Warren Duggan died on August 26, 1945, after the war had ended. Albert Dietrich, another CPSer, was severely injured in November 1945 when he was thrown from a work truck. A discussion of his injury and efforts to get the government to cover his surgery appears in Frank Dietrich and Albert Dietrich, *Army GI, Pacifist CO: The World War II Correspondence of Frank and Albert Dietrich*, ed. Scott H. Bennett (New York: Fordham University Press, 2005), 43–44; for the letters see 271–72, 283–84, 292.

133 Minutes of the NSBRO, December 12–13, 1941, and October 3, 1941, both in box IX-7-10, folder 3, Mennonite archives.

When Caleb Foote emerged from prison after serving a sentence for refusing to report to CPS, he sent a scathing open letter to NSBRO, which said, in part, "The handful of men who initiated the CPS program have given the stamp of pacifist approval to the totalitarian force of conscription.... A handful of religious leaders, who themselves have not been directly affected, were so anxious to control the program that they committed thousands of c.o.s, many of whom they did not represent, to a compulsory program of labor without pay."[134]

CPS men repeatedly petitioned NSBRO and the Selective Service for pay. One such petition, in October 1941, initiated at the camp in San Dimas, California, and circulated nationally, called for COs to get equal pay with soldiers. Another petition seeking pay, circulated in February 1943, was signed by nine hundred CPS men. Eight hundred CPS men signed a petition originating in Middletown, Connecticut, demanding pay; the petition was sent to NSBRO and the Selective Service.[135]

Some CPS men, including Art Emery, walked out of camp mainly or exclusively as a protest against the lack of pay. On August 23, 1943, Armenag James Manoukian was arrested at the Mancos (Colorado) CPS camp for refusing to work without pay, which he called "slave labor." Jack Hemelcke, a Mancos walkout, referred to CPS as "slave labor." When he was sentenced by US District Judge John Foster Symes, the judge said, "If I had my way this young man would be dragged in front of the German machine guns in Italy [*sic*] so he would see how he would like taking what other American boys are having to go through. Since that can't be done, I sentence him to three years in the penitentiary."[136]

Wally Nelson, who was Black, linked the lack of pay to the history of slavery when he walked out of CPS Camp 23 in Coshocton, Ohio, on July 1, 1943: "Is there any significant difference between the treatment given a man who is classified 4E [a CO] under the Draft and that meted out to the slaves? Being a Negro, coming out of slave background, I will never consciously allow myself to be made a slave. Whenever I am in a situation which I feel is forced labor or slavery, I will [resist], regardless of consequences."[137]

---

134 Caleb Foote, "An Open Letter to NSBRO," *Pacifica Views* 2, no. 4 (June 30, 1944): 2–3.

135 Bennett, *Radical Pacifism*, 104; Baird Petition, WRL Records, A/24, A/29, SCPC; *Reporter* 2, no. 21 (May 1, 1944): 1; *Conscientious Objector* 6, no. 5 (May 1944): 3. Both newsletters are in the SCPC's Center on Conscience and War Records.

136 United States v. Emery, 168 F.2d 454 (2d Cir. 1948); "ACLU Backs Manoukian," *Pacifica Views* 1, no. 18 (October 8, 1943): 3.

137 Wallace Nelson, "My Reasons for Leaving Civilian Public Service," and "Summary of Case," Nelson Papers, box 6, folder "Draft Resistance—CPS," SCPC; John W. Wood III, "Wally and Juanita Nelson and the Struggle for Peace, Equality, and Social Justice: 1935–1975" (PhD diss., Morgan State University, 2008).

Dozens of other CPS protesters (including both men who walked out and men who stopped work) were motivated by the lack of pay and other issues, such as the heavy role of the military men who headed the Selective Service in running the CPS, which by law was supposed to be under civilian control. Despite years of protests, CPS men were never paid for their years of labor.[138]

In 1943, a bill introduced in Congress would have provided some compensation for CPS men injured in the line of duty. (The bill did not become law.) During the hearings on the bill, a number of senators were obviously quite surprised that the CPS men—who, after all, had been drafted by the government and were working on government projects—were not being paid by the government. In response to a question from Senator John Chandler Gurney (R-SD), Hershey simply lied and pretended that CPS men didn't want to be paid. Hershey said, "It is our opinion that the great mass of conscientious objectors believe that it is not good public relations to pay them anything." In fact, it was Hershey's own often-expressed opinion that COs should not be paid for their work for reasons of public relations, that is making the government's treatment of COs palatable to the public.[139]

∴

The perceived problems with NSBRO—especially NSBRO's close cooperation with the Selective Service—led to resignations from NSBRO's board as well as from NSBRO's staff. On March 22, 1943, at a meeting of its executive committee expressly called for that purpose, the War Resisters League (WRL) voted to withdraw from NSBRO's Consultative Council because of the "obvious injustice" perpetrated by NSBRO, including the lack of pay for CPS men and NSBRO's cooperation with Selective Service rules for punishment of CPS men.[140] In December 1944, the Fellowship of Reconciliation withdrew from NSBRO. In November 1945, the Association of Catholic Conscientious Objectors also withdrew. Finally, on March 2, 1946, the AFSC and the Episcopal Pacifist Fellowship (affiliated with FOR) withdrew from NSBRO.[141]

---

138 Atherton v. United States, 176 F.2d 835 (9th Cir. 1949); Kramer v. United States, 147 F.2d 756 (6th Cir. 1945); and Roodenko v. United States, 147 F.2d 752 (10th Cir. 1944), involved 31 COs who walked out of camp, refused to return from leave or furlough, or refused to work. For a thoughtful exposition of the thinking of men who sought and obtained IV-E classification and then felt compelled to walk out of camp, see George Reeves, "Why They Went to Prison," *Pacifica Views* 1, no. 44 (April 7, 1944): 1, 4.

139 *Hearing Before the Committee on Military Affairs on S-675*, 17.

140 "WRL Resigns from NSBRO; League Quits over Broad Issues," *Conscientious Objector* 5, no. 4 (April 1943): 1, 6.

141 French to NSBRO no. 420, December 14, 1944, box IX-12-1, folder 2; Sheehan to French, November 15, 1945 and French to NSBRO no. 564, October 30, 1945, both in box IX-12-1,

In November 1943, Russell Freeman, who had been the director of CPS Camp 52 in Powellsville, Maryland, then run by the AFSC, resigned his post, saying, "I am increasingly uneasy because we as Friends have become confused about our values and consequently have allowed Selective Service to assume excessive control of" CPS.[142]

In November 1942, two senior staff people—R. Boland Brooks, a lawyer who had been the head of NSBRO's Appeals Section, and George Reeves, the head of its Placement Division—resigned from NSBRO as a matter of principle. In separate resignation letters, the two men said that they differed with NSBRO policy on such matters as pay and discipline and that NSBRO had sacrificed the rights of CPS campers in order have smoothly running camps.[143] Brooks and Reeves immediately set up a new organization to aid COs, Legal Service to Conscientious Objectors (LSCO). The new group's governing committee read like a Who's Who of the peace movement: A. J. Muste (then executive secretary of FOR) was the chairman; Evan Thomas (then chairman of WRL) was the treasurer; Abe Kaufman and Frank Olmstead (of WRL) were committee members, as were Norman Thomas and Richard Gregg, a leading twentieth-century author and theorist of nonviolence; Roger Nash Baldwin (of the ACLU); and John Nevin Sayre (of FOR).[144]

Almost immediately, the new entity was absorbed by the ACLU and LSCO became the National Committee on Conscientious Objectors (NCCO) of the ACLU. In its statement of purpose, the LSCO, and subsequently the NCCO, said that it would provide counsel and legal aid to all COs in their dealings with draft boards and appeal boards, provide representation for men in CPS regarding problems of camp life, provide representation for camp walkouts, and maintain a bail fund for resisters awaiting trial. The NCCO continued to operate as a separate entity from NSBRO for the remainder of the war and into the postwar period.[145]

On December 2, 1942, Roger Nash Baldwin wrote to NSBRO on NCCO letterhead and discussed what he viewed as the differences between NSBRO and the newly formed NCCO. NSBRO cooperated closely with the Selective Service,

---

folder 2; French to NSBRO no. 609, February 20, 1946, box IX-12 -1, folder 1, all in Mennonite archives.

142 "Three Directors Leave CPS Jobs," *Conscientious Objector* 5, no. 12 (December 1943): 6.

143 Lee Nichols, "Move Launched to Form New CO Aid Group," *Conscientious Objector* 4, no. 12 (December 1944): 1. The full texts of the resignation letters of both Brooks on November 21, 1942, and Reeves on November 24 are in the WRL Records, ser. A, 1923–49, subser. VI, box 22, folder marked "Legal Service to COs, 1942," SCPC.

144 Legal Service to Conscientious Objectors (LSCO) Statement of Purpose, December 14, 1942, WRL Records, ser. A, 1923–49, subser. VII, box 33, folder marked "NCCO," SCPC.

145 "New CO Legal Group Headed by Brooks, Reeves / Muste Heads Committee of 14 Sponsors," *Conscientious Objector* 5, no. 1 (January 1943): 1–2; "ACLU Takes Direction of Brooks Group," *Conscientious Objector* 5, no. 2 (February 1943): 1, 4, both in Center on Conscience and War Records, SCPC; LSCO Statement of Purpose.

he said, and that cooperation made it impossible for NSBRO to do two other things that Brooks, Reeves, and the NCCO thought essential. One was to provide legal representation to COs and CPS men who were suing the Selective Service. The other was to agitate for political changes to the entire Selective Service apparatus, including getting parole for men who went to prison when their CO claims were denied, creating civilian rather than military control of CPS, and creating non-CPS jobs for men classified IV-E.[146]

6 Resisters

Just as the government wasted no time in implementing conscription once the conscription statute had been enacted, the peace movement, too, wasted no time in organizing opposition to the draft. Well before the first registration day (October 16, 1940), eight divinity students at Union Theological Seminary announced that, for reasons of religious belief and conscience, they would publicly refuse to register for the draft.[147] They were Donald Benedict, Joseph Bevilacqua, Meredith Dallas, David Dellinger, George Houser, William Lovell, Howard Spragg, and Richard Wichlei, a group that came to be known as the Union Eight. On registration day, the Union Eight appeared together at the registration location at the seminary and presented their statement to the clerk; because of the advance publicity, the government was ready for the Union Eight, and someone from the US Attorney's Office handed each of the men a subpoena to appear before a grand jury.[148]

In their public statement, "Why We Refused to Register," the Union Eight said that their opposition to conscription grew from their religious pacifism and explained why they drew the line at the act of registration: "If we register under the act, even as conscientious objectors, we are becoming part of the act. The fact that we as conscientious objectors may gain personal exemption from the most crassly un-Christian requirements of the act does not compensate for the fact that we are complying with it and accepting its protection."[149]

146 Baldwin to French, December 2, 1943, WRL Records, ser. A, 1923–49, subser. VI, box 22, folder marked "Legal Service to COs, 1942," SCPC.

147 Don Benedict et al., "Why We Refused to Register," in Staughton Lynd, ed., *Nonviolence in America: A Documentary History* (New York: Bobbs-Merrill, 1966), 296–305.

148 George Houser, "Reflections of a Religious War Objector," in Larry Gara and Lenna Mae Gara, eds., *A Few Small Candles: War Resisters of World War II Tell Their Stories* (Kent, OH: Kent State University Press, 1999), 134–35; "First Draft Objectors Are Called to Federal Inquiry Here Today," *New York Times*, October 17, 1940; David Dellinger, *From Yale to Jail: The Life Story of a Moral Dissenter* (New York: Pantheon Books, 1993), 73–80.

149 Benedict et al., "Why We Refused to Register," 297.

FIGURE 5.2 Eligible men were obligated to register for the draft on October 16, 1940. Eight seminary students at Union Theological Seminary in New York were among those who publicly refused to register. This photograph shows the Union Eight that day.
CREDIT: GETTY IMAGES / BETTMANN / CONTRIBUTOR

On October 16, 1940, draft registration was big news and so was opposition to the draft. The next day, the *New York Times* ran three front-page stories about registration and about resistance. Along with a separate article about the Union Eight nonregistrants, the *Times* ran a photo of the eight seminarians, all appropriately well groomed and besuited (see Figure 5.2). Also on October 16, WRL and FOR held a religious service in opposition to conscription at Tabernacle Church at Broadway and Fifty-Sixth Street, in New York, followed by an antidraft demonstration.[150]

The Union Eight were promptly indicted, and all pleaded guilty; sentencing was set for November 14. On that date, antidraft picketers marched outside the

150 "Millions Register in City and Nation; Draft Rolls Completed without Hitch; Figure Here Is Estimated at 991,000," *New York Times*, October 17, 1940; "First Draft Objectors Are Called to Federal Inquiry Here Today," *New York Times*, October 17, 1940; Charles Hurd, "President Speaks: Draft Imposed on Us by Those Who Threaten War," *New York Times*, October 17, 1940.

FIGURE 5.3 The Union Eight were tried, convicted, and sentenced to prison. Here they are shown in a police wagon after their conviction.
CREDIT: GETTY IMAGES / BETTMANN / CONTRIBUTOR

courthouse, and friends, relatives, and supporters of the resisters packed the courtroom and overflowed into the halls. The court had been inundated with letters supporting the resisters. Each of the resisters was sentenced to a year and a day in prison. The sentencing made the front page of the *New York Times* again, along with another photograph of the formally dressed Union Eight, this time in a police wagon (see Figure 5.3).[151]

The Union Eight may have been the most widely publicized of the nonregistrants of the World War II era, but they were not the only ones. A total of 271 men were criminally prosecuted because they publicly refused to register for the draft for religious, political, or moral reasons (or for a combination of those).[152] For example, Winslow Wilson, a twenty-nine-year-old Methodist minister from Brownsdale, Minnesota, with two young children, refused to register.[153] Of the wider group of resisters, these nonregistrants were the men

151 "8 Draft Objectors Get Prison Terms," *New York Times*, November 15, 1940; Houser, "Reflections of a Religious War Objector," 130–39.

152 US Selective Service System, *Conscientious Objection*, 264.

153 *Conscientious Objector* 3, no. 4 (August–September 1941).

who believed that any cooperation with or participation in the conscription system was wrong.

The nonregistrants always made their protests public. For example, when it was time for John Griffith to register, he sent a certified letter to the General H. B. Springs, the head of Selective Service in South Carolina, that said, "I am conscientiously opposed to war, for any cause whatever, and shall refuse to comply with this act ... which I feel to be a contradiction of Christian teachings."[154]

Many nonregistrants would have been eligible for complete exemption as divinity students or clergy or for classification as COs. In discussing their position, they explained that they did not want to avail themselves of special privileges that might not be available to others. For example, Joe Bevilacqua, one of the Union Eight, said at his sentencing, "Because we are theological students does not mean that we should have privileges which other people do not have. We want to identify ourselves, not with a selected group of men or class of people, but with the vast majority of people."[155]

In addition to the 271 men prosecuted for nonregistration, a group of twenty "overage" men publicly refused to register for the draft but were not prosecuted. Recall that the amendment to the draft law enacted on December 20, 1941, had extended mandatory draft registration to men as old as sixty-five. Although men older than forty-five never were drafted, the law still required them to register in the fourth registration, on April 27, 1942. This requirement created an opportunity for older men to share some of the risk younger draft resisters were taking by refusing to register, thereby opening themselves up to prosecution.

These overage public nonregistrants were keenly aware of the risk they were taking. Six had been imprisoned as resisters during World War I, including Max Sandin, who had originally been sentenced to death; Ammon Hennacy, who had been imprisoned twice and threatened with a death sentence; and Howard Moore, who had been in the last group of resisters released after the end of the war. Julius Eichel and Evan Thomas were both leaders of WRL at the time; Harold Gray was the sixth ex-con from World War I.[156]

In a joint statement, three of these World War I resisters (Moore, Eichel, and Evan Thomas) wrote, "We refused to be conscripted during the last World War, and we cannot sanction nor submit to conscription in this war." In a separate, letter to Hershey, Eichel elaborated: "War resistance is not a new attitude on my part, for I was imprisoned for twenty-six months ... in various lock-ups, jails, cantonments, guardhouses and prisons because I refused to participate

154 John H. Griffith, "War Resistance in World War II," in Gara and Gara, *A Few Small Candles*, 98.

155 Houser, "Reflections of a Religious War Objector," 138.

156 Hassler, *Conscripts of Conscience*, 24–36.

in the First World War.... I have had twenty-five years in which to think it over, and I am as firmly against regimentation, conscription, and war as ever."[157]

Two other overage nonregistrants knew the risk they were running because, when the men publicly refused to register, they had nonregistrant sons in prison. Edward Richards's older son, Ned, had been a student at Haverford College when he was sentenced to a year and a day; his younger son, William, had been a student at Antioch College when he was sentenced to five years. Ralph Templin's son, Lawrence, had been a student at Bethel College when he was sentenced to three years for nonregistration. The elder Templin had enlisted in the Army Air Corps during World War I but had later become a pacifist. By 1942, he was a minister and would have been exempt from conscription had he registered.[158] Although these twenty overage nonregistrants had all clearly violated the law, the government chose not to prosecute them.[159]

A counterpart to the situation of the overage draft violators whom the government chose not to prosecute was the situation of resisters who were prosecuted multiple times. The legal obligation to register was a continuing one, and 203 resisters were prosecuted a second time for not registering, and two were prosecuted a third time.[160] For example, when Dave Dellinger, of the Union Eight, got out of prison for refusing to register, he was told that he must now register; naturally, he again refused. He was prosecuted a second time and sent right back to prison.[161] The ACLU at the time, and Dellinger later, denounced these second prosecutions as a violation of the constitutional prohibition on double jeopardy, that is, being prosecuted twice for the same offense.[162] But the ACLU and Dellinger were mistaken. These resisters were not prosecuted twice for the *same* offense; the second prosecution was for a *new*, different offense. Interestingly, if these resisters had registered after being released from prison, they should have been exempt from being drafted because felons were classified as IV-F (morally unfit for military service), but these resisters simply would not register.

World War II era draft resisters, however, were a much larger group than the 271 public nonregistrants. The government prosecuted 6,086 draft resisters and sent them to prison during the World War II era. According to the Selective Service, which kept careful records, 3,721 of these resisters were convicted of

157 Joint Statement of Howard W. Moore, Julius Eichel, and Evan W. Thomas, and Julius Eichel to Brigadier General Lewis B. Hershey, April 13, 1942, both in Eichel Family Collection, RG 131, box 2, folder marked "General Correspondence, 1942," SCPC.

158 Hassler, *Conscripts of Conscience*, 37–42.

159 Sibley and Jacob, *Conscription of Conscience*, 335; Wittner, *Rebels against War*, 51.

160 US Selective Service System, *Conscientious Objection*, 264.

161 Dellinger, *From Yale to Jail*, 113–18.

162 ACLU, *Conscience and the War*, 35; Dellinger, *From Yale to Jail*, 101, 115.

refusing induction. While technically accurate, this information is not especially helpful, because it conflates men who viewed their resistance differently and who consequently followed different paths.[163]

One category of resister who failed to report for induction comprised men like Jim Peck, who knew from the beginning that they would refuse to be inducted but nevertheless registered. Unlike the nonregistrants, these men saw no harm in merely registering and may even have appreciated the delay in prosecution that registering afforded. They often did little or nothing beyond registration and simply waited for their lottery number to be called, at which time they refused induction. In Peck's case, for example, it took almost two years for that to happen; he registered on October 16, 1940, but, because he drew a high lottery number, he was not drafted until the autumn of 1942; he was convicted and sent to prison on November 26, 1942. But for all the "cooperating" Peck did with the Selective Service, he might as well have been a nonregistrant. After the war, Peck went on to a full-time career in the peace and civil rights movements, participating in the Freedom Rides and working on the staffs of both WRL and the Congress of Racial Equality (CORE).[164]

Another group of resisters who refused induction were men like Ralph DiGia, who were pacifists—conscientiously opposed to war in any form—but did not base their objection on religion and therefore did not qualify as COs under the statute. DiGia registered on October 16, 1940, and filled out Selective Service Form 47 for COs; DiGia wanted to be a CO, but he knew he did not meet the statutory requirement and so was not surprised when his application was turned down. DiGia was drafted, refused induction, and went to prison. After the war, DiGia became a mainstay staff person of WRL for decades.[165]

A closely related group of men were those who registered and had their CO applications turned down—but who (unlike DiGia) actually did qualify under the statute but were turned down anyway. Bronson Clark was initially in this category; his draft board in Cleveland Heights, Ohio, had never granted anyone CO status. Because of public scorn for COs and the wide discretion allowed local draft boards, it was not unusual for draft boards to deny meritorious CO applications, and many men who satisfied the legal requirements to be a CO ended up in prison instead.[166]

163 US Selective Service System, *Conscientious Objection*, 264. See generally Mulford Sibley and Ada Wardlaw, *Conscientious Objectors in Prison* (Philadelphia: Pacifist Research Bureau, 1945).

164 Peck, *We Who Would Not Kill*, 1–70.

165 Ralph DiGia, "My Resistance to World War II," in Gara and Gara, *A Few Small Candles*, 38–52.

166 Bronson P. Clark, "Prison Memoir," in Gara and Gara, *A Few Small Candles*, 1–6.

All these categories of men—those who registered but otherwise refused to cooperate, those who wanted to be COs but did not qualify under the law, and those who did qualify but were denied CO status anyway—got lumped together by the Selective Service as merely refusing induction.

In a related category were the COs who, after initially accepting assignment at a CPS camp, later grew disenchanted with CPS and walked away. In all, 320 CPS men walked out.[167]

In postwar memoirs, both Dave Dellinger and Jim Peck recount the story of the draft resister at the old federal West Street jail in Manhattan who crossed paths with Louis Lepke, the notorious Mafia hit man later executed in the electric chair. The resister was having some difficulty getting Lepke to understand his (the resister's) offense. As Dellinger tells the story, Lepke said, "I am in jail for killing people, but you are here for refusing to kill anyone. It doesn't make sense."[168]

∴

When the 6,086 idealistic World War II draft resisters arrived at the various federal prisons, they found much that was not to their liking. As a result, they organized hunger strikes, work stoppages, and other protests against a variety of injustices, including racial segregation, censorship of incoming and outgoing mail, restrictions on publications the prisoners could receive, parole restrictions, and other matters. The best known of these prisoner protests—but by no means the only one—was the 135-day work strike, from August 11 to December 22, 1943, at the federal prison in Danbury, Connecticut, undertaken by a racially integrated group of twenty-three resisters to desegregate the prison dining hall. The Danbury strike may be prominent in the historiography of imprisoned World War II draft resisters for several reasons. The strike was stunningly successful; as a result of the strike, Danbury was the first US prison in history to have a desegregated mess hall—at time when the US Army remained totally segregated. In addition, some participants wrote memoirs that described the strike. Although many people have written about the Danbury strike, including participants Ralph DiGia and Lowell Naeve, and scholars such as Scott Bennett, Nora Duffy, Mulford Sibley and Philip Jacob, James Tracy, and Lawrence Wittner, the most comprehensive and engaging account remains the discussion in Jim Peck's memoir, *We Who Would Not Kill*.[169]

---

167 US Selective Service System, *Conscientious Objection*, 264.

168 James Tracy, *Direct Action: Radical Pacifism from the Union Eight to the Chicago Seven* (Chicago: University of Chicago Press: 1996), 37.

169 Ibid. For participant accounts see DiGia, "My Resistance to World War II," 43–45; Lowell Naeve, *Field of Broken Stones* (Glen Gardner, NJ: Libertarian Press, 1950). For scholarly

Before going on strike, the Danbury strikers engaged in weeks of negotiations with the prison warden over their demand to desegregate the mess hall. They called a strike only when those negotiations failed. When the strike started, the strikers refused to eat in the segregated mess hall and also refused to work. The strikers knew that they faced a potential raft of serious consequences: they could be put in "the Hole," an isolation cell; lose exercise and other so-called privileges; and lose "good time" taken off their sentences. The last could result in additional months in prison for the strikers.[170]

The prison administration wanted to put the strikers in solitary confinement, but the prison did not have enough solitary confinement cells for so large a group. Instead, the prison converted the second floor of one wing of the prison into solitary cells and put the strikers together in adjacent cells. The prisoners were not supposed to be able to communicate with each other, but they quickly developed a system for clandestinely circulating notes and, later, their own publication. In addition to being segregated from the main prison population, the strikers were also given reduced rations and reduced exercise time.

The strikers immediately set up an internal democratic decision-making structure to address strike-related matters as they came up. The first thing the strikers decided was that they would refuse to go to the main mess hall, even during off-hours, as long as segregation prevailed at the prison. This decision could have forced the men into a prolonged fast, and they viewed it as a small victory when the prison authorities almost immediately started to deliver meals to them in their new quarters. Another small interim victory came for the strikers when they went on a hunger strike to protest prison censorship of incoming books and periodicals. After three days, the censorship on books, newspapers, and magazines was substantially lessened.

An amusing incident during the strike revealed prison authorities' abysmal lack of understanding of what animated the strikers' actions. Prison barber shops were, like everything else in the prison system, strictly segregated: Black barbers, using scissors designated for Blacks only, cut the hair of Black prisoners sitting in chairs designated for Blacks only. White barbers, white scissors, and white chairs were for were white prisoners. At one point, the prison sent two sets of barbers with two sets of scissors to cut the prisoners' hair—and

---

accounts see Bennett, *Radical Pacifism*, 113–19; Nora Duffy, "Jim Crow behind Bars: Benjamin Davis and Early Prison Desegregation Efforts," *Columbia Undergraduate Journal of History* 2, no. 1 (Spring 2017): 11–13; Sibley and Jacob, *Conscription of Conscience*, 374; Tracy, *Direct Action*, 19, 37, and Wittner, *Rebels against War*, 87–88; Peck, *We Who Would Not Kill*, chaps. 10–12.

170 This account that follows is drawn primarily from Peck, *We Who Would Not Kill*, chaps. 10–12.

were shocked that the strikers refused to submit to segregated haircuts. The prisoners refused to have their hair cut until the prison authorities allowed the strikers to have desegregated haircuts. Another small victory for the strikers.

The strikers coordinated their efforts to get publicity for their action with WRL and FOR activists on the outside. As outside publicity increased later in the autumn, the Danbury administration started giving the strikers additional privileges, most notably that their cell doors were open eight hours a day so that they could congregate and socialize with each other.

On December 22, the Danbury administration caved in to the strikers' demands.

Although certainly the best known, the successful desegregation strike at Danbury in 1943 was not the first protest at Danbury by war resisters. When the Union Eight arrived at Danbury in the autumn of 1940, well before the United States had entered the war, Dave Dellinger was immediately put into solitary confinement for sitting in the Black section at the Saturday night movie showing. This was not a planned or organized protest; it was just an individual action by Dellinger.[171]

Nor was Danbury the only prison affected by protests and organizing efforts of imprisoned draft resisters. At the federal prison in Lewisburg, Pennsylvania, a work strike began in May 1943 (while the Danbury strike was still in progress). The Lewisburg strike began as a protest against segregation, when the Black resister Bill Sutherland was not allowed to sit with white resisters in the mess hall. By the autumn, the strike had morphed into a protest demanding an end to the use of "the Hole" as a punishment tool and an end to prison censorship of incoming and outgoing prisoner mail. Dellinger, now at Lewisburg for his second sentence for nonregistration, was one of the participants. Some strikers also fasted, and the prison administration force-fed them through nasal feeding tubes for the last month of their fast. The protest ended on December 1, in a partial victory for the strikers in which the prison administration agreed to modify the censorship rules.[172]

In June 1944, three resisters in the federal prison in Petersburg, Virginia, conducted a work strike to protest segregation.[173]

171 Dellinger, *From Yale to Jail*, 82.

172 Ibid., 120–22; David Dellinger, "Why I Refused to Register in the October 1940 Draft and a Little of What It Led To," in Gara and Gara, *A Few Small Candles*, 35; "Hunger Strike at Lewisburg," *Pacifica Views* 1, no. 24 (November 19, 1943): 2; "Lewisburg Strike Over," *Pacifica Views* 1, no. 28 (December 17, 1943): 2; "Six COs Fast in Lewisburg," *Conscientious Objector* 5, no. 11 (November 1943): 1–2; Bennett, *Radical Pacifism*, 126–30; Sibley and Jacob, *Conscription of Conscience*, 374–76.

173 "In Prison," *NSBRO Reporter* 3, no. 2 (July 15, 1944): 1.

In the autumn of 1944, resisters, including Bayard Rustin and Larry Gara, conducted a work strike in the federal prison in Ashland, Kentucky, to protest segregation.[174] Earlier, when Rustin had been held temporarily at the federal detention facility in New York, he had been placed in solitary for refusing to accept segregation in the dining hall. He was released when the jail's choir, which he had organized, spontaneously refused to sing at Sunday services unless Rustin was released.[175] In June 1945, a larger strike against segregation at Ashland involved fourteen men.[176]

In September and October 1944, seven resisters, including Ralph DiGia and Matt Kauten, held another hunger strike at Lewisburg; the men were protesting the harsher parole conditions meted out to paroled COs than to "white slavers, dope smugglers, post office robbers, and other ordinary prisoners."[177]

Not all the protests by imprisoned resisters were successful, but in August 1945, a protest by a group of resisters, including Igal Roodenko, who later chaired WRL, successfully integrated a dormitory and dining hall at the federal prison at Sandstone, Minnesota. Also in 1945, a twelve-day hunger strike by a single imprisoned draft resister, Manuel Talley, achieved desegregation of the dining hall at the federal prison in Tucson, Arizona.[178]

∴

The decision by the War Department in early 1944 to draft Japanese internees sparked widespread resistance. As Frank S. Emi, one of the resisters, later wrote, "Drafting Nisei [US citizens of Japanese descent] from concentration camps was morally wrong and legally indefensible." On the moral issue, Emi was certainly correct; he was also correct on the legal issue. Selective Service regulations stated that men in any sort of an asylum, jail, penitentiary, "or similar institution" were not even eligible for draft registration until "the day of

---

174 Bayard Rustin (1912–1987) was a lifelong peace and civil rights activist associated with Fellowship of Reconciliation, War Resisters League, and *Liberation* magazine. Recent biographies of Rustin include John D'Emilio, *Lost Prophet: The Life and Times of Bayard Rustin* (New York: Free Press, 2003), and Jervis Anderson, *Bayard Rustin: Trouble I've Seen* (Berkeley: University of California Press, 1998). Rustin's "Letter to the Draft Board" appears in Devon W. Carbado and Donald Wiese, *Time on Two Crosses: The Collected Writings of Bayard Rustin* (San Francisco: Cleis Press, 2003), 11. See also Sibley and Wardlaw, *Conscientious Objectors in Prison*, 47, and Bennett, *Radical Pacifism*, 121.

175 "Rustin in Prison Fights Jim Crow," *Conscientious Objector* 6, no. 5 (May 1944): 2.

176 "Pacifist Cons Fight Jim Crow," *Pacifica Views* 3, no. 14 (July 27, 1945); Bennett, *Radical Pacifism*, 121; Larry Gara, "My War on War," in Gara and Gara, *A Few Small Candles*, 86–87.

177 "Hunger Strike at Lewisburg," *Pacifica Views* 2, no. 21 (October 27, 1944): 3.

178 "Hunger Strike Ends Race Segregation," *Conscientious Objector* 7, no. 10 (October 1945): 2.

[their] discharge." The ten concentration camps in which the Nisei were held certainly qualified as such institutions. Also writing later about his World War II draft experience, another of the Nisei resisters, Yosh Kuromiya, wrote that when he received his draft notice, "I thought 'NO! ... I refuse to comply."[179]

Both Emi and Kuromiya were prisoners at the Heart Mountain camp in Wyoming, where eighty-five men openly resisted the draft. At the Tule Lake camp in California, twenty-seven men were public resisters. At the Minidoka camp in Idaho, thirty-eight men openly resisted conscription.[180] In all, 315 Nisei in ten camps resisted the draft, were criminally prosecuted, and went to prison.[181] These figures are remarkable in at least two ways. First, the resisters' families and friends, who were often eager to use their public support of the US war effort as evidence of their patriotic bona fides, often strongly opposed these individual acts of resistance.[182] It stung when the Japanese American Citizens League, the organized voice of the Japanese American community, referred to the resisters as "dim-witted cowards" and provocateurs and denounced them as traitors.[183] In this context, the strength of mind and purpose shown by the Nisei resisters was remarkable.

Also, the Nisei internees resisted the draft at a rate that was eight times higher than the rate for all American men. Moreover, this occurred in the context of the virulent anti-Japanese racism endemic during the war. The range of sentences that could be, and was, meted out to resisters was wide, and the internee resisters faced not only opprobrium from their families and communities but a serious risk of long imprisonment.[184]

At Heart Mountain, the internees formed the Fair Play Committee to oppose conscription of camp internees, in January 1944. Frank Emi was one of the lead organizers. "We felt that we had to challenge the legality of conscripting inmates held in concentration camps," he later explained. The committee held nightly camp meetings often attended by hundreds of people. On March 4,

---

179 Frank Emi, "Protest and Resistance, An American Tradition," in Mackey, *A Matter of Conscience*, 53; Selective Service Regulations, 5 Fed. Reg. 3779 (1940), § 14, ¶ 246; Yosh Kuromiya, "The Fourth Option," in Mackey, *A Matter of Conscience*, 78.

180 Muller, *Free to Die for Their Country*, 76–77, 65–66, 67, 74.

181 William Hohri, "Nesei Resistance: How We Remember," in Mackey, *A Matter of Conscience*, 121.

182 Muller, *Free to Die for Their Country*, 6.

183 Emi, "Protest and Resistance, An American Tradition," 57.

184 The 6,086 open resisters who went to prison represented 0.13 percent of the 46.8 million men registered. US Selective Service System, *Conscientious Objection*, 264. In contrast, the 315 open Nisei resisters represented nearly one full percentage of the 32,230 registered Nisei who were living in the ten camps. Muller, *Free to Die for Their Country*, 4.

FIGURE 5.4 The sixty-three US citizens of Japanese descent who were tried together for draft resistance in June 1944 in a single courtroom in Cheyenne, Wyoming, the largest mass trial of draft resisters in US history.
PHOTOGRAPH COURTESY OF FRANK ABE/CONSCIENCE AND THE CONSTITUTION

1944, it circulated a flyer in the camp citing the Thirteenth Amendment's bar against involuntary servitude and urging men to resist.[185]

The flyer, and Fair Play Committee's organizing efforts at Heart Mountain, had an immediate effect. On March 6, two men refused to go; on March 7, three more men refused; on March 8, seven men refused to go. Draft resistance at Heart Mountain continued to grow, and by June sixty-three resisters had been arrested at the camp and were in jails and prisons throughout Wyoming. The sixty-three resisters were tried together in a mass trial in the federal courthouse in Cheyenne beginning on June 12, 1944 (see Figure 5.4). The US district judge, T. Blake Kennedy, was described by law professor Eric L. Muller as "a racist, an anti-Semite, and a xenophobe," who addressed the defendants in open court—with a stenographer present—as "you Jap boys." All were convicted and, on June 26, 1944, all were sentenced to three years in prison.[186] This was the largest

185 Emi, "Protest and Resistance, An American Tradition," 54; Muller, *Free to Die for Their Country*, 82–84.

186 Muller, *Free to Die for Their Country*, 84, 99, 102, 104, 111.

mass trial of resisters ever held during World War II (or since). Nevertheless, the *New York Times*—which had run two front-page stories (both with photographs) about the eight white Union Seminary students—printed no stories at all about the Japanese American resisters: not when they were arrested, not when they were tried, not when they were sentenced, and not at any other time.

7 The End of the World War II Draft

World War II ended in August 1945 with Japan's surrender, following the atomic bombings of Hiroshima on August 6 and Nagasaki on August 9. The rapid demobilization of US military forces followed. The conscription statute was set to expire on May 16, 1946. By the time the new president, Harry Truman, sent his first State of the Union message to Congress on January 21, 1946, half of the 12 million Americans who had been in the armed forces at the end of the war had already returned to civilian life.

On May 14, in the midst of a debate about whether to continue the Selective Service, Congress passed a stop-gap measure extending the draft for forty-five days, until July 1.[187] On June 25, Congress extended the World War II draft once more, for nine months, until March 31, 1947.[188]

That was the last extension of the World War II draft. On March 3, 1947, Truman sent a message to Congress recommending no further extensions of the law. Congress followed Truman's recommendation, and on March 31, 1947, the World War II draft law expired. Also on March 31, 1947, Congress passed a law creating the Office of Selective Service Records to hold the draft files of the previous six and a half years, and Truman named Hershey to stay on as its director (see Figure 5.5).[189]

8 Legal Cases Pertaining to the Conscription Statute

Three important US Supreme Court cases addressed conscription during the World War II era, and a number of decisions in US District Court and Circuit Court cases also are relevant.

187 Joint Resolution to Extend the Selective Service Act, Pub. L. No. 79-379, 60 Stat. 181 (1940).

188 An Act To extend the Selective Training and Service Act of 1940, as amended, Pub. L. No. 79-473, 58 Stat. 798 (1940). For the text of this statute, see *ss 4th*, 285, 290–92.

189 Harold Hinton, "Truman Requests Selective Service End on March 31," *New York Times*, March 3, 1947 (the full text of the presidential message appeared on p. 15); *ss 4th*, 288; An Act To establish an Office of Selective Service Records, Pub. L. No. 80-26, 61 Stat. 31 (1947); "Selective Service Law Expires; 10,200,367 Inducted under It," *New York Times*, April 1, 1947.

FIGURE 5.5 President Harry Truman and General Lewis B. Hershey in 1947. Hershey, the director of the Selective Service during World War II, once again became its director in 1948 and kept the post until the Nixon administration.
CREDIT: GETTY IMAGES / BETTMANN / CONTRIBUTOR

Consider first the Supreme Court case of *Billings v. Truesdale*. The facts of the case are the essence of simplicity. Arthur Billings registered and claimed to be a CO. His draft board disagreed and classified him I-A. Billings received an induction order, went to the induction center, and there refused induction. An army officer read the oath of induction to Billings, told him that his refusal to take the oath himself made no difference at all, and said, "You are in the army now." Now considered a soldier, Billings refused to obey military orders. He was court-martialed, convicted, and sentenced under the military Articles of War. Billings sued, claiming that—because he had refused induction—he had never entered the army, was therefore not subject to military jurisdiction, and must instead be prosecuted in a civilian court for the civilian crime of refusing induction.[190]

The Supreme Court agreed with Billings. The court's opinion, written by Justice William O. Douglas, was short and direct. In the first sentence Douglas

190 Billings v. Truesdale, 321 U.S. 542 (1944).

quoted the dispositive section of the statute: "Section 11 of the Selective Training and Service Act of 1940 provides that 'No person shall be tried by any military or naval court-martial in any case arising under this Act.'" He said that Congress had said what it meant and meant what it said: Congress wanted to, and did, vest civil courts with exclusive jurisdiction over all violations of the draft law. Douglas observed that Congress knew perfectly well how to write a draft law that gave the army jurisdiction over people who refused induction, and it had done just that with the World War I draft law. "But Congress did not choose that course in the present emergency."[191]

Despite the simplicity of the issue presented, and the brevity of the Supreme Court's disposition of the matter, the case has at least three shocking aspects. First, both the trial court and the Tenth Circuit Court of Appeals had ruled against Billings.[192] Indeed, the trial court judge thought the case was simple: "There can be no question but that the petitioner is subject to military law. The question is not close enough to even be debatable." The judge also called Billings, who held a master's degree from Harvard, "an over-educated, egotistical, scholastic slacker."[193] Equally shocking is that, on appeal, the Court of Appeals affirmed the ruling against Billings without citing—or even mentioning—the relevant section of the underlying statute that required draft violators to be tried in civilian courts.

Second, so clear was the congressional mandate that civilian courts had sole jurisdiction over draft violators that multiple judges in multiple courts had already recognized that fact and issued multiple reported decisions based on it. As soon as conscription became law, there were *civilian* prosecutions of men who refused to register. In one such case in New York, the judge, presaging what Douglas would write four years later in *Billings*, said, "Section 11 [of the draft law] specifically provides that any person who violates any provision of the Act ... is to be tried in the United States District Court and that no person is to be tried by court-martial until he has been actually inducted." In an Oregon case at about the same time, the judge said much the same thing: the defendant cannot "be subjected to military law nor liable to court-martial until after induction, which includes swearing allegiance. Prior to that time he is still entitled to protection of any rights he may have by the civil tribunals." Other reported cases say exactly the same thing.[194]

---

191 Ibid., 556.

192 *Ex parte* Billings, 46 F. Supp. 663 (D. Kan. 1942); Billings v. Truesdale, 135 F.2d 505 (10th Cir. 1943).

193 *Ex parte* Billings, 46 F. Supp. at 667, 668.

194 United States v. Rappeport, 36 F. Supp. 915, 917 (S.D.N.Y. 1940); Stone v. Christensen, 36 F. Supp. 739, 741–42 (D. Ore. 1940); United States v. Bowles, 131 F.2d 818 (3d Cir. 1942); United States v. Grieme, 128 F.2d 811 (3d Cir. 1942).

Third, it is shocking that the Supreme Court's ruling in *Billings* came on March 27, 1944, three and a half years after Roosevelt had signed the unambiguous conscription statute into law (and after Billings had been imprisoned for two full years in the US Disciplinary Barracks at Fort Leavenworth). Moreover, what happened to Arthur Billings was not unusual. The case reports of the era show that draft violators were often tried under military law, and when the Supreme Court decision in *Billings* was announced, the NSBRO staff recognized that the result would be that many of the men then being held in army disciplinary barracks would have to be released.[195] Indeed, after *Billings*, a single army prison, Army Detention Barracks Stormville in Green Lawn, New York, had to release ten COs.[196]

That is, notwithstanding the unambiguous language of the statute to the contrary and notwithstanding the earlier correct rulings of multiple judges, until March 1944 American men were being wrongly tried and sentenced under military law and imprisoned in army disciplinary barracks, sometimes for years, for civilian violations of the draft law.

When Billings and others in the same position were released from military prisons after the Supreme Court ruling, they still faced federal (civilian) charges of violating the conscription statute, charges that were tried in federal (civilian) courts. Billings himself was tried before US District Judge J. Foster Symes in Kansas City and sentenced to two additional years in civilian prison. The Justice Department was keen to make an example out of Billings because of the precedent his case had set.[197]

∴

The other two World War II era US Supreme Court cases of note dealt with an identical issue: whether a Selective Service registrant who disagrees with a decision of his local draft board can appeal to a federal court. The two cases came to opposite conclusions. In *Falbo v. United States*, decided in January 1944, the court held that the disgruntled registrant (or draftee) has no right to appeal to a federal court. In *Estep v. United States*, decided in February 1946, the court addressed the identical question and came to the opposite conclusion.[198]

---

195 Bowles, 131 F.2d 818; United States *ex rel.* Bergdoll v. Drum, 107 F.2d 897 (2d Cir. 1939), *cert. denied*, 310 U.S. 648 (1940); Winslow E. Osborne to NSBRO, March 29, 1944, box IX-12-1, folder 2, Mennonite archives.

196 "C.O.s and the Law," *NSBRO Reporter* 3, no. 2 (September 15, 1944).

197 "C.O.s and the Law," *NSBRO Reporter* 3, no. 8 (October 15, 1944): 2–3.

198 Falbo v. United States, 320 U.S. 549 (1944); Estep v. United States, 327 U.S. 114 (1946).

Recall that the conscription statute gave local draft boards the "power ... to hear and determine ... all questions or claims with respect to ... exemption or deferment from training and service." Congress added this explicit command: "The decisions of such local boards shall be final" (except for the *administrative* appeals set forth in Selective Service regulations, first to a regional Selective Service appeal board, and then to "the President," but never to a court).

Since Congress had explicitly stated that draft board decisions were final, the decisions of the lower federal courts are not surprising. The first courts to receive appeals from disgruntled draftees were the US district courts, the trial courts of the federal system. In one typical early example, a US district judge in Ohio wrote, "An examination of the Selective Training and Service Act reveals that not only was no authority given for judicial review, but the action of [the Selective Service] was definitely made final: 'The decision of such boards shall be final.' ... This Court might disagree with the interpretation of the law made by [the Selective Service] in this case ... but it would be powerless to change [the] decision."[199] Soon enough the issue was presented to the federal circuit courts, the appellate courts of the federal system, and the courts for the Third, Fourth, Fifth, Seventh, and Eighth Circuits soon announced the same ruling against the defendants.[200] Many of these circuit court decisions were brief and simply affirmed that draft board "decisions are final and may not be disturbed by the courts." The Seventh Circuit, in *United States v. Sauler*, provided a succinct public policy rationale for the rule: allowing registrants who were dissatisfied with a draft board ruling to appeal to the courts "would lead to disastrous results in the administration of the Selective Service Act. It would mean that any registrant not satisfied with his rating [classification] could in effect obtain a review thereof in court by openly defying the order directing him to report for induction. Such a procedure would produce chaos."[201]

To be sure, some courts recognized the problems that could be presented by the rule that draft board decisions could not be subjected to judicial review. Even leaving aside the passions excited by war, draft board members were human and so might err. Some draft board members had never heard

---

199 Shimola v. Local Board No. 42 for Cuyahoga County, 40 F. Supp. 808, 810, 811 (N.D. Ohio 1941). Another case that came to the same conclusion is *Petition of Soberman*, 37 F. Supp. 522 (E.D.N.Y. 1941).

200 United States v. Grieme, 128 F.2d 811 (3d Cir. 1942); Goff v. United States, 135 F.2d 610 (4th Cir. 1943); Fletcher v. United States, 129 F.2d 262 (5th Cir. 1942); United States v. Sauler, 139 F.2d 173 (7th Cir. 1943); United States v. Mroz, 136 F.2d 221 (7th Cir. 1943); Bronemann v. United States, 138 F.2d 333 (8th Cir. 1943).

201 Sauler, 139 F.2d at 174.

of conscientious objection, and some who had heard of the concept boasted that they would never grant such status to anyone. Any of these circumstances could lead to mistakes by the Selective Service. Such was the case of Whitney Bowles. Bowles claimed to be a CO and, under the statute, qualified as a CO, but his draft board classified him I-A and drafted him into the army. Bowles refused induction, was prosecuted in civilian court, convicted, and sentenced to prison. On appeal, the appellate court agreed that Bowles qualified as a CO and that Bowles's draft board had been wrong—but also that the court could do nothing in the circumstance: "The case thus discloses … a gross violation of [Bowles's] rights which has plainly thwarted the intention of Congress. It is, however, settled in this court that the action of the selective service authorities in making an erroneous classification, even though arbitrary and capricious," is final.[202] In other words, Bowles was right, his draft board was wrong, a grave injustice had been done, and there was not a thing the court could do about it, because the law was clear: Selective Service rulings are final.

It is also not surprising, then, that when the US Supreme Court considered this question in late 1943 in the *Falbo* case, the court ruled that decisions of local draft boards are indeed final and that unhappy registrants and draftees have no recourse to the federal courts to change those rulings. Justice Hugo Black wrote the court's majority opinion. (Note that the court's decision in *Falbo* is not inconsistent with the court's ruling in *Billings*, because the two cases dealt with different, though related, questions. *Billings* held that draft law violators had to be tried in civilian courts; *Falbo* held that, when that happens, the draftee does not have a right to have a federal court second-guess a local draft board determination.)

Justice William Francis Murphy wrote a strong dissent in *Falbo*. Murphy relied on long-standing notions of "basic principles of justice," procedural due process, and fundamental fair play in arguing that Selective Service decisions must be subject to judicial review: "Common sense and justice dictate that a citizen accused of a crime should have the fullest hearing possible, plus the opportunity to present every reasonable defense. Only an unenlightened jurisprudence condemns an individual without according him those rights. Such a denial is especially oppressive where a full hearing might disclose that the administrative action underlying the prosecution is the product of excessive wartime emotions." Today, it is widely accepted that procedural due process requires that decisions by administrative agencies must be subject to judicial review (although the scope of that review can often be quite constrained), but

202 Bowles, 131 F.2d at 819, *aff'd per curiam*, Bowles v. United States, 319 U.S. 33 (1943).

during World War II the rule was generally that agency decisions *may* be subject to judicial review if, but only if, the statute creating the agency so directed. In that context, Murphy's dissent was unusually enlightened for the time.[203]

Two years later, in *Estep*, the Supreme Court addressed the same issue and this time reached the opposite conclusion. The *Estep* opinion, written by Douglas, basically adopted the views of Murphy in his *Falbo* dissent. In *Estep*, Murphy wrote a lengthy concurring opinion, setting forth his views more fulsomely than he had in *Falbo*. Not to allow appeals of draft board decisions "would require adherence to the proposition that a person may be criminally punished without ever being accorded the opportunity to prove that the prosecution is based on an invalid administrative order. That is a proposition to which I cannot subscribe. It violates the most elementary and fundamental concepts of due process of law. It condemns a man without a full hearing and a consideration of his alleged defenses." Murphy acknowledged the government's argument that "Congress [had] so designed the Selective Training and Service Act of 1940 as to preclude courts from inquiring into the validity of an induction order," and added, "But if that is true, the Act is unconstitutional."[204]

One of the arguments used against the Burke-Wadsworth bill during the debate in Congress was that it was an unprecedented and unconstitutional peacetime draft. In the period after the draft was signed into law in September 1940 and before the United States entered the war in December 1941, draft resisters plied this argument in courts.[205]

It is not surprising that the argument that a peacetime draft was unconstitutional was not successful in the courts. It is, however, notable and unfortunate that the courts did not even treat the argument seriously. At the least, the constitutional question about peacetime conscription was an unsettled issue in 1940. After all, what was, both then and now, regarded as the landmark US Supreme Court case *upholding* the right of Congress to impose conscription, *Arver* (1918), had expressly linked the power to conscript to the power to make war. The opening lines of *Arver* stated, point-blank, that the power to conscript "is found in the power to declare war." In that context, there was, in 1940, at least a serious argument that a peacetime draft might be unconstitutional.

203 Falbo, 320 U.S. at 556–57 (Murphy, dissenting).
204 Estep, 327 U.S. at 125–26 (Murphy, concurring).
205 Rappeport, 36 F. Supp. 915; United States v. Herling, 120 F.2d 236 (2d Cir. 1941). *Rappeport* concerned several men who had refused to register under the law.

But the courts of the time did not treat it as such. In one case, *United States v. Herling*, the Second Circuit Court of Appeals used exactly one paragraph to rule that a peacetime draft is constitutional.[206] To support that holding, the Second Circuit Court cited three US Supreme Court cases as precedents. Two were World War I cases that upheld a wartime draft: *Arver* and *Cox v. Wood*. Obviously, neither of these addressed the question in *Herling* about a peacetime draft. The third case cited as a precedent by the Second Circuit was even further afield, *Hamilton v. Regents of the University of California*. This was a 1934 case (that is, during peacetime) that challenged the constitutionality of mandatory participation (for able-bodied male students) in the Reserve Officers' Training Corp at California land grant colleges. In *Hamilton* the Supreme Court made a narrow ruling that held that the statute creating land grant colleges contained language pertaining to mandatory ROTC for students who choose to attend such colleges.[207] *Hamilton* was not about conscription—it was about college ROTC—and it was certainly not about peacetime conscription. The Second Circuit's 1941 ruling in *Herling* was an embarrassment to thoughtful jurisprudence.

The World War II era conscription statute defined COs as those "who, by religious training and belief, [are] conscientiously opposed to participation in war in any form." This was broader language than that in the World War I statute; that law had exempted only members of the peace churches. The change brought two kinds of problems for the Selective Service.

The first set of problems went to the nature of religion: What is "religious belief"? Does one have to believe in a deity to be religious? Does one have to hold formal membership in a church or religious institution? Does one have to attend religious services? What is the nature of religion, anyway?

The second, related, set of problems presented by the new language was epistemological. How does the Selective Service determine whether a person is religious? The inquiry during World War I had been relatively simple and often relied on membership in a Mennonite Church or Quaker Meeting. But during World War II, the inquiry was much harder. The law no longer required membership in a specific sect, so membership records were no longer dispositive.

206 Herling, 120 F.2d 236. Other cases that concluded that the 1940 peacetime draft was constitutional include *United States v. Lambert*, 123 F.2d 395 (3d Cir. 1941); *United States v. Garst*, 39 F. Supp. 367 (E.D. Pa. 1941); and *United States v. Cornell*, 36 F. Supp. 81 (D. Idaho 1940).

207 Cox v. Wood, 247 U.S. 3 (1918); Hamilton v. Regents of the University of California, 293 U.S. 245 (1934).

As a result of both sets of questions, draft boards often faced the thorny problem of how an outsider could determine whether a man was, or was not, religious within the meaning of the law. As I noted earlier, there was wide discrepancy in how draft boards addressed this set of questions, and it is not surprising that different courts also came up with different answers.

As previous scholars have noted, two different lines of cases developed in the courts: a rather more liberal interpretation by the Second Circuit of what religion is and a narrower interpretation in the Ninth Circuit.[208] An early case, *United States v. Kauten*, from the Second Circuit, involved Matt Kauten, who claimed to be a CO but whose draft board disagreed.[209] It classified Kauten I-A and sent him an induction order. Kauten refused induction, was indicted, prosecuted, and convicted. Although the court ruled that it lacked jurisdiction, it nonetheless went on to opine at some length on what religion is.[210] The court stated that religion "may justly be regarded as a response of the individual to an inward mentor, call it conscience or God, that is for many persons at the present time the equivalent of what has always been thought of as a religious impulse." The court said that religion "is a belief finding expression in conscience" and cited as historical examples of religious people Socrates, Martin Luther, Wordsworth, and the Greek poet Menander, who said, "Conscience is a God to all mortals."[211] Saying that religion is nothing more than an inward mentor or conscience is a broad definition of religion indeed. Despite the fact that this definition of religion in *Kauten* appeared only in dictum (that is, language that was not necessary for the court's central holding), it was consistently cited in later Second Circuit cases as being controlling.[212]

The Ninth Circuit's view of religion was rather the opposite. In *Berman v. United States*, the Ninth Circuit said: "It is our opinion that the expression 'by reason of religious training and belief' is plain language, and was written into the statute for the specific purpose of distinguishing between a conscientious social belief, or a sincere devotion to a high moral philosophy, and one based on an individual's belief in his responsibility to an authority higher and beyond any worldly one." The court went on to say, "It would be quite ridiculous to argue that the use of the word 'religion' could have ... [the] meaning to be

208 Sibley and Jacob, *Conscription of Conscience*, 428–31.

209 United States v. Kauten, 133 F.2d 703 (2d Cir. 1943).

210 The Second Circuit cited the previous circuit court rulings in *Grieme* and *Bowles*—as well as the plain language of the conscription statute—in support of its lack of jurisdiction.

211 Kauten, 133 F.2d at 708.

212 United States *ex rel.* Reel v. Badt, 141 F.2d 845 (2d Cir. 1944); United States *ex rel.* Phillips v. Downer, 135 F.2d 521 (2d Cir. 1943).

inclusive of morals or of devotion to human welfare.... Philosophy and morals and social policy without the concept of a deity cannot be said to be religion."[213]

∴

During World War II, several constitutional challenges to conscription were brought. One challenge was based on the Thirteenth Amendment's ban on involuntary servitude. The argument was no more successful in World War II than it had been in World War I.[214]

Another constitutional argument against conscription during World War II was that it violated the free exercise of religion clause of the First Amendment. This recalled Walter Nelles's argument in the ACLU's amicus brief in the *Arver* case during World War I. Multiple cases during the World War II draft, including *Commers v. United States*, used this argument but with no more success than Nelles had had a generation earlier.[215]

∴

In only one case, *United States v. Masaaki Kuwabara*, did a group prosecution of Japanese American draft resisters not end in conviction.[216] The case was tried in Eureka, California, which, as Eric Muller discusses, was a town with an especially virulent strain of anti-Asian racism. Before and during the trial, the local newspaper railed against the "Japs" (the defendants), and on the eve of the trial one of the defendants' two court-appointed lawyers made a public speech denigrating "the Japs" and mocking his own clients.[217]

The *Kuwabara* case involved twenty-six resisters from the Tule Lake Relocation Center in California who had refused preinduction physicals. The court's decision to throw out the criminal prosecution came in response to a defense motion presented on the eve of trial. The motion sought dismissal of the indictment both on both statutory grounds (including that the draft board had erred in classifying the defendants I-A) and constitutional grounds (including that the defendants' placement in Tule Lake Relocation Center and subsequent criminal prosecution for draft resistance deprived the defendants of their liberty without due process of law).

---

213 Berman v. United States, 156 F.2d 377, 380 (9th Cir. 1946).

214 United States v. Ryals, 56 F. Supp. 772 (N.D. Ga. 1944); Commers v. United States, 66 F. Supp. 943 (D. Mont. 1946).

215 Commers, 66 F. Supp. 943.

216 United States v. Masaaki Kuwabara, 56 F. Supp. 716 (N.D. Cal. 1944).

217 Muller, *Free to Die for Their Country*, 135–40.

In his ruling, US District Judge Louis Goodman first addressed the statutory defense (error by the draft board). The judge cited *Falbo* for the proposition that, even if it were true that the defendants had been wrongly classified I-A, they could not assert that defense and the court could not hear the defense. Then, remarkably, the judge went right on to consider the defendants' constitutional due process argument. This was remarkable because *Falbo* had made no distinction about the types of arguments that a court could or could not entertain in Selective Service cases; *Falbo* had said that the decisions of draft boards are final and could not be reviewed in court.

In addressing the defendants' due process argument, Goodman sidestepped the issue of whether the forced relocation of American citizens of Japanese descent into concentration camps was constitutional, saying that the issue was not for him to decide. (The US Supreme Court later upheld the constitutionality of forced relocation in the now-infamous case of *Korematsu v. United States.*)[218]

But Goodman did not sidestep the related issue of whether it was constitutional to draft American citizens of Japanese descent out of those concentration camps and into the army. Instead, Goodman addressed that issue head-on: "It is shocking to the conscience that an American citizen be confined on the ground of disloyalty, and then, while so under duress and restraint, be compelled to serve in the armed forces, or be prosecuted for not yielding to such compulsion.... Certainly 'fair and just' compulsory military training in a free society is wholly inconsistent with [this] proceeding." With that, Goodman threw out the indictments of twenty-six Japanese American draft resisters.[219]

Muller recounts that residents of Eureka were so reactionary that Goodman and his law clerk, Eleanor Jackson Piel, thought that they might be lynched when the decision was announced: "Judge Goodman read it from the bench as a car containing his packed suitcase idled outside the courthouse, waiting to take him and his clerk back to San Francisco."[220]

∴

At the end of the war, the US Supreme Court ruled that Illinois was within its rights to preclude from the practice of law men who had been COs. The case involved Clyde Summers who, although otherwise fit to be a lawyer, was excluded from the bar because he had been a CO. As the Supreme Court recognized, "The so-called 'misconduct' for which [Summers] could be reproached

218 Korematsu v. United States, 232 U.S. 214 (1944).
219 Masaaki Kuwabara, 56 F. Supp. at 719.
220 Muller, *Free to Die for Their Country*, 141–42.

for was his taking the New Testament too seriously."[221] The holding of the court was not that COs could not be lawyers; the court merely held that Illinois, if it wished to do so, could choose to exclude COs without violating the First Amendment's right of freedom of religion. (Here again was an echo of Walter Nelles's argument in *Arver.*) Black dissented: "I cannot believe that a state statute would be consistent with our constitutional guarantee of freedom of religion if it specifically denied the right to practice law to all members of one of our great religious groups, Protestant, Catholic, or Jewish.... I am not ready to say that a mere profession of belief in the Gospel is sufficient reason to keep otherwise qualified men out of the legal profession."[222]

9 Conclusion

In addition to pulling off an enormous undertaking, the World War II Selective Service operated extremely efficiently. In a period of only a few weeks, starting in late 1940, the Selective Service set up a decentralized system of more six thousand local draft boards, registered and classified tens of millions of men, and started drafting millions of men into the army. The classification system was complex, with more than a dozen different classifications, and the classifications were grounded in multiple factors, based on detailed information about a registrant's private life as reflected in the classification forms filled out by the men. All this was done before the advent of electronic records or electronic communication.

Inevitably, this efficiency came at the expense of fairness. At least since the enactment of the Fourteenth Amendment in 1868, equal protection under the law has been a central aspiration of American jurisprudence. The US Supreme Court said as early as 1920 that the essential command of the equal protection clause of the Fourteenth Amendment is "the direction that all persons similarly situated should be treated alike."[223] But the World War II draft did not treat people who were similarly situated in a similar manner. Often, different local draft boards treated identically situated men in different ways—and, needless to say, during wartime these differences often meant the difference between life and death for the men involved. The situation may

221 *In re* Summers, 325 U.S. 561 (1945).

222 Ibid., 576 (Black, dissenting).

223 City of Cleburne v. Cleburne Living Center, 473 U.S. 432, 439 (1985), citing Plyler v. Doe, 457 U.S. 202, 216 (1982), which quotes F. S. Royster Guano Co. v. Virginia, 253 U.S. 412, 415 (1920). The decision in *Royster Guano* asserts the proposition that "all persons similarly circumstanced shall be treated alike."

have been particular acute for COs because many draft board members had never heard of pacifism and others viewed COs as slackers, but the problem was by no means limited to COs. Unequal treatment was common for all registrants who went through the system.

There were additional problems as well. In September 1940, Congress had made a careful, deliberate decision that violators of Selective Service law were to be tried in civilian courts and, if convicted and sentenced, serve their time in civilian prisons. The legislative mandate was unequivocal, and multiple courts recognized it early—before the United States even entered the war. But the government ignored the mandate in many cases, and it took a Supreme Court decision in April 1943 (*Billings*) to set matters right.

The more than six thousand local draft boards making tens of millions of individual decisions constituted an enormous administrative bureaucracy. Today, a fundamental notion is that procedural due process requires that almost any decision by an administrative agency—let alone decisions involving life or death—be reviewable by a court. But the US Supreme Court case that made Selective Service decisions reviewable by courts (*Estep*) came in 1946, after the war had ended and the army was well along in demobilizing millions of draftees. This left no judicial avenue of appeal for tens of thousands of men who disagreed with Selective Service rulings during the entire period that the country was at war.

During World War II, the entire US military was racially segregated. While this was both legally and morally deplorable, the problem cannot fairly be laid on the doorstep of the Selective Service, which merely reflected the Jim Crow norms of the country at large. But the way the draft treated the Nisei can be laid at the doorstep of the Selective Service. The first decision—to classify American citizens of Japanese descent as aliens (III-C)—was not just morally repugnant, it was also nonsensical: both the legal and dictionary definitions of an alien is a noncitizen. The next decision—to draft men out of concentration camps to make cannon fodder of them—was even worse. And that only a single judge, Louis Goodman, had the decency to throw out the criminal charges against the Nisei resisters is shameful.

Supporters of conscription might argue that some unfairness in the draft was an inevitable trade-off in order to achieve the efficiency of operation that was absolutely necessary in light of the global emergency that the war presented. However, opponents of conscription might counter that one reason that conscription must never be countenanced is precisely because rank unfairness is an inevitable concomitant of a draft.

∴

On the first page of a 285-page post–World War II publication, the Selective Service describes "the twofold objective of Selective Service" thus:

> The Selective Service System during World War II had *two related objectives*. The first was to furnish enough men having acceptable qualifications to meet the requisitions of the armed forces; the second was to leave enough men in civilian status to provide for the greatest productive accomplishment of farm and factory in the nation's history, and to support the national interest....
>
> From its beginning the Selective Service System visualized its two objectives as equal parts of a twofold objective. This objective was to examine carefully the suitability of *each man* for military service and for retention in his civilian employment and to select him for military service or defer him for civilian work in accordance with the best interests of the entire nation's war effort.[224]

Several governmental regulations pertaining to conscription that otherwise make no sense at all are easily understandable when viewed in the context of the aspiration of universal channeling. The most obvious was the presidential order that prohibited the recruitment of volunteers during the war. Another example is the Selective Service regulation that provided that men who were deferred because they worked in an essential defense industry would be reclassified I-A if they changed jobs without previous Selective Service approval *even if their new job would also qualify for an occupational deferment*. The requirement that civilians obtain governmental approval before switching jobs—even switching to a new job that is government approved—smacks more of totalitarianism than of a free society.

∴

Lawrence Wittner, in *Rebels against War*, compares the respective percentages of inducted COs and imprisoned resisters in World Wars I and II (as a function of overall draft numbers). But it is unclear how useful such comparisons can be. Because of the stark differences between the conscription statutes during the First and Second World Wars, direct, apple-to-apples comparisons of the two eras are impossible, and scholars should be cautious about trying to make these comparisons. In World War II, 12,000 men classified IV-E were legally

---

224 US Selective Service System, *Quotas, Calls and Induction: Special Monograph No. 12* (Washington, DC: Government Printing Office, 1948), 1 (emphasis added).

and actually inducted but into alternative civilian service for which there was no World War I analogue. In World War II, more than 6,000 draft violators were in civilian prisons, but civilian prisons held almost no draft violators during World War I (because of the difference in the two statutes that was the subject of *Billings*). Wittner compares the 500 pacifists in military prisons during World War I to the 6,000 draft violators in civilian prisons during World War II, but that may not be a fair comparison. Many or most of the World War I prisoners would have been happy to accept alternative civilian service had it been available (and did accept it when it was belatedly offered to them), but many of the World War II prisoners were in prison because they refused to accept alternative civilian service. As a result, direct comparisons of the two experiences are difficult to make.[225]

What can we say about CPS in World War II and about the churches and individuals that invented it, served in it, and administered it? Here, the answers are fraught with ambiguity.

Certainly the churches and individuals who invented CPS were wise. It is deeply revealing of the government's attitude toward COs that, even though the Joint Army Navy Selective Service Committee had been preparing since 1926 for the potential resumption of conscription, the original version of the Burke-Wadsworth bill contained the identical, deeply flawed, language pertaining to COs that had been such an unfortunate feature of the World War I draft. In this context, the inventors of CPS were especially wise to have had the foresight to prepare for a resumption of conscription before the law was enacted; as a result of this, they were able, on short notice, to step into the vacuum created by the lack of government planning and offer concrete and reasonable proposals.

The men drafted into CPS and the people who ran the program were also diligent. Many worked eighty- or one hundred–hour weeks under difficult conditions. Although COs, too, had been drafted and many had dependents, including children, no COs received pay, health insurance, or death benefits, although all the GIs did. Indeed, CPS men had to pay their own way for the privilege of working on projects that the government believed served the national interest.

225 Wittner, *Rebels against War*, 41.

The peace churches that, for the most part, ran CPS were also generous. The churches, which were often poor themselves, donated the money to run the program. Mennonite churches collected "suggested assessments" of twenty-five cents per month per member to fund the program. Brethren sold bonds and stamps. Quakers made assessments of their yearly meetings. Northern Baptists established a wartime emergency fund that, in part, went to support COs. Presbyterians, Methodists, Episcopalians, and Unitarians contributed through their denominational entities.[226]

Certainly the creators of CPS were also well-meaning. Many had firsthand recollections of the calamities that had befallen COs during the First World War. Coming out of that experience, they were determined to fashion a better system for COs in the Second World War. Despite the shortcomings of the CPS system, their intentions in creating that system were nothing but good.

Nevertheless, the entire CPS system was morally compromised. This was an inevitable consequence of the exquisitely close entanglement of CPS with the Selective Service. The Treasury Department's ruling that CPS was exempt from federal taxes because CPS was doing the work of the Selective Service was merely an exemplar of that inextricable interconnection.

Especially troubling was the decision made by a relatively small group of Quakers, Mennonites, and Brethren to agree, on behalf of others, that CPS men would work without pay. Even in the context of the last-minute bait-and-switch perpetrated by the government, that decision was wrong, because those church representatives were making, on behalf of other people, a decision with important ethical implications.

And the choice made by the peace churches to have Paul Comly French head the NSBRO was particularly unfortunate. French was a martinet. As events showed, he was temperamentally well suited to work hand-in-glove with the career military brass who were in charge of the Selective Service, but he was a poor fit for a group of iconoclastic, individualist COs, a great many of whom were not members of the peace churches. His shameful eagerness to punish CPS men for perceived infractions, his petty turf guarding when non-NSBRO peace activists met with officials to discuss draft-related issues, and his shocking offers to have the NSBRO staff do Selective Service work at Selective Service headquarters—on one occasion without the prior knowledge or consent of the NSBRO board—were an embarrassment even then, and they look much worse when viewed with the benefit of hindsight.

226 *Reporter* 1, no. 13 (March 15, 1943): 4, and *Conscientious Objector* 3, no. 2 (April–May 1941): 7, both in the Center on Conscience and War Records, SCPC.

CHAPTER 6

The Cold War and the Korean War

Two diametrically opposite tendencies characterized the zeitgeist of the United States in the immediate post–World War II years. On the one hand, Americans felt tremendous relief that the war was over and a strong desire to bring the armed forces home, demobilize the troops, and return to civilian life. They placed great hopes for world peace on the newly created United Nations. And their traditional skepticism of standing armies, an important part of the American psyche since colonial days, had a significant resurgence. On the other hand, the immediate postwar years saw the birth of the Cold War, an era of strong, sometimes pathological, anti-Communist sentiment in the United States that shaped many aspects of US foreign and domestic policy.[1]

The early Cold War years gave rise to two distinct ideas among the military, politicians, and the general public about the future of American military manpower procurement. One was universal military training (UMT); the other was selective service. In the years following World War II—and especially during 1948, when Congress was considering proposals for conscription—the same people often supported and opposed UMT and selective service: supported them to maintain the armed forces and opposed them as an un-American step toward militarism and war.

Both UMT and selective service were forms of conscription; both could be enacted and implemented either as standalone policies or in tandem. Nevertheless, there was an important distinction between the two. UMT, as the name suggests, required universal service: every man would be drafted for six to twelve months of basic training (proposals varied) and then serve for additional years in the reserves, to be recalled in the event of another war or national emergency. In a way, UMT was viewed as a throwback to the colonial era state militias in which every man was deemed to be a citizen-soldier to be called forth during periods of need. In contrast, selective service was predicated

1 For general background on the Cold War, see John Lewis Gaddis, *The United States and the Origins of the Cold War, 1941–1947* (New York: Columbia University Press, 1972). For a strictly conventional perspective, see Odd Arne Westad, *The Cold War: A World History* (New York: Basic Books, 2017). For an approach that deals with cultural changes during the Cold War from a liberal—not anti-imperialist—perspective, see Louis Menand, *The Free World: Art and Thought in the Cold War* (New York: Picador, 2021). For a pacifist perspective, see Joseph Gerson, *With Hiroshima Eyes: Atomic War, Nuclear Extortion and Moral Imagination* (Philadelphia: New Society, 1995).

 | DOI:10.1163/9789004546684_008

on the idea of *selective* service. In keeping with the thinking of Provost Marshal Enoch H. Crowder (during World War I) and General Lewis B. Hershey (during World War II), the function of a selective service would be to decide which (usually smaller) group of men would best aid the nation's war-making capacity by entering the armed forces and which other men should serve a broader war effort by remaining civilians. While postwar discussions of UMT and selective service were often intertwined, it is important to understand that these were distinct ideas.

On January 6, 1947, in his State of the Union address, President Harry S. Truman announced the creation of the President's Advisory Commission on Universal Training to study the idea of UMT. The World War II draft was still in effect, and Truman was coy in his speech about what ought to be done about the Selective Service System, saying only that he expected the War Department to give some recommendations.[2]

On March 3, in a written message to Congress, Truman recommended that the Selective Service, which was scheduled to lapse on March 31, not be extended (see chapter 5). Truman's message made no mention of UMT. On March 31, the Selective Service System went out of business.[3]

On March 29, the president received the report of his President's Advisory Commission, entitled *A Program for National Security.* The commission had heard from more than two hundred witnesses, and its report ran to ninety-five pages (448 pages with appendixes).[4] The commission came to a unanimous, and strong, recommendation in favor of UMT, saying that it "is a matter of urgent military necessity." The commission recommended that UMT be universal ("That is the essence of democracy") but of course should exclude women. UMT should start with a mandatory six months of basic military training for all eighteen-year-olds. After this, men could complete their service in one of several ways, such as by enlisting in the regular military for the normal term of enlistment, joining the reserves, entering a service academy, or joining the

2 "The Text of President Truman's Call on Congress," *New York Times*, January 7, 1947; President's Advisory Commission on Universal Training, *A Program for National Security: Report of the President's Advisory Commission on Universal Training* (Washington, DC: Government Printing Office, May 29, 1947), 99.

3 Harold Hinton, "Truman Requests Selective Service End on March 31," *New York Times*, March 4, 1947; Lewis B. Hershey, *Report of the Director of the Office of Selective Service Records* (Washington, DC: Government Printing Office, 1950); US Selective Service System, *Selective Service under the 1948 Act* (Washington, DC: Government Printing Office, 1951).

4 Among the two hundred witnesses was a three-person delegation—Clarence Pickett, Harold Evans, and Ray Wilson—from the American Friends Service Committee that met with the President's Advisory Commission on January 24, 1947, at the White House.

National Guard. No one's military obligation would be complete with just six months of basic training.[5]

A year later, on March 17, 1948, Truman appeared before a joint session of Congress. Declaring that "the Soviet Union and its agents have destroyed the independence and democratic character of a whole series of nations in Central and Eastern Europe," Truman called on Congress to enact both UMT and a draft under a reconstituted Selective Service System. "There is no conflict between the requirements of selective service for the regular forces and universal training for the reserve components. Selective service is necessary until the solid foundation of universal training can be established," he said.[6]

During the spring and summer of 1948, multiple bills providing for UMT and/or selective service were introduced in Congress. Hearings on the various bills were combined. The United States had nearly 1.4 million men in uniform, but the Pentagon sought a combined armed forces of 2 million, an increase of 45 percent. At 540,000 men, the army was 129,000 below its authorized strength of 669,000 men, but the Pentagon wanted an authorization for 837,000, an increase of 55 percent from its actual size.[7]

1 The 1948 Congressional Debate

The Senate and House Armed Services Committees held hearings on the various conscription bills between March and May 1948.[8] The administration's case for renewing the Selective Service System and introducing UMT was well organized and well coordinated; indeed, hearings on the Senate side commenced later in the day on March 17, the same day that Truman delivered his address to Congress. Many senior administration officials testified, and all sang from the same hymnal. Both selective service and UMT were needed, they said, and complemented each other. Selective service was necessary for the immediate, short-term ramping up of forces; UMT was needed for maintaining military

5 President's Advisory Commission, *A Program for National Security*, 40–81, 89–95.

6 US Selective Service System, *Selective Service under the 1948 Act*, 5, 13; "The Text of President Truman's Address to the Joint Session of Congress," *New York Times*, March 18, 1948.

7 Hershey, *Report of the Director*, 97–98.

8 *Hearings Before the Committee on Armed Services, United States Senate on Universal Military Training*, 80th Cong., 2d sess. (1948) (hereafter, *Senate Hearings on Universal Military Training*); *Hearings Before the Committee on Armed Services, House of Representatives on H.R. 6274 and H.R. 6401*, 80th Cong., 2d sess. (1948) (hereafter, *House Hearings on H.R. 6274 and H.R. 6401*).

strength for the long term.[9] The essence of the government's position was that Soviet imperialism created an urgent need for conscription.

The retired Supreme Court justice Owen J. Roberts testified in support of conscription by saying that "the organized core of opposition to universal military training centers [is] in the Communist Party." To support that argument, Owens placed into the record two newspaper articles from the *Washington Daily News*. The first bore a headline that said, "Commies Plot District of Columbia UMT Demonstration"; its first sentence read, "As their latest move to strengthen Russia at the United States' expense, the Communists are preparing a grandstand play to capture the sentiment in this country against universal military training." The headline on the second article was "Commies Woo Big Names to Back Anti-UMT Demonstration Here"; the first sentence of that article said, "Prominent Americans are being asked to back a Communist-organized plan to corral into the pro-Soviet camp sentiment in this country against universal military training."[10]

Testimony against the conscription bills came from expected sources, including peace groups, student associations, and religious bodies, including the peace churches. But opposition also came from less expected sources as well, including farmers. The National Grange testified against the bills, as did the National Farmers Union (conscription "is futile for defense, dangerous to peace, damaging to youth, and directly opposed to our national traditions"), and the American Farm Bureau Federation ("the only real security in this day, with guided missiles and atomic bombs, lies in the abolition of war").[11] Both the American Federation of Labor and the Congress of Industrial Organizations (they had not yet merged) were opposed, as were the Catholic War Veterans Association, the Southern Baptist Convention, and the National Congress of Parents and Teachers.[12] Even the Mormon Church came out against conscription, saying that "by the creation of a great war machine we shall

9 James Forrestal to House of Representatives, *House Hearings on H.R. 6274 and H.R. 6401*, 6078, and statement of James Forrestal, US secretary of defense, April 12, 1948, 6081, 6083, 6122.

10 *Senate Hearings on Universal Military Training*, 97, 104, 105.

11 *House Hearings on H.R. 6274 and H.R. 6401*, statement by J. T. Sanders, legislative counsel to the National Grange, April 19, 1948, 6476–78; *Senate Hearings on Universal Military Training*, statements by Herbert W. Voorhees for the National Farm Bureau, March 23, 1948, 138–40; Russell Smith, legislative secretary, National Farmers Union, March 23, 1948, 144–45; J. T. Sanders, legislative counsel to the National Grange, March 23, 1948, 151–55.

12 *Senate Hearings on Universal Military Training*, statement by Seymour Linfield for the Progressive Citizens of America, March 24, 1948, 254–55. Linfield named the organizations cited in text, as well as a long list of others opposed to universal military training.

invite and tempt the waging of war against foreign countries, upon little or no provocation; for the possession of great military power always breeds thirst for domination, for empire, and for a rule of might not right."[13]

One controversial subject during the congressional hearings was segregation. Jesse O. Dedmon, Jr., speaking on behalf of the NAACP, spent exactly five words stating the organization's official position ("We oppose peacetime military conscription") and then went on at considerable length to argue that, if conscription were to be enacted, it must clearly prohibit racial segregation in recruiting, drafting, and service in the military. He said, "We gain little if now and in the future the military protects us from Hitler's brand of racism, yet itself exemplifies ... a domestic brand of racism equally at variance with democratic ideals." As I will show, the civil rights leader A. Philip Randolph went considerably further.[14]

Because the bill reestablishing the Selective Service would have allowed objectors to noncombatant military service to be excused even from alternative civilian service (unlike the World War II requirement for service in Civilian Public Service camps), the peace churches were much less involved in 1948 than they had been in 1940 in discussing and negotiating the fine points of treatment for COs. Representative James Wadsworth (R-NY), who had been one of the lead sponsors of the 1940 conscription statute, testified that providing for alternative civilian service in his bill had been a mistake. Groups such as the Friends Committee on National Legislation and the War Resisters League (WRL) mostly restricted their comments to opposition to conscription and said little or nothing about the minutiæ of how COs ought to be treated under the new law.[15]

One matter discussed during the 1948 hearings that had, for obvious reasons, not arisen in 1940 was whether, in the age of nuclear weapons, conscription was any use at all. Some anticonscription witnesses said that Hiroshima and Nagasaki had rendered large armies obsolete. The Reverend Dutton Peterson of the New York Council of Churches said, "In prospect of atomic warfare our

---

13 *Senate Hearings on Universal Military Training*, "Statement on Universal Compulsory Military Training by the First Presidency of the Church of Jesus Christ of Latter-Day Saints," by George Albert Smith, J. Reuben Clark, Jr., and David O. McKay, first secretaries, March 20, 1948, 122–23.

14 *Senate Hearings on Universal Military Training*, statement by Jesse O. Dedmon, Jr., secretary, Veterans' Affairs, National Association for the Advancement of Colored People, March 31, 1948, 662–68.

15 *House Hearings on H.R. 6274 and H.R. 6401*, statement by Samuel R. Levering for Friends Committee on National Legislation, April 16, 1948, 6278–87, and statement by Edward C.M. Richards for the War Resisters League, April 17, 1948, 6348–52.

scientists tell us that the use of large standing armies is now obsolete." The Reverend Donald Harrington of the New Jersey Council Against Conscription said something similar. Government witnesses were at pains to refute the assertion. Said Secretary of Defense James Forrestal, "The comfortable assurance of a pushbutton war is an illusion."[16] Given what we know today about the decades of intervention by the United States, using large armies in Vietnam, Iraq, Afghanistan, and other nations, Forrestal had the better argument.

Both the Senate and House Armed Services Committees heard proposals to bar racial discrimination by the Selective Service, but none gained traction. On June 7, 1948, when the bill that would become the Selective Service Act was debated on the floor of the Senate, Senator William Langer (R-ND) introduced seven civil rights amendments. One would have barred racial discrimination in the operation of the Selective Service and in the armed forces. The amendment was defeated overwhelmingly, 67 to 7. The opposition was broadly bipartisan. It wasn't just ardent segregationist southerners like Richard Russell (D-GA), John Stennis (D-MI), James O. Eastland (D-MI), and John Sparkman (D-AL) in opposition. The chair of the Democratic National Committee, northerner J. Howard McGrath (D-RI), opposed the amendments. So did many Republicans, like Majority Leader Wallace White (R-ME), Leverett Saltonstall (R-MA), and George Aiken (R-VT). Senator John Chandler Gurney (R-SD), who chaired the Armed Services Committee, led the opposition to the civil rights amendments, saying that they were unnecessary and would create "a morale problem of the first magnitude."

One of the amendments Langer proposed would have made it illegal to lynch a member of the armed forces. Another would have prohibited racial discrimination against soldiers by government officials or in public accommodations like hotels and restaurants off base. One amendment would have barred employment discrimination by companies that sell products to the government for use by the armed forces. All were soundly defeated. One civil rights amendment passed; it barred charging a poll tax to members of the armed services.[17]

---

16 *House Hearings on H.R. 6274 and H.R. 6401*, statements by Dutton Peterson, vice president of the New York Council of Churches, April 16, 1948, 6287–88; Donald Harrington for the New Jersey Council Against Conscription, April 16, 1948, 6329; James Forrestal, April 12, 1948, 6081.

17 94 Cong. Rec. 7220–81 (1948); C. P. Trussell, "Most 'Rights' Kept Out of Draft Bill by Senate Votes," *New York Times*, June 8, 1948.

2 The 1948 Conscription Statute

Truman signed the Selective Service Act of 1948 into law on June 24, 1948.[18] Depending on one's view of the 1940 conscription law, this was, or was not, the first peacetime draft law in US history. It provided for the increases in maximum troop levels that the military had sought, gave the president legal authority to call up the reserves, and, by its own terms, had a duration of only two years.[19]

The Selective Service Act of 1948 re-created the entire Selective Service System that had disappeared in 1947, merging the existing Office of Selective Service Records into the reconstituted Selective Service System. Fully reinstated was the old setup of local draft boards, composed of civilians in each county; appeal boards in each judicial district; and the conventional rule-making authority of an ordinary administrative agency.[20] Registration was required for men aged eighteen to twenty-six, and the law imposed the possibility of conscription for those aged nineteen to twenty-six for twenty-one months of active-duty service, to be followed by five years in the reserves.[21]

Congress thought that the 1948 law's channeling function was so important that, for the first time in US history, it included that function in the preliminary declarations in the first section of the law: "The Congress ... declares that adequate provision for national security requires maximum effort in the fields of scientific research and development, and the fullest possible utilization of the Nation's technological, scientific, and other critical manpower resources."[22] A lengthy section in the middle of the statute provided for deferment for "study, research, medical, scientific or other endeavors" necessary for defense. This included, for the first time, the possibility of indefinite student deferments for those in college.[23]

Once again, clergy and ministerial students were exempt; high school students were deferred until graduation or age twenty; and "those persons found to be physically, mentally, or morally deficient or defective" (i.e., missing a leg or gay) also were exempt. Deferments were permitted but not required for men with dependents (wives and/or minor children). A sole surviving son, all of

18 An Act to provide for the common defense by increasing the strength of the armed forces of the United States, including the reserve components thereof, and for other purposes, Pub. L. No. 80-759, 62 Stat. 604 (1948) (hereafter, Conscription Statute).

19 Conscription Statute, §§ 2, 7, 17.

20 Ibid., § 10.

21 Ibid., §§ 3, 4.

22 Ibid., § 1.

23 Ibid., § 6.

whose siblings had been killed in combat, was exempt, as were men who had served during World War II.[24]

Once again, the law set out two categories of COs. Those who objected only to carrying arms were obligated to serve in the military in noncombatant roles. However, for the first time, those who objected to even noncombatant service were entirely excused from service. (This exemption even extended to alternative civilian service but did not extend to other obligations under the law, including registration, filling out forms, and carrying draft cards.) Congress also addressed the issue that had arisen under the 1940 statute about the meaning of "religious training and belief" (see chapter 5) by narrowing the definition. The 1948 act declared: "Religious training and belief in this connection means an individual's belief in relation to a Supreme Being involving duties superior to those arising from any human relation, but does not include essentially political, sociological, or philosophical views or a merely personal moral code." (This sentence was extensively litigated by draft opponents during the 1960s; see chapter 7.)

For CO applicants who were denied recognition by local draft boards, the 1948 act provided for a review by the Department of Justice; however, that department would make a merely advisory recommendation, with the ultimate decision resting with the Selective Service Appeal Board. This provision also led to a significant amount of litigation.[25]

One other noteworthy provision in the 1948 act was entitled "Utilization of Industry" and permitted the president, if he found that "it is in the interest of national security," to seize any materials the government wanted or needed and/or to seize any plant, mine, or other facility that manufactured items needed for defense.[26] Although perhaps novel, this provision was neither outrageous nor necessarily unconstitutional. If, in time of war or other national crisis, the government has the power to conscript people and send them

24 Ibid.

25 Ibid.

26 Section 18 of the Conscription Statute allowed the government to indefinitely operate a plant, mine, or other facility it had seized to advance a war effort. In particular, steel mills and nuclear facilities were thought most likely to be seized. The statute did not define either "national security" or "the interest of national security." It did have a requirement that "fair and just compensation" be paid in the event of government seizure of materials or facilities. This was necessary because the last clause of the Fifth Amendment says "private property [shall not] be taken for public use without just compensation." But the 1948 statute did not explain how just compensation would be determined, nor by whom, nor did it contain any appeal right for a property owner who believed that the compensation the government proffered for seized property was unfair.

into battle where they may be maimed or killed, there is no reason why the government should not also be able to "conscript" a ton of steel (or the plant that can turn out a ton of steel). Indeed, the Constitution suggests that this *can* be done—so long as just compensation is paid for the seized goods. The government never invoked the Utilization of Industry provision in the 1948 act. When, during the Korean War, Truman did seize the nation's steel mills—and the seizure was challenged in the courts—the administration was careful to state that it had relied on the president's implied powers under the Constitution to do so, and not the 1948 act (or any other statute).[27]

3 The 1950 Congressional Debate and Conscription Statute

The 1948 statute was set to expire on June 26, 1950. During the spring of 1950, bills were introduced into Congress to extend the Selective Service System or to create a new program of UMT. The House also entertained a bill to extend the Selective Service's registration and classification functions but to suspend involuntary conscription entirely.[28] Secretary of Defense Louis Johnson, Chairman of the Joint Chiefs of Staff Omar Bradley, Secretary of State Dean Acheson, and Hershey all testified in favor of extending conscription.[29]

Once again, a number of groups opposed extension of the draft: labor groups, including the American Federation of Labor; farm groups, including the National Grange; a wide range of Jewish and Christian religious groups; a college president; and the National Council Against Conscription.[30] In his

---

27 In the case of the steel mills Truman seized during the Korean War, the Supreme Court ruled that the president does not have an implied constitutional power to seize steel mills. Youngstown Sheet & Tube Co. v. Sawyer (Steel Seizure Case), 343 U.S. 579 (1952).

28 US Selective Service System, *Selective Service under the 1948 Act Extended, July 9, 1950–June 19, 1951* (Washington, DC: Government Printing Office, 1953) (hereafter, *Annual Report*, 1951), 18. This was the first annual report that Congress required the Selective Service to make each year from 1951 to 1967. From 1952 on, all the detailed annual reports to Congress were published under the title *Annual Report of the Director of Selective Service for the Fiscal Year 19__ to the Congress of the United States Pursuant to the Universal Military Training and Service Act as Amended.* Each report covered one government fiscal year (July 1 to June 30) and is named for the calendar year in which the fiscal year ended. These are cited in subsequent notes as *Annual Report*, followed by the year.

29 *Hearings Before the Committee on Armed Services of the House of Representatives on H.R. 6826*, 81st Cong., 2d sess. (1950) (hereafter, *Hearings on H.R. 6826*), statements of Louis Johnson, January 20, 1950, 5099; Omar Bradley, January 25, 1950, 5159; Dean Acheson, January 25, 1950, 5164; and Lewis Hershey, January 26, 1950, 5186.

30 *Hearings on H.R. 6826*, statements of the American Federation of Labor, January 30, 1950, 5233; National Grange, January 30, 1950, 5246; Central Conference of American Rabbis,

testimony on behalf of the War Resisters League, Executive Secretary Roy Kepler linked the WRL's opposition to conscription to its opposition to all war: "Conscription ... has shown itself to be an integral part of war. It is not simply an effect of modern war-making, but also a cause. Conscription, in our view, is not a peace measure."[31]

Once again, in 1950, when Congress was considering extending the 1948 act, it also considered a series of proposed civil rights amendments. These were substantially similar to the Langer amendments that had been rejected in 1948, including to prohibit discrimination in conscription and segregation in the armed forces and to make it illegal to lynch a draftee. Once again, the amendments failed.[32]

In addition, in 1950 there was a new argument against the Selective Service: it was simply unnecessary and, in fact—despite the existence of the 1948 act—the Selective Service System just wasn't drafting many people (see Table 6.1). In the two years since the 1948 act became law, there had been few draft calls; most months, no one was drafted at all. Only ten thousand people were called for the entire period of January to June 1949. And when the draft was set to expire in June 1950, not one person had been drafted for more than a year.

Draft boards had little to do. Forty-one percent of the 3,659 local draft boards were open only eight hours per week or fewer; perhaps partly as a result of that, men turning eighteen were getting careless about registering. All 3,659 local draft boards and the fifty-four state headquarters had a combined total of only 1,234 full-time employees, that is, about one full-time employee for every three draft boards.[33] The argument was that the Selective Service System wasn't necessary—why extend a law that isn't being used? In his congressional testimony, Hershey countered this argument by likening the Selective Service to an insurance policy. No one criticizes his life insurance company if, after two years, he hasn't collected on the policy because he is still alive, Hershey noted.[34]

January 30, 1950, 5259; Presbyterian Church (USA), January 30, 1950, 5261; Northern Baptist Convention, February 1, 1950, 5308; Church of the Brethren, January 31, 1950, 5272; John Elliott, president of Alderson Broaddus College, January 31, 1950, 5265; Alonzo Myers, chair of the National Council Against Conscription, January 31, 1950, 5269.

31 *Hearings on H.R. 6826*, statement of Roy Kepler, January 31, 1950, 5277. For a biography of Kepler, including his World War II activism and antidraft protests during the Vietnam War, see Michael Doyle, *Radical Chapters: Pacifist Bookseller Roy Kepler and the Paperback Revolution* (Syracuse, NY: Syracuse University Press, 2012).

32 *Hearings on H.R. 6826*, statement of US Rep. Isidore Dollinger, D-NY, February 1, 1950, 5297–98.

33 *Annual Report*, 1951, 4, 18; *Hearings on H.R.* 6826, statement of Lewis Hershey, director of Selective Service, January 26, 1950, 5189. The fifty-four state headquarters included those for territories, such as Alaska and Hawaii.

34 *Hearings on H.R. 6826*, 5188–89.

TABLE 6.1 Calls and inductions under the 1948 conscription act, July 1948–August 1950

Month	Calls	Inductions
July 1948	0	0
August 1948	0	0
September 1948	0	0
October 1948	0	0
November 1948	10,000	7,839
December 1948	15,000	12,509
January–June 1949	10,000	9,781
July 1949	0	0
August 1949	0	0
September 1949	0	0
October 1949	0	0
November 1949	0	0
December 1949	0	0
January 1950	0	0
February 1950	0	0
March 1950	0	0
April 1950	0	0
May 1950	0	0
June 1950	0	0
July 1950	0	0
August 1950	0	0
Total	**35,000**	**30,129**

SOURCE: US SELECTIVE SERVICE, *ANNUAL REPORT OF THE DIRECTOR OF SELECTIVE SERVICE FOR THE FISCAL YEAR 1957 TO THE CONGRESS OF THE UNITED STATES PURSUANT TO THE UNIVERSAL MILITARY TRAINING AND SERVICE ACT AS AMENDED* (WASHINGTON, DC: GOVERNMENT PRINTING OFFICE, 1958), APP. 19, 95.

Note: Between January and June 1949, the Selective Service tabulated the number of calls and inductions only as an aggregate figure.

On May 4, 1950, the House, by a vote of 216 to 11, passed the Manpower Registration Act, which would have permitted continued registration and classification for two years but barred conscription absent a national emergency (so declared by Congress, not by the president). Although the bill did not become law (because the Korean War intervened), it was nevertheless a fair representation of the deep skepticism in that chamber of peacetime conscription.[35]

On June 8, 1950, the Senate Armed Services Committee sent to the Senate floor a bill that would have extended the 1948 act for three years—and allowed every drafted man to state whether he wanted to serve in a segregated unit. Floor debate on the measure began on June 21.[36]

The next day, because it was clear that the Senate and House bills could not be reconciled before the 1948 act was set to expire on June 26, Congress passed a stop-gap bill extending the 1948 act for fifteen days. On June 23, Truman signed the fifteen-day extension into law.[37]

On June 24, North Korean troops invaded South Korea. As a result of the new situation, Congress quickly passed the four-sentence Selective Service Extension Act of 1950, which extended the (entire unedited) 1948 act for one year and authorized the president to call up the reserves. On June 30, 1950, Truman signed the bill into law.[38]

4 The Doctors' Draft (1950) and Other Draft Laws, 1951–63

At the beginning of the Korean War, the United States had a perceived shortage of doctors and dentists in the military that was about to be exacerbated by the sudden increase in the size of the armed forces. In response, Congress passed a bill establishing a doctors' draft that became law on September 9, 1950. The doctors' draft provided for separate registration, classification, and drafting of doctors and dentists. If drafted, their length of active duty was the same as for regular draftees (twenty-one months), but their liability continued up to age fifty (rather than twenty-six). Doctors who were so drafted entered the military as commissioned officers. The statute also provided that doctors and dentists

35 *Annual Report*, 1951, 4.

36 Ibid.

37 Ibid., 4–5; Joint Resolution Extending the period of effectiveness of the Selective Service Act of 1948 for fifteen days, Pub. L. No. 81-572, 64 Stat. 254 (1950).

38 An Act To extend the Selective Service Act of 1948, as amended, for a period of one year, and for other purposes, Pub. L. No. 81-599, 64 Stat. 318 (1950). With passage of the Selective Service Extension Act, the entire body of existing Selective Service regulations, enacted under the 1948 Conscription Statute, was carried forward in full force and effect.

who had received their medical training at the expense of the army or navy and had never served would be drafted first.[39]

In June 1953, Congress extended the doctors' draft for two years, until July 1, 1955, while shortening the time of active duty for doctors and dentists to seventeen months.[40] In the first approximately four years that the doctors' draft was in effect (September 9, 1950, to June 30, 1954), about twelve thousand men were drafted under it.[41]

Another significant change in US conscription law occurred with the enactment of the Universal Military Training and Service Act (UMTSA) on June 19, 1951. The UMTSA effected several specific changes to the Selective Service Act of 1948.[42]

First, the authority to draft, which was set to expire on July 9, 1951, was extended until July 1, 1955. Second, the UMTSA changed the term of active duty for conscripted men to twenty-four months. Third, the UMTSA provided that, if a man had ever received a deferment before his twenty-sixth birthday (say, as a student or on the basis of a job he no longer held), his period of liability for the draft would extend until his thirty-fifth birthday (rather than until his twenty-sixth birthday, as under the 1948 act). As the Selective Service correctly observed, the purpose of the last change was to increase the size of the pool of draft-eligible men.[43]

The UMTSA also significantly changed the requirement for COs who were opposed to noncombatant service. Instead of being excused from service entirely, these COs were obligated to perform two years of alternative civilian service. However, unlike the situation during World War II (with the creation of Civilian Public Service camps), each CO had to find his own alternative service; there would be no organized program for these COs. Also, unlike the World War II experience, these noncombatant COs would be paid (by their employers).

---

39 An Act to amend the Selective Service Act of 1948, as amended, so as to provide for special registration, classification, and induction of certain medical, dental, and allied specialist categories, and for other purposes, Pub. L. No. 81-779, 64 Stat. 826 (1950).

40 An Act To amend the Universal Military Training and Service Act, as amended, so as to provide special registration, classification, and induction of certain medical, dental, and allied specialist categories, and for other purposes, Pub. L. No. 83-84, 67 Stat. 86 (1953) § 6; *Annual Report*, 1953, app. 19, 93.

41 *Annual Report*, 1954, 41.

42 An Act To provide for the common defense and security of the United States and to permit the more effective utilization of manpower resources of the United States by authorizing universal military training and service, and for other purposes, Pub. L. No. 82-51, 65 Stat. 75 (1951).

43 *Annual Report*, 1951, 2.

Recall that three of the major irritants for some CPS men during World War II had been the Selective Service's control of CPS, the lack of meaningful work, and the lack of pay. All three were now removed. Men could look for work as meaningful as they wished, they would not be supervised by military men, and they would be paid the going rate for their work. Other objections, such as the involuntary nature of the service, remained.

One more thing must be said about the UMTSA. The statute contemplated, in some considerable detail, a national security training corps, in which all eighteen-year-old men, with no exceptions, would serve for six months. But under the UMTSA, this proposed, theoretical national security training corps would have required further congressional action before it actually came into being, and that further congressional action never happened. Thus, the lengthy, highly elaborated plans for a national security training corps contained in the 1951 statute never came to fruition. The name change stuck, however, and for years—indeed, well into the Vietnam War period—the controlling conscription law in the United States was the Universal Military Training and Service Act. But the name change that seemingly specified universal military training was entirely without substance, and the Selective Service remained a *selective* service system that self-consciously saw its role as channeling, by means of deferments, the entire male population of the country.

On June 30, 1955, the day before conscription authority was set to expire, Congress extended the draft for four years, until July 1, 1959, and extended the doctors' draft for two years until July 1, 1957. This statute made no other major changes in the Selective Service.[44] On June 27, 1957, just before the doctors' draft was set to lapse, Congress discontinued it and merged several of its key provisions into the basic draft law, the UMTSA.[45]

In March 1959, in a statute only five sentences long, Congress extended the draft for four years until July 1, 1963, making no other changes in the underlying statute. Then, in March 1963, in a similarly terse statute, Congress extended the draft for another four years, until July 1, 1967, again, making no other substantive changes in the law.[46]

44 An Act to further amend the Universal Military and Training Act, Pub. L. No. 84-118, 69 Stat. 223 (1955).

45 An Act To Amend the Universal Military Training and Service Act, as amended, Pub. L. No. 85-62, 71 Stat. 206 (1957).

46 An Act to extend the induction provisions of the Universal Military Training and Service Act, as amended, Pub. L. No. 86-4, 73 Stat. 13 (1959); An Act To extend the induction provi-

5 The Draft in Operation

With the Korean War on, monthly draft calls quickly shot up (see Table 6.2). In September 1950, 50,923 draftees were inducted; in October, there were 55,263 draftees; in November, 71,390; and in December, there were 42,189 draftees.[47]

After the Korean War ended in July 1953, monthly draft calls dropped immediately by about 50 percent, from roughly 50,000 per month to about 25,000 per month. Neither monthly nor annual inductions again fell to their pre–Korean War levels; annual inductions hit lows of just more than 61,000 men in 1961 and just fewer than 72,000 men in 1963 in the immediate pre–Vietnam War years.

TABLE 6.2 Calls and inductions by Fiscal year, 1951–63

Year	Calls	Inductions
1951	550,397	587,444
1952	367,288	381,006
1953	523,000	560,798
1954	251,000	268,018
1955	211,000	213,716
1956	136,000	136,580
1957	175,000	179,321
1958	124,958	126,369
1959	109,000	111,889
1960	89,500	90,549
1961	58,000	61,070
1962	147,500	157,465
1963	70,000	71,744
Total	**2,847,643**	**2,976,098**

SOURCE: US SELECTIVE SERVICE SYSTEM, *ANNUAL REPORT OF THE DIRECTOR OF SELECTIVE SERVICE FOR THE FISCAL YEAR 1963 TO THE CONGRESS OF THE UNITED STATES PURSUANT TO THE UNIVERSAL MILITARY TRAINING AND SERVICE ACT AS AMENDED* (WASHINGTON, DC: GOVERNMENT PRINTING OFFICE, CA. 1964), 20.

sions of the Universal Military Training and Service Act, Pub. L. No. 88-2, 77 Stat. 4 (1963).

47 *Annual Report*, 1957, app. 19, 95. Of these 50,923 September inductees, 1,577 technically entered the army in August but in response to the September call.

The number of local draft boards grew slowly but steadily from 3,659 in 1950 to 4,046 in 1963.[48] Similarly, the number of men newly registering per annum—mostly men turning eighteen—rose steadily from 1.1 million new registrants in the year ending June 30, 1952, to 1.5 million new registrants in the year ending June 30, 1963.[49]

Several significant changes in the Selective Service System's classifying of registrants were made during this period as well. The first two of these changes occurred in 1951.

On September 25, 1951, the Selective Service created the II-S classification (student deferment). The statute did not require a separate deferment for students; the Selective Service took this action on its own initiative by issuing a regulation under permissive provisions of the statute that allowed the agency to create deferments in the interest of national defense as part of its broader mission of channeling. At first, in order to qualify for the deferment, a student had to either be in graduate school or be an undergraduate who achieved a certain score on a new Selective Service examination (developed for the Selective Service by the then-new Educational Testing Service in Princeton, New Jersey) or be ranked in the upper two-thirds of his class. The examination was not rigorous. By 1963, 91 percent of students taking the exam achieved a high enough score to get a deferment, and the requirement pertaining to examination or class rank was eliminated. Thereafter, every graduate and undergraduate student who was making satisfactory academic progress received a student deferment.[50]

That same year, the Selective Service made a second significant change in classification, this one pertaining to those COs who were opposed to both combatant and noncombatant military service. It eliminated the IV-E classification for these COs (which had existed since the 1940 draft) and replaced it with a new I-O classification. COs who were willing to do noncombatant military service continued to be classified I-A-O; now COs who would do only alternative civilian service were classified I-O.[51]

The Selective Service made two additional significant classification changes in the early 1960s. Throughout the 1950s Selective Service officials had been concerned about the relatively large pool of men who were classified IV-F, that is, completely unfit for military service because of physical, mental, or moral deficiencies. On January 5, 1962, a new classification, I-Y, was created for

---

48 The 1963 figures are taken from *Annual Report*, 1963, 2.

49 *Annual Report*, 1952, 11; 1963, 1.

50 Ibid., 1951, 9–21, 27; 1963, 2, 14.

51 Ibid., 1951, 40. The classification of IV-E changed to the new I-O classification on September 25, 1951, pursuant to Executive Order No. 10292, 16 Fed. Reg. 9843 (September 25, 1951).

previously IV-F men whose problems were sufficiently minor that they would be acceptable for military service in time of war or other national emergency. The I-Y classification later became useful as a category for men whose physical issue, while severe, was not permanent. Thus, a man with a broken leg might be classified I-Y, while a man with a missing leg would be classified IV-F.[52]

On March 14, 1963, a presidential order specified that all, not merely some, fathers would be eligible for the III-A classification. Table 6.3 shows what the Selective Service classifications looked like as a result of these changes.[53]

TABLE 6.3 Selective service classifications as of June 30, 1963

Classification	Description
I-A	Available for military service
I-A-O	Conscientious Objector available for noncombatant military service only
I-O	Conscientious Objector available for alternative civilian service only
I-Y	Qualified for military service in time of war or national emergency
II-A	Deferred due to occupation (nonagricultural)
II-C	Deferred due to occupation (agricultural)
II-S	Student deferment
III-A	Registrant with child or children or extreme hardship to dependent(s)
IV-B	Officials exempt by statute
IV-C	Aliens
IV-D	Clergy (or divinity school student)
IV-F	Not qualified for physical, mental, or moral reason
IV-A	Registrant who has completed military service or sole surviving son
I-W	CO who is currently doing, or has completed, alternative civilian service
I-C	Current member of the military

SOURCE: US SELECTIVE SERVICE SYSTEM, *ANNUAL REPORT OF THE DIRECTOR OF SELECTIVE SERVICE FOR THE FISCAL YEAR 1963 TO THE CONGRESS OF THE UNITED STATES PURSUANT TO THE UNIVERSAL MILITARY TRAINING AND SERVICE ACT AS AMENDED* (WASHINGTON, DC: GOVERNMENT PRINTING OFFICE, CA. 1964), APP. 9, 50.

52 *Annual Report*, 1961, 19–20; 1963, 1.

53 Exec. Order No. 11098, 28 Fed. Reg. 2615 (March 14, 1963); *Annual Report*, 1963, 35.

Channeling continued to be a prime function of the Selective Service System. In 1958, Hershey lamented that "there is evidence that part of the general public still seems to have the idea that Selective Service is merely an apparatus for drafting men into the Armed Forces. Actually, this function is a relatively small fraction of the system's total operation. The great changes brought about by today's air and space age have caused Selective Service to evolve into an organization which is far beyond the narrow concept of the draft."[54]

In 1960, Hershey elaborated on this point in his annual report to Congress:

> One of the basic functions of Selective Service is the channeling of registrants into endeavors, occupations and activities that are in the national interest. This function arises in connection with the deferment operations of Selective Service.
>
> The process of channeling by not taking men from certain activities who are otherwise liable for military service, or by giving deferment to qualified men in certain occupations, is the counterpart to procuring manpower for civilian activities.
>
> Many young men would have not pursued higher education had there not been a Selective Service program of student deferment. Young scientists, engineers, tool-and-die makers and other possessors of scarce skills would not remain in their jobs in the defense effort were it not for occupational deferments.[55]

6 Conscientious Objectors

Under the UMTSA, signed into law in June 1951, COs who objected to noncombatant military service became liable for alternative civilian service.[56] However, the regulations pertaining to how that alternative service was to be managed were not issued until February 20, 1952, and the new program was not up and running until July 1, 1952. The regulations issued on February 20, 1952, provided that COs had to "volunteer" in order to serve their two years of alternative service at a nonprofit organization involved in charitable, health, welfare, educational, or scientific work. The entity for which the CO worked had to be approved by the Selective Service. To ensure that COs were not too

---

54 *Annual Report*, 1958, 51.

55 Ibid., 1960, 25 (in chap. entitled "Channeling").

56 An Act to … to permit the more effective utilization of manpower resources, Pub. L. No. 82-51, 65 Stat. 75 (1951), Title I, § 1(q); *Annual Report*, 1951, 27.

comfortable, the regulations also specified that, except in rare circumstances, such work could not be in the CO's home community.[57]

In the event, some COs worked for projects operated by the service committees of the peace churches, the American Friends Service Committee, Mennonite Central Committee, and Brethren Service Committee. Other COs served in such places as Veterans Administration hospitals; Goodwill Industries; a wide range of religious, state, and other nonprofit hospitals; so-called training schools for delinquent youths; the US Public Health Service; US Department of Agriculture; and a variety of universities and colleges.[58]

During this period (unlike during World War II), the Selective Service did not provide statistics for the numbers of men who were classified I-A-O (subject to noncombatant military service). The numbers of men classified IV-E (and, later, I-O), however, are known. In January 1949—that is, during the first seven months that the 1948 conscription statute was in effect—4,903 men were classified IV-E.[59] By September 1949, that number had more than doubled, to 10,516.[60]

Between July 1, 1952, when the requirement that COs perform alternative civilian service began, and January 1963, 14,727 men performed alternative service, though the number of men claiming to be COs during this period was vastly higher. The largest category of men claiming to be COs that is not included in the 14,727 number are men whose CO claims were never processed or addressed because they were immediately placed in a higher classification by the Selective Service because of a physical ailment, job deferment, farm work, or other factor.[61]

---

57 Exec. Order No. 10328, 17 Fed. Reg. 1645 (Feb. 20, 1952); *Annual Report*, 1952, 6–8; "CO Regulations Signed by President," *NSBRO Reporter* 9, no. 1 (January–February 1952): 1–4.

58 "Where They Serve," *NSBRO Reporter* 10, no. 5 (May 1953): 3–4.

59 "Nearly Five Thousand Classified IV-E," *NSBRO Reporter* 6, no. 6 (March 1949): 1. Although Selective Service provided statistical enumerations of COs in its annual reports to Congress, the figures are essentially meaningless, because they lump together men who were newly classified IV-E (or, later, I-O), men who were currently performing alternative civilian service, and COs who had completed their alternative service. Thus many COs were double- or triple-counted in the Selective Service tabulations: once when they became IV-E or I-O, again when they began alternative service, and may have been counted yet a third time when they completed that service. NSBRO kept much more accurate statistics on the numbers of COs, and those are the figures used in this section.

60 "Number of IV-Es Soars," *NSBRO Reporter* 7, no. 1 (November 1949): 4.

61 J. Harold Sherk, "A Report from the NSBRO," January 16, 1963, Center on Conscience and War Records, RG 025, pt. II, ser. A, box 16, folder labeled "Reports by NSBRO," Swarthmore College Peace Collection (SCPC), Swarthmore College, Swarthmore, PA.

During this period (again, unlike World War II), the Selective Service did not record the religious affiliations of men who performed alternative service. However, NSBRO maintained a tally that, while not covering all COs, did account for about 70 percent of the men who performed alternative service. Once again, the overwhelming majority (87 percent) were members of the peace churches.[62]

Being a CO during this period was not entirely without risk. During the 1950s, some states had statutes prohibiting or restricting the hiring of men who had been COs. An Iowa state law, for example, prohibited the hiring of former COs by any city or town; this included positions as teachers, firefighters, librarians, and so forth. During the 1950s, various organizations argued that such bans were poor public policy. However, it is doubtful that legal challenges to such laws, including on constitutional grounds, would have been successful. And such laws, where they existed, were enforced. For example, in June 1958, Donald E. Laughlin, a high school science teacher in West Branch, Iowa, was forced to resign his position because he had been a CO. On May 30, 1962, Daniel Gerber, a Mennonite CO performing his alternative civilian service at a Christian leprosarium in Ban Me Thuot, Vietnam, faced the most dire risk: he was kidnapped by guerrillas and presumed killed.[63]

7 Resisters

Resistance to the new draft started while Congress was still discussing and debating draft proposals. Eighteen months before Truman signed the new peacetime draft bill into law, opponents called for a national day of draft resistance called Break With Conscription (see Figure 6.1). They circulated "An Appeal for Action" that said, in part: "We the undersigned propose to break with conscription now. [On February 12, 1947] we shall get rid of our draft credentials publicly, by returning them to the President, accompanied by letters of explanation, or disposing of them in open demonstrations. We hope to see demonstrations organized in our local communities to dramatize the importance of this event." The "Appeal for Action" was signed by World War

62 "Few Churches Account for Bulk of I-Ws," *NSBRO Reporter* 17, no. 9 (September 1960): 1.

63 "Iowa State Law Restricts Employment for CO's," *NSBRO Reporter* 15, no. 2 (February 1958): 2; "Teacher Ousted Because of CO Convictions," *NSBRO Reporter* 15, no. 7 (July 1958): 1; Daniel Gerber, "Daniel Gerber Summary (2017 Luke Martin Update)," ASIA 1943—Ongoing, Vietnam 1954—2021, Mennonite Central Committee Archives, Akron, PA. See also *NSBRO Reporter* 19, no. 7 (July 1962): 1.

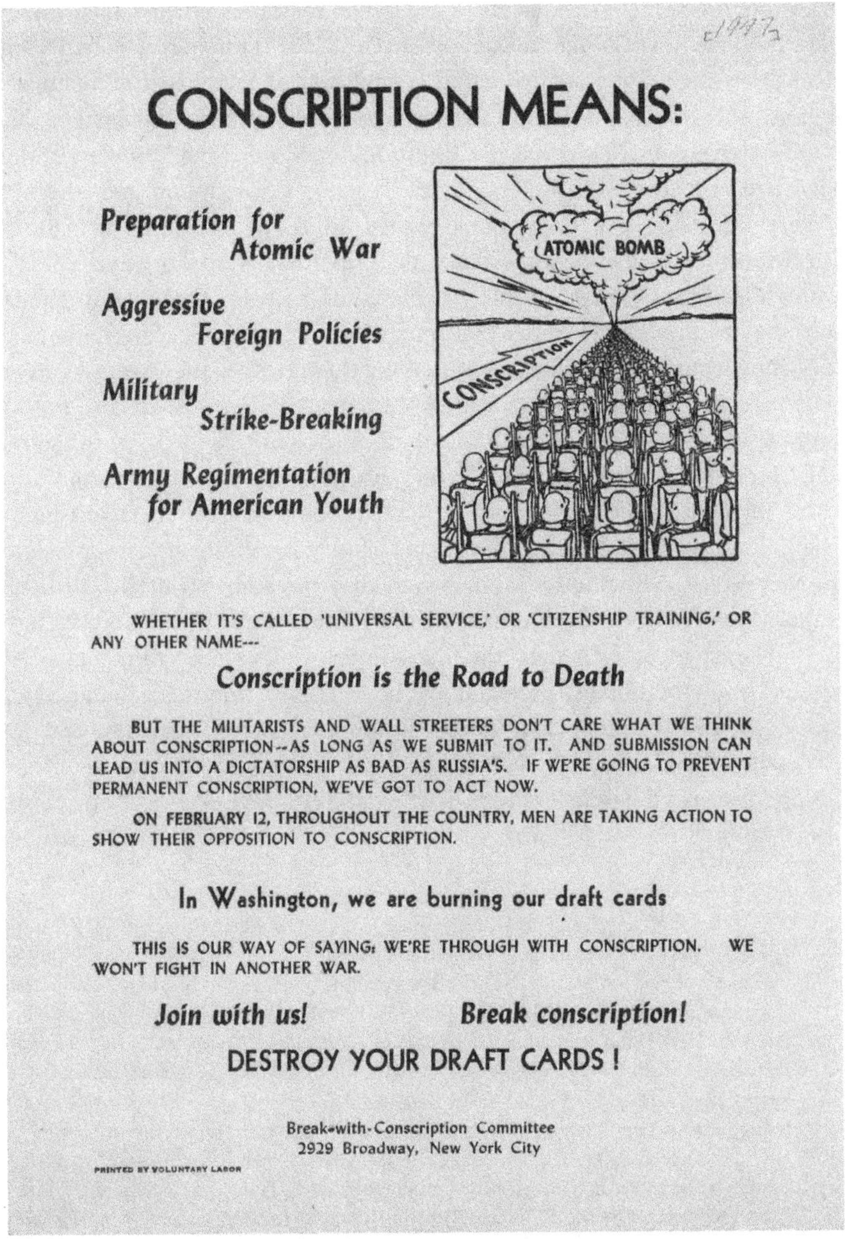

FIGURE 6.1 One of the publicity flyers for the nationally coordinated Break with Conscription demonstrations on February 12, 1947
COURTESY OF THE SWARTHMORE COLLEGE PEACE COLLECTION ARCHIVES

I resisters Julius Eichel and Scott Nearing; two members of the Union Eight, Dave Dellinger and George Houser; other World War II resisters who had been recently released from prison, including Bent Andresen and Bayard Rustin; as well as by Frank Olmstead and George Reeves of the War Resisters League, Donald Harrington, A. J. Muste, the author Milton Mayer, John Swomley, Jr., the new acting director of the National Council Against Conscription, and others.[64]

Three hundred seventeen people signed up in advance to participate in the civil disobedience. George Houser, the principal national organizer of the event, hoped that perhaps as many as five hundred people would participate and was delighted when more than one thousand people broke the law by publicly burning their draft cards or mailing them back to the government, at demonstrations on February 12 in thirty-seven states. Demonstrations and/or draft card burnings took place in major cities such as New York, Washington, DC, Chicago, Philadelphia, and Los Angeles, as well as in smaller cities and towns, including Wichita, Kansas; Flemington, New Jersey; Akron, Ohio; Champaign, Illinois; and Galesville, Wisconsin.[65] Today, historians underappreciate the 1947 Break With Conscription (when they remember it at all). In their authoritative history of the Vietnam era draft, Lawrence M. Baskir and William A. Strauss say incorrectly that "the first known" draft-card burning "occurred in 1964." In their otherwise excellent 1971 book *The Resistance*, Staughton Lynd and Michael Ferber report that "four or five hundred Americans in several cities destroyed their draft cards or mailed them to President Truman." Andrew E. Hunt also uses the latter figure in his biography of Dave Dellinger, one of the participants.[66]

---

64 "An Appeal for Action," *Politics* 4, no. 1 (January 1947): 31. Abraham Johannes (A. J.) Muste (1885–1967) was a minister, labor activist, and leading light of the pacifist wing of the US peace movement in the twentieth century. An early biography of Muste is Nat Hentoff, *Peace Agitator: The Story of A. J. Muste* (New York: Macmillan, 1963). A more recent biography is Leilah Danielson, *American Gandhi: A. J. Muste and the History of Radicalism in the Twentieth Century* (Philadelphia: University of Pennsylvania Press, 2014). See also A. J. Muste, *The Essays of A.J. Muste*, ed. Nat Hentoff (New York: Simon & Schuster, 1967).

65 George Houser, "Brief Report," March 5, 1947, Central Committee for Conscientious Objectors (CCCO) Records, RG 073, box 10, folder labeled "Break With Conscription Committee"; George Houser, "Break with Conscription Press Release," Collected Document Group–A (CDGA) Collection Box, folder labeled "Boston Alliance Against Registration and the Draft through Bristol, James, 'Break With Conscription Committee'"; Joseph W. Reilly, Report on February 12th Demonstration (Confidential—Not for Publication), CDGA, folder labeled "Boston Alliance Against Registration," all in SCPC.

66 Lawrence M. Baskir and William A. Strauss, *Chance and Circumstance: The Draft, the War, and the Vietnam Generation* (New York: Alfred A. Knopf, 1978), 65; Michael Ferber and Staughton Lynd, *The Resistance* (Boston: Beacon Press, 1971), 3; Andrew E. Hunt, *David Dellinger: The Life and Times of a Nonviolent Revolutionary* (New York: New York University Press, 2006), 89–90.

In New York, a demonstration was held at the Labor Temple, 242 East Fourteenth Street, where sixty-three men burned their draft cards. The men had planned to burn their draft cards inside the Labor Temple but were headed off by the fire marshal, who said that the indoor fire was not permitted. Thus, the well-behaved resisters—who meant to violate only federal draft laws, not local fire codes—tore up their draft cards indoors and later burned them outside. One of the speakers at the event, which was attended by 350 people, was Dwight Macdonald, then publisher of *Politics*, who said, "We have decided to attack conscription by the simplest and most direct way possible: that is, by refusing, as individuals, to recognize the authority of the State in this matter."[67]

In Washington, DC, twelve men burned their draft cards outside the White House. In a secret memorandum, which I obtained through a Freedom of Information Act request to the FBI, then-FBI director J. Edgar Hoover advised then-attorney general Tom Clark that "to try to stop this action would make martyrs out of those responsible, which is just what they want."[68] In Philadelphia, demonstrators marched past the Liberty Bell in Independence Hall and then—again in compliance with local fire codes—burned their draft cards outdoors. The FBI file on the Break With Conscription demonstrations also contains one of the letters sent to Truman, which said, "This is not only a protest against the present Selective Service Act but against any proposal for compulsory military training.... My draft card is a symbol of acquiescence to the conscriptive system when in my possession; I will not even passively submit to the act."[69]

After the 1948 conscription statute was enacted, the official attitude of the peace churches underwent a notable change in regard to out-and-out draft resistance. During World War II, the historic peace churches had been wary and skeptical of nonregistrants and other resisters. Instead, the peace churches worked closely with the Selective Service to administer the Civilian Public Service program for COs. According to some disgruntled CPS men, this involved complicity by the churches in carrying out improper Selective Service directives or policies. But in 1948, all three of the historic peace churches adopted formal, denominational positions supporting both conscientious objection (through cooperation with the new draft law) and resistance (noncooperation).

67 Houser, "Brief Report"; "Group Here Burns Many Draft Cards," *New York Times*, February 13, 1947. The *Times* reported, incorrectly, that they had burned forty-two draft cards. See also "Dwight Macdonald speech," reprinted in *Politics* 1, no. 3 (March–April): 54–55. For an account of this demonstration, see Jerry Elmer, *Felon for Peace* (Nashville: Vanderbilt University Press, 2005), 56–57, based on FBI files obtained by the author through FOIA request. For the War Resisters League, the Break With Conscription demonstration, and the immediate aftermath, see Bennett, *Radical Pacifism*, 178–80.

68 Houser, "Brief Report"; Elmer, *Felon for Peace*, 57.

69 Elmer, *Felon for Peace*, 57.

At a special "called meeting" of Friends in the United States in July, a decision was made "to support Young Friends and others who express their opposition to conscription either by nonregistration or by registration as conscientious objectors." Just to be clear, the statement added, "We warmly approve civil disobedience under Divine Compulsion as an honorable testimony fully in keeping with the history and practices of Friends." Similarly, the General Brotherhood Board of the Church of the Brethren gave support to both conscientious objection and noncooperation. The Mennonite Central Committee encouraged Mennonites to register as COs but also said that they would respect those whose loyalty to God led them to resist.[70] Both the Mennonites (through the MCC) and the Brethren (through the Brethren Service Committee) also provided bail for their young men who had been charged with Selective Service violations.[71]

During this period hundreds of men resisted the draft. Between 1948 and 1953, about one hundred men were prosecuted for refusing to register and another 250 were prosecuted for refusing induction.[72] More than a dozen of the imprisoned draft resisters were World War II military veterans (these men were exempt from being drafted but were not exempt from other obligations under the law such as registering and filling out forms), and more than two dozen had been World War II resisters or in CPS.[73] Eight members of the Nationalist Party of Puerto Rico, which refused to recognize US sovereignty over the island, refused to register, were prosecuted, and received sentences of one or two years. One of the eight Puerto Rican resisters was a World War II veteran.[74] Harold Flasker, another (draft-exempt) World War II veteran, refused to register and wrote to the attorney general to inform him of the fact.[75]

---

70 "Mennonites," *NSBRO Reporter* 6, no. 2 (November 1948): 3.

71 "Mennonites and Brethren Furnish Bail for COs," *NSBRO Reporter* 6, no. 7 (April 1949): 4.

72 "About 350 COs Arrested since Passage of 1948 Draft Law; Well over 200 Imprisoned; Majority of Sentences over Two Years," *NSBRO Reporter* 10, no. 9 (September 1953): 2–3. For an extended examination of two such cases, those of David McReynolds and Vern Davidson, see Scott H. Bennett, "Conscience, Comrades, and the Cold War: The Korean War Draft Resistance Cases of Socialist Pacifists David McReynolds and Vern Davidson," *Peace & Change* 38, no. 1 (January 2013): 83–120.

73 Minutes of NSBRO, March 30, 1949, Center on Conscience and War Records, pt. II, ser. A, box 1, folder labeled "Meeting Minutes, Board of Directors," SCPC.

74 Non-Registrants in Puerto Rico, August 3, 1940, CCCO Records, box 8, folder labeled "Special Releases 1949–1951," SCPC. The eight Puerto Rican nationalist draft resisters sentenced were Reinaldo Trilla Martinez, Luis Manuel O'Neill Rosario, Dario Berrios Cruz, Dimas Matos Nieves, Miguel Angel Ruiz Alicea, Rafael Cancel Miranda, Ramon Heraclio Medina Maisonave, and Jose Miguel Castillo Vega. Vega was the World War II veteran.

75 "Statement of Harold Flasker—Non-Registrant for the Draft, November 1948"; "Statement of Harold Flasker Non-Registrant Before Judge Leo F. Rayfiel, United States District Judge,"

Second prosecutions (after a resister had been released from prison after serving a first sentence) were not unusual. One such case involved four brothers, Joel, Orin, Paul, and Sid Doty, of Bruno, Minnesota. For the Dotys, draft resistance was something of a family tradition. Their father, William Doty, had been imprisoned during World War I for refusing induction. In April 1951, the four Doty brothers were prosecuted for refusing to register. Joel was sentenced to two years; Orin, Paul, and Sid each received eighteen-month sentences. If they had registered after their release, they should have been classified 4-F as convicted felons, but they again refused to register.[76] All were prosecuted a second time and were resentenced to two years each. Another man, Amos Brokaw, who served time during World War II as a CPS walkout, was prosecuted twice more under the new peacetime draft. His first sentence was 120 days; his second sentence was two years. Gilbert McFadden was prosecuted twice (first sentence, three years; second sentence, probation), as were the brothers David and Paul Seaver (first sentence, eighteen months; second sentence, one day).[77] Robert Michener (first sentence, one year and one day) was prosecuted a second time in February 1951 and sentenced to two consecutive five-year terms; his sentence was later changed to two concurrent five-year terms.[78]

In the first half-year after the 1948 conscription statute was in effect, more than a dozen people were prosecuted for counseling others to resist the draft. One such case involved Larry Gara, a history professor at Bluffton College, a small Mennonite school in Ohio, who had served a three-year sentence for publicly refusing to register during World War II. The chronology of the *Gara* case shows its complete absurdity.

December 16, 1948. Both documents are archived in the CCCO Records, box 8, "Special Releases 1949–1951" folder, SCPC.

76 Section 6(m) of the Conscription Statute precluded the drafting of felons. Section 1622.44 of the Selective Service regulations required classification of felons as 4-F.

77 "Brief on Second Prosecutions Presented to General Hershey," *NSBRO Reporter* 11, no. 4 (April 1954): 1–2. With regard to Brokaw, see Third Prosecution Prepared for Indiana Conscientious Objector, April 17, 1950, CCCO Records, box 8, "Special Releases 1949–1951" folder, SCPC. The Dotys argued that they were ineligible by reason of their earlier felony draft convictions but were unsuccessful. Doty v. United States, 218 F.2d 93 (8th Cir. 1955). Thomas Bouziden was another resister who was prosecuted twice and, in his second prosecution, unsuccessfully raised the fact that he was ineligible for the draft by reason of an earlier conviction. See *United States v. Bouziden*, 108 F. Supp. 395 (W.D. Okla. 1952), in which the defendant was acquitted on other grounds. See also *United States v. Palmer*, 223 F.2d 893 (3d Cir. 1955), which affirmed a second conviction.

78 "Brief on Second Prosecutions," *NSBRO Reporter* 11, no. 4 (April 1954): 1–2; "Conscientious Objector Gets 10-Year Sentence," February 27, 1951, CCCO Records, box 3, folder labeled "Form/Appeal Letters, 1950–1954," SCPC. See stapled "Special Note" in which the sentence was changed to five years on April 17, 1951.

In June 1948, when Truman signed the conscription statute into law, Charles Rickert, a student at Bluffton, decided that, for religious reasons, he could not register; Rickert informed his father of his decision. In September 1948, on registration day, Rickert followed through with his decision and did not register. When Rickert committed his so-called crime, he had not yet met Gara. The following month, in October 1948, Rickert learned that Gara was a faculty member and went to talk to him. Like most resisters would have done in a similar position, Gara gave young Rickert two pieces of advice. Gara counseled that Rickert could face serious consequences, including prison, for civil disobedience; Gara urged Rickert not to break the law unless he was thoroughly comfortable with the potential consequences. But Gara also told Rickert that if he had thought through the consequences, and was comfortable with them, he should not allow himself to be bullied by the government.

On November 8, Rickert was arrested for refusing to register. When that happened, Gara and his wife, Lenna Mae Gara, wrote to Attorney General Clark and to the local US attorney about the case: "We have recently learned of the arrest of Charles Rickert, a senior at Bluffton College who, as a conscientious objector to war and conscription, has refused to register for the draft. We have been very much concerned about the imprisonment of young war objectors.... We have openly urged young men to take this position and shall do all in our power to further the cause of civil disobedience to conscription in this country. If men like Charlie are guilty then we, too, are guilty." Note that the Garas' letter to the government did not say that they had counseled (nor even met) Rickert. Note, too, that the government would have had no idea of the Garas' views had they not written.

On January 4, 1949, Larry Gara (but not Lenna Mae) was indicted in Toledo for counseling Rickert to resist the draft. In March, Gara was tried and convicted. On May 6, 1949, US District Judge Frank Kloeb, saying that he believed Gara to be mentally unbalanced, sentenced him to eighteen months in prison. At the time, there was considerable concern among certain segments of the religious community about the Gara prosecution (see Figure 6.2). Gara had counseled a young man to stand by his religious principles; if this were a prosecutable criminal offense, many clergy might be guilty of the same crime. On July 24, 1949, in a show of support for Gara, four hundred prominent clergy issued a public statement of support for Gara. The signers included John C. Bennett, Reinhold Niebuhr, Elton Trueblood, Paul Tillich, and Henry Cadbury.[79]

79 For the *Gara* case discussed in the preceding paragraphs, see multiple statements, flyers, and press releases, all in CCCO Records, box 8, "Special Releases 1949–1951" folder, SCPC; "400 Assail Jailing of Draft Objector," *New York Times*, July 25, 1949; Bennett, *Radical Pacifism*, 179–80; and Larry Gara, "My War on War," in Gara and Gara, *A Few Small Candles*, 91–93.

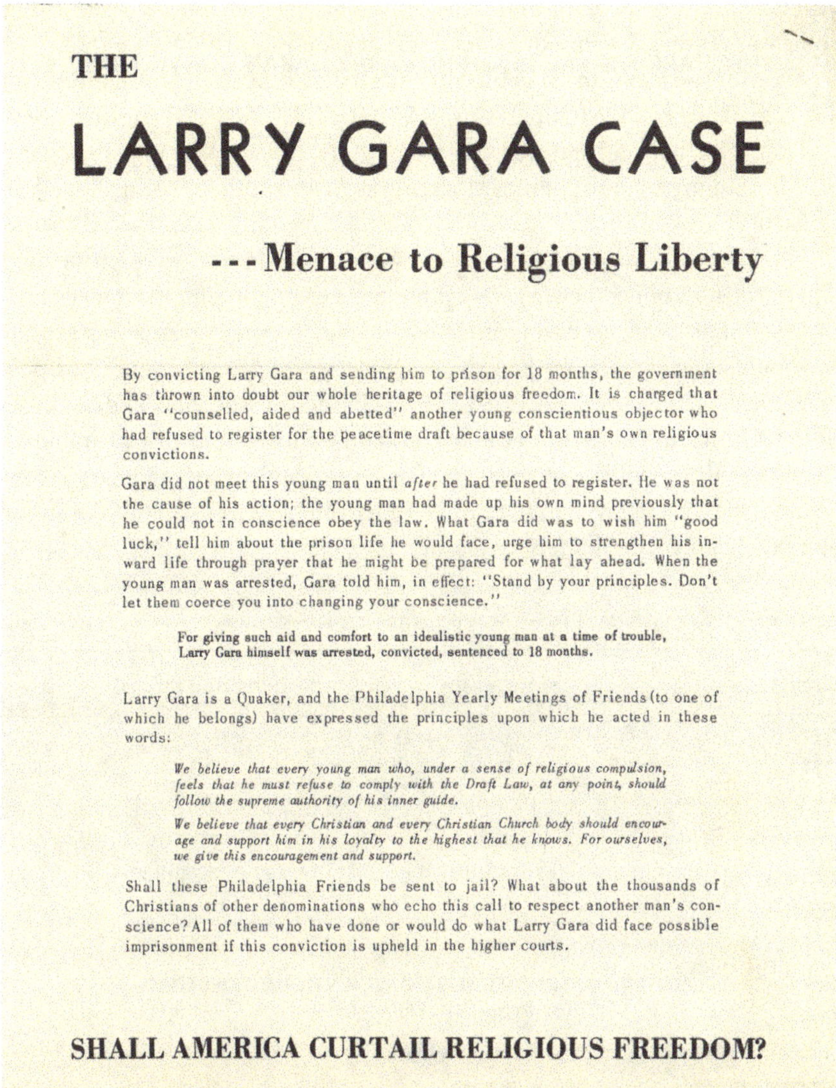

THE

LARRY GARA CASE

---Menace to Religious Liberty

By convicting Larry Gara and sending him to prison for 18 months, the government has thrown into doubt our whole heritage of religious freedom. It is charged that Gara "counselled, aided and abetted" another young conscientious objector who had refused to register for the peacetime draft because of that man's own religious convictions.

Gara did not meet this young man until *after* he had refused to register. He was not the cause of his action; the young man had made up his own mind previously that he could not in conscience obey the law. What Gara did was to wish him "good luck," tell him about the prison life he would face, urge him to strengthen his inward life through prayer that he might be prepared for what lay ahead. When the young man was arrested, Gara told him, in effect: "Stand by your principles. Don't let them coerce you into changing your conscience."

For giving such aid and comfort to an idealistic young man at a time of trouble, Larry Gara himself was arrested, convicted, sentenced to 18 months.

Larry Gara is a Quaker, and the Philadelphia Yearly Meetings of Friends (to one of which he belongs) have expressed the principles upon which he acted in these words:

We believe that every young man who, under a sense of religious compulsion, feels that he must refuse to comply with the Draft Law, at any point, should follow the supreme authority of his inner guide.

We believe that every Christian and every Christian Church body should encourage and support him in his loyalty to the highest that he knows. For ourselves, we give this encouragement and support.

Shall these Philadelphia Friends be sent to jail? What about the thousands of Christians of other denominations who echo this call to respect another man's conscience? All of them who have done or would do what Larry Gara did face possible imprisonment if this conviction is upheld in the higher courts.

SHALL AMERICA CURTAIL RELIGIOUS FREEDOM?

FIGURE 6.2 A flyer produced and distributed by the Central Committee for Conscientious Objectors to publicize the case of Larry Gara, a college professor prosecuted for counseling a student to resist the draft
COURTESY OF THE SWARTHMORE COLLEGE PEACE COLLECTION ARCHIVES

Gara appealed to the US Court of Appeals for the Sixth Circuit. That court affirmed the conviction and sentence, citing the World War I case of *Schenck v. United States* as the relevant draft-related precedent that supported the proposition that the government can send an antidraft activist to prison for espousing and expressing antidraft sentiments (see chapter 4). The circuit court was correct: *Schenck* did stand for that proposition.[80]

Another prosecution for counseling draft resistance involved a physician, Dr. Wirt Warren, in Wichita, Kansas. Warren had urged his stepson not to register. The lad had ignored his stepfather's advice and registered anyway. In January 1949 the senior Warren was prosecuted for counseling draft refusal.[81]

In addition to the hundreds of open draft resisters who were prosecuted, and the dozens of people who were prosecuted for counseling draft resistance, this period also saw an enormous level of quiet, covert evasion. This evasion was not classic civil disobedience, because it was not done openly and publicly. Nevertheless, the Selective Service referred tens of thousands of cases of quiet noncompliance with the law to the Department of Justice. Table 6.4 shows the total number of cases (by fiscal year) of noncompliance that the Selective Service referred to the Justice Department and breaks out the subset of cases dealing with induction refusal.

TABLE 6.4 Cases referred to the justice department for prosecution

Fiscal year	Total cases	Induction refusal
1952	20,072	12,542
1953	n/a	11,730
1954	17,600	7,500
1955	20,362	6,546
1956	10,430	3,572
1957	12,003	4,098
1958	13,848	4,156
1959	14,195	7,768
1960	9,708	2,920
Average (annual)	14,778	6,759

SOURCES: US SELECTIVE SERVICE SYSTEM, *ANNUAL REPORT OF THE DIRECTOR OF SELECTIVE SERVICE FOR THE FISCAL YEAR 19__ TO THE CONGRESS OF THE UNITED STATES PURSUANT TO THE UNIVERSAL MILITARY TRAINING AND SERVICE ACT AS AMENDED* (WASHINGTON, DC: GOVERNMENT PRINTING OFFICE), 1952, 37; 1953, 28; 1954, 30; 1955, 34; 1956, 32; 1957, 30; 1958, 25; 1959, 28.

80 Gara v. United States, 178 F.2d 38 (6th Cir. 1949).

81 "Arrests for Counseling Non-Registration," January 6, 1949, CCCO Records, box 8, "Special Releases 1949–1951" folder, SCPC.

Part of what makes these figures remarkable is that in nearly every case, referral to the Justice Department came only after both the Selective Service and the FBI had been unsuccessful in cajoling a man to cooperate. These figures, then, represent men who were fairly determined, even after being confronted by governmental authority, not to provide their cooperation with the requirements of the law.

8 National Service Board for Religious Objectors, the Central Committee for Conscientious Objectors, and the National Council against Conscription

During this period, two organizations worked to service the needs of men facing the draft: the National Service Board for Religious Objectors (NSBRO), which had existed since 1940, and the newly created Central Committee for Conscientious Objectors (CCCO). Another new organization, the National Council Against Conscription (NCAC), did not provide services to COs or resisters but fought proposals in Congress for selective service and UMT.

8.1 *National Service Board for Religious Objectors*

Between March 31, 1947, when the World War II draft ended, and June 24, 1948, when Truman signed the new 1948 conscription statute into law, NSBRO maintained a small staff to close out matters pertaining to the wartime CPS program and lobby Congress for broad CO provisions in the new law.[82] When Truman signed the new law in June 1948, the organization dramatically ramped up its efforts. During and following the Korean War, NSBRO provided information to men, especially members of the peace churches, about how to fill out the special Selective Service form for COs (Form 150); maintained an extensive literature service of materials pertaining to conscientious objection and Selective Service regulations; worked with a panel of attorneys knowledgeable about Selective Service matters who could advise both NSBRO and individuals on matters pertaining to the draft; advocated for fair wages for COs performing alternative service; visited resisters in federal prisons; and lobbied the Selective Service and the Department of Justice against second prosecutions of resisters coming out of prison.[83] After the law changed in 1951 to require that COs

82 "Between the Drafts," *NSBRO Reporter* 6, no. 10 (October 1948): 2.

83 "Dear Reporter Reader," mimeograph inserted in *NSBRO Reporter* 10, no. 7 (July 23, 1953); "How to Fill in Form 150 for Conscientious Objectors," *NSBRO Reporter* 7, no. 10 (November–December 1950): 1, 3–4, including a sample form 150; "Literature on

classified I-O perform alternative service, NSBRO maintained a job referral service for these men to help them find suitable positions. The NSBRO board was made up mainly of men from the historic peace churches, and a large percentage of the young men with whom NSBRO worked were Quakers, Mennonites, or Brethren.[84]

Throughout this period, NSBRO maintained the same close and cordial relations with the Selective Service and its personnel that had been so controversial during World War II. NSBRO staff met frequently with Hershey and other Selective Service officials.[85] In fact, on September 21, 1956, Hershey spoke at a NSBRO luncheon and praised NSBRO's "special contribution to the Nation's health, safety, and interest." The organization was so pleased with the general's praise that it touted the honor on the front page of the *NSBRO Reporter*, its newsletter; the article included a picture of General Hershey with NSBRO board members (see Figure 6.3).[86] While NSBRO was willing to advise all men, including resisters, who had conscientious scruples about war and conscription, a broad segment of that population clearly did not believe itself to be well served by NSBRO.

Conscientious Objectors Available from National Service Board," *NSBRO Reporter* 9, no. 2 (April 16, 1951): 3; "Robert B. Myers, Attorney, Serving NSB Cases," *NSBRO Reporter* 8, no. 6 (September–October 1951): 1; "Memo on Wages, CPS Service, Legal Procedures, Presented to Officials," *NSBRO Reporter* 10, no. 1 (January 1953): 1; "Prison Visitation Policy of World War II Continued by NSBRO Directors," *NSBRO Reporter* 10, no. 2 (February 1953): 1; "Brief on Second Prosecutions Presented to General Hershey," *NSBRO Reporter* 11, no. 4 (April 1954): 1.

84 "Board of Directors and Guests Plan Job Referral Service for COs," *NSBRO Reporter* 8, no. 7 (November–December 1951): 5; "Job Referral Services for COs Set Up by NSB and Constituent Agencies," *NSBRO Reporter* 9, no. 2 (March–April 1952): 5; "Information on Job Opportunities for CO's in Many States Now Available," *NSBRO Reporter* 9, no. 10 (December 1952): 2–4; "Job Opportunities for C.O.s," *NSBRO Reporter* 10, no. 10 (October 1953): 2.

85 "Brief on Second Prosecutions Presented to General Hershey," *NSBRO Reporter* 11, no 4 (April 1954): 1; "NSBRO Constituents Interview Presidential Appeal Board, Justice Department, Pentagon Officials," *NSBRO Reporter* 11, no. 11 (November 1954): 1; "NSBRO Convenes for Annual Meeting—General Hershey Addresses Luncheon," *NSBRO Reporter* 19, no. 11 (December 1962): 1; "Confidential Memorandum on Informal Conversation with General Hershey, General Renfrow, and Col. Kosch," by NSBRO's Orie Miller, Ora Huston, and others about their chat with military men on June 7, 1951; "Dear Board of Directors," December 17, 1951, a confidential letter describing a September 24, 1951, meeting; "Memo to the Files," March 4, 1953, describing a meeting of multiple NSBRO officers, including Orie Miller, Harold Row, Stauffer Curry, and others, with Hershey; a confidential letter dated May 28, 1954, describing a meeting between NSBRO and Hershey. All these documents are archived in Center on Conscience and War Records, pt. II, ser. A, box 16, folder labeled "Reports of Interviews and Meetings with Selective Service Officials," SCPC.

86 "NSBRO Given Special Tribute by Two Top Gov't Officials," *NSBRO Reporter* 13, no. 10 (October 1956): 1.

Left to right: Dr. C. N. Hostetter, President of Messiah College; General Lewis B. Hershey, Director of Selective Service; Dr. Arthur S. Flemming, Director of the Department of Defense Mobilization; Dr. W. Harold Rowe, Executive Secretary, Brethren Service Commission.

FIGURE 6.3 This photo of General Lewis B. Hershey, the Selective Service director (second from left), with NSBRO board members, September 21, 1956, at a luncheon at which the Selective Service thanked NSBRO for its "splendid job" of cooperating with the Selective Service.
COURTESY OF THE SWARTHMORE COLLEGE PEACE COLLECTION ARCHIVES

8.2 *The Central Committee for Conscientious Objectors*

The perception that NSBRO and the Selective Service were too cozy led directly to the creation of the Central Committee for Conscientious Objectors on August 29, 1948. CCCO's first press release explained that the organization had been created for "assisting those men who may violate the law by refusing to register for the draft." The organizing committee for the new entity included Albon Man and Abe Kaufman of the War Resisters League (WRL); A. J. Muste of the Fellowship of Reconciliation; and Mildred Olmstead of the Women's International League for Peace and Freedom.

The CCCO's initial press release also addressed the obvious question of the new entity's relationship to NSBRO and described a cooperative relationship, with the two groups having complementary, yet distinct, functions: "While promising to work closely with the already existing National Service Board for Religious Objectors in order to prevent duplication in function, the new

Central Committee feels that the NSBRO cannot adequately represent" resisters. NSBRO, for its part recognized the same distinction; NSBRO was for COs who cooperated with the law; the CCCO was for the noncooperators. In an article in the *NSBRO Reporter*, the organization explained: "The jurisdictional line between [NSBRO] and [CCCO] appears to be drawn on the differences in attitude toward the draft. [NSBRO] will assist groups and individuals who want recognition of their conscientious objection under the law through administrative procedures.... The CCCO will assist objectors who refuse to cooperate with the law, who do not qualify as conscientious objectors under the law and who refuse to cooperate [with] certain existing and prevailing practices."[87]

The distinction, at least at first, was accurate. The NSBRO board was composed of elders of the peace churches who were alumni of the wartime CPS program. In contrast, the executive committee of CCCO was heavily weighted with draft noncooperators from World War I (Julius Eichel) and World War II (Bent Andresen, Jim Bristol, Ralph DiGia, Albon Man, Dave Dellinger, and Igal Roodenko), as well as other radicals like A. J. Muste. (The CCCO board also met every other month at the WRL office in New York.)[88] One early CCCO publication was entitled *The Non-Cooperator and the Draft*. This twelve-page document went through sections headed "Some Reasons for Non-Cooperation," "Consequences of Civil Disobedience," the risk of "Repeated and Multiple Prosecutions," and "Prison Life." It contained brief quotations and explanatory statements from some World War II resisters like Dave Dellinger and Wally Nelson and many statements from contemporary noncooperators. Another publication, *Positions of Conscientious Objectors*, described in even-handed terms the arguments in favor of entering the army as a noncombatant (Selective Service classification I-A-O), doing alternative civilian service (as a I-O), and resistance. Each section was written by a man who had taken that particular position.[89] These were not publications that could have come from NSBRO at the time.

87 CCCO press release, CCCO Records, box 3, folder labeled "Publicity: Press Releases," SCPC; "N.S.B. Continues," *NSBRO Reporter* 6, no. 1 (October 1948): 3.

88 Minutes of Meeting of Organizing Committee, CCCO, August 17, 1948; Meeting of the Executive Committee, CCCO, October 7, 1948; Minutes of Meeting, CCCO, November 1, 1948, all in CCCO Records, box 1, folder labeled "Meeting Minutes, 1948," SCPC.

89 *The Non-Cooperator and the Draft* and *Positions of Conscientious Objectors* [November 1955], both in CCCO Records, box 3, "Publicity: Press Releases" folder. Note that the term *conscientious objector* continues to be used even for men who were draft resisters. Indeed, that an organization explicitly founded to help draft *resisters* called itself the Central Committee for *Conscientious Objectors* is telling.

Like NSBRO, the Central Committee for Conscientious Objectors maintained an extensive list of publications related to conscription, visited resisters in prison, and had a panel of attorneys available to both draft counselors and CCCO staff.[90] Unlike NSBRO, the CCCO worked extensively with local draft counselors and counseling agencies across the country, providing them with materials for advising young men about the draft. Its publication, *CCCO News Notes* provided up-to-date information about important changes in Selective Service regulations and procedures pertaining to COs.[91] The CCCO was at the forefront of publicizing the Larry Gara and Wirt Warren cases, and it organized the public statement by four hundred religious leaders in support of Gara.[92]

Starting in 1952, the CCCO produced the *Handbook for Conscientious Objectors*. The manual provided detailed, comprehensive, legally accurate information about applying for CO status and engaging in resistance. The handbook soon became the Bible for men considering conscientious objection and/or resistance. It clearly filled an important need; the first printing of eight thousand copies sold out quickly.[93] The CCCO updated the handbook with new editions (and new printings) as required by changes in the draft law, and thousands of copies sold throughout the 1950s and into the 1970s.[94]

While the distinction between NSBRO and the CCCO was clear in 1948 (NSBRO served COs who obeyed the law; the CCCO served resisters who violated the law), the distinction had largely disappeared by 1950. NSBRO became gradually more willing to work with resisters, and the CCCO got heavily involved with COs and draft counseling (within the constraints of the law). Each organization sent representatives to sit in on meetings of the other, the

90 "Materials for Counseling Conscientious Objectors," including Basic Materials, Periodicals, Pacifist Materials, and Supplementary Literature List and materials specifically for draft counselors, as well as other lists of general materials, materials for COs, information on deferments and exemptions, and emigration, all in box 3, folder labeled "Literature Lists, Bibliography." For multiple flyers and press releases regarding visits to imprisoned resisters, see box 1, folder labeled "Series A, Organization, Flyers, History, and Purpose"; for multiple letters to CCCO panel of attorneys and CO counseling agencies, see box 3, folder labeled "Literature Lists, Bibliography," all in CCCO Records, SCPC.

91 "CCCO: An Agency for Military and Draft Counseling," CCCO Records, box 1, "Series A, Organization, Flyers, History, and Purpose" folder.

92 Multiple flyers, statements, and press releases in CCCO Records, "Special Releases 1949–1951" folder.

93 Minutes of the Executive Committee Meeting, January 26, 1951, and December 12, 1952, both in CCCO Records, box 1, folder labeled "Meeting Minutes, 1950–1952."

94 "Dear Friend," November 1958, CCCO Records, box 3, folder labeled "Appeal Letters 1955–1959." This letter appealing for donations was signed by co-chairs A. J. Muste and Lyle Tatum.

staffs of both organizations kept in touch with each other, and coordinating their efforts was the rule. This arrangement spoke well of the people associated with both organizations, for, given the similarity in functions, each could have viewed the other as a rival or competitor.[95]

In the early 1960s, the CCCO took on a trio of lower-court cases that squarely questioned the constitutionality of the statutory requirement that a man must believe in a Supreme Being to qualify as a CO. All three men—Daniel Seeger, Arno Jakobson, and Britt Peter—claimed to be COs even though they did not believe in a god. All three men were denied CO status by their draft boards. All three refused induction, were tried, and were convicted. All three, represented by CCCO lawyers, appealed. Seeger and Jakobson won their appeals in the US Circuit Courts of Appeal; Peter lost. The losing sides in all three cases appealed to the US Supreme Court, which took the cases and consolidated them into a single case (since the issue presented in each case was the same). In 1965 those three consolidated cases became the most important draft case of the Vietnam era—and perhaps of all time (see chapter 7).[96]

8.3 *The National Council against Conscription*

The National Council Against Conscription came into being on November 9, 1945, at a meeting in Philadelphia. The thirty-one people in attendance adopted a statement of purpose for the new organization that reflected the widely held and highly idealized hopes for disarmament and peace that constituted one part of the American zeitgeist in the immediate aftermath of World War II: "In the name of God and humanity there must never be another war. Every nation seeks security. Our only hope of security is the same as that of every nation—namely, total peace together. Our world demands one cooperative

---

95 See, for example, "Relationship with Other Organizations," in Minutes of Executive Committee, September 25, 1950, 2, and "NSBRO Report," in Minutes of Executive Committee, March 19, 1954, 1, both in CCCO Records, box 1, "Meeting Minutes, 1950–1952" folder. See also a "Dear Reporter Reader," mimeograph dated July 22, 1950; it reports that, with the onset of the Korean War, NSBRO would provide "advisory services to all COs" and is archived in an SCPC box labeled "The Reporter—1950." See as well "COs Testify through Non-cooperation," *NSBRO Reporter* 14, no. 9 (September 1957): 1–2, and Minutes of Directors of NSBRO and CCCO, March 30, 1949, reporting that staff of both organizations had been attending each other's meetings to facilitate communication, both in a collection of some *NSBRO Reporter* issues and other materials archived in the Center on Conscience and War Records, SCPC.

96 Minutes of the Executive Committee, CCCO, February 3, 1961, CCCO Records, box 1, folder labeled "Meeting Minutes, 1961–1962"; United States v. Seeger, 326 F.2d 864 (2d Cir. 1964); United States v. Jakobson, 325 F.2d 409 (2d Cir. 1963); Peter v. United States, 324 F.2d 173 (9th Cir. 1963).

endeavor based on international morality.... In as much as we covet for our nation such moral leadership we oppose conscriptive universal military training in peacetime."[97]

At a second meeting, on December 13, administrative details for the newly created organization were worked out: the founders formed an executive group to develop policy; decided to solicit names for a list of "more prominent and influential citizens who will lend their names and influence but who are not expected to give time"; and agreed to hire a salaried staff. The NCAC would seek to defeat proposals in Congress for selective service and UMT by issuing public statements and holding public meetings. In particular, the NCAC saw its role as working with local anticonscription committees at the city and state levels, of which there were about eighty nationwide at the time. They also decided that a separate group, the National Council Against Peacetime Conscription Now, would be merged into the NCAC.[98]

During 1946, the National Council had temporary staff. In December 1946, John Swomley, Jr., who had been working for the Fellowship of Reconciliation, was hired as the acting director of NCAC at the suggestion of A. J. Muste. In 1947, Swomley became the permanent director, a position he held until the organization dissolved at the end of 1959. At FOR, Swomley had been producing a twice-monthly, four-page typescript newsletter, *Conscription News*, that he brought with him to the NCAC. By 1947, *Conscription News* had six thousand subscribers nationwide.[99]

In almost everything it did, NCAC filled niches and carried out work that no other organization was doing. As envisioned at its founding meeting, the NCAC consistently organized statements and publicity by "more prominent and influential citizens" who were opposed to conscription. Its letterhead and brochures carried long lists of these names, including those of Professor Zechariah Chafee of Harvard Law School; William Faulkner; Harry Emerson Fosdick, a prominent liberal minister; Mrs. Allan Knight Chalmers, a civil

---

97 "Proceedings leading to the organization of the National Council Against Conscription," November 9, 1945, National Council Against Conscription (NCAC) Records, RG 052, box 1, folder labeled "History and Purpose," SCPC.

98 Minutes of the Meeting of the Executive Group, NCAC, December 13, 1945, NCAC Records, box 1, "History and Purpose" folder, SCPC.

99 For NCAC's leadership in this period, see NCAC Records, 1944–60, box 1, "History and Purpose" folder, esp. "Historical Introduction"; Minutes of the Meeting of the Executive Group, NCAC, December 13, 1945; Minutes of Meeting, Russian Inn, Philadelphia, February 5, 1946; Minutes of Meeting, Broadway Tabernacle Church, December 20, 1946; and Minutes of the Meeting of the NCAC, October 29, 1947.

rights activist who was married to a prominent clergyman; Colonel Roscoe S. Conkling, the noted Black journalist and acclaimed orator, and John Dewey.

On January 18, 1947, while Congress was debating UMT, the NCAC released a public statement against militarism and UMT signed by Albert Einstein; Ray Lyman Wilbur, a former cabinet secretary and Stanford University president; Robert Wilson, former president of the Federation of American Scientists; and other prominent people.[100] On June 4, 1947, the NCAC released a petition signed by 652 clergy and religious leaders calling on Truman and Congress to defeat any proposal for UMT or reconstituting the Selective Service System.[101]

On July 6, three months after Truman's advisory commission issued its report endorsing UMT, the NCAC released its own twenty-eight-page analysis rejecting the conclusions of the advisory commission point by point and concluding that UMT would "drastically change the character of our democracy and virtually destroy the chances of preventing a third world war." Signing the document were Josephus Daniels, US secretary of the navy during World War I; Senator Edwin Johnson (D-CO); Robert Hutchins, chancellor of the University of Chicago; and other luminaries. The Associated Press and Scripps Howard newspapers picked up the story, so articles about the NCAC analysis appeared across the country. The *New York Times* ran its own story about the statement and listed many of the prominent signers—but failed to mention that the NCAC had organized and published the analysis.[102]

On March 22, 1948, less than a week after Truman's address to a joint session of Congress calling for both UMT and reestablishing the Selective Service, the NCAC released a statement rejecting conscription that was signed by prominent educators and clergy. "The President, in his message to Congress last week, failed to make a case for universal military training or for Selective Service," the statement said.[103]

Two weeks after the statement by educators, the NCAC released a statement on April 4 signed by twenty-six prominent historians who declared that conscription would bring "neither peace nor security." A week later, on April 11, the organization announced it was sending a letter opposing conscription to every senator; the letter was signed by Albert Einstein and eight religious leaders. On May 9, the NCAC released an antidraft statement by fifteen prominent women,

100 "Press Release: Public Statement against Militarism, January 18, 1947," NCAC Records, box 5, folder labeled "Publicity/Press Releases—1946–1948".

101 "Petition Opposes Military Training," *New York Times*, June 5, 1947.

102 "For release, July 7, 1947," NCAC Records, box 5, folder labeled "Publicity/Press Releases—1946–1948"; "Military Training Is Held Outmoded," *New York Times*, July 7, 1947.

103 "NCAC News Release," March 22, 1948, NCAC Records, box 5, folder labeled "Publicity/Press Releases—1946–1948".

including the author Pearl Buck and the Nobel laureate Emily Greene Balch; on May 16, it released an antidraft statement signed by eighty-three college presidents.[104]

The enactment of conscription in June 1948 did nothing to slow the flow of statements from the NCAC. In a press release dated August 8, 1948, the NCAC said, "More than a hundred clergymen, including some of the nation's leading Protestant ministers, today called upon the churches and people of America to observe Sunday, August 22—a week before draft registration begins—as a Day of Mourning and Repentance." On February 13, 1949, it released a sixty-four-page report, *New Evidence of the Militarism of America*, in which it denounced the Selective Service, which, the report said, could not have been re-created by law without "an inspired war scare." The report bore the names of Albert Einstein, Pearl Buck, Ray Lyman Wilbur, Victor Reuther, and others. Once again, the statement received press coverage across the country, including in the *New York Times* and *Herald Tribune*.[105]

The NCAC kept at it throughout the 1950s. For example, on January 25, 1959, it issued a statement signed by "58 prominent Americans, including former Senator Edwin C. Johnson, Brig. Gen. Hugh B. Hester, Harold Case, President of Boston University, Rev. Harry Emerson Fosdick… . call[ing] upon Congress to stop conscripting Americans. We have had conscription long enough."[106]

Another aspect of the NCAC's work was coordinating and supporting the efforts of scores of small, grassroots anticonscription organizations at the state and municipal levels. Some of these groups, including those in Philadelphia, Los Angeles, New York, and Denver, had large memberships and produced fancy printed letterhead and flyers with their own local address. Other local groups, including the state organizations in Minnesota, Illinois, and Ohio, were smaller, less sophisticated, and produced mimeographed (or even dittoed) flyers. The Ohio group was run by Gene Sharp, who would later become among the world's most influential advocates of active nonviolence. Some smaller

---

104 NCAC news releases, April 5 and April 12, 1948, both in NCAC Records, box 5, folder labeled "Publicity/Press Releases—1946–1948," SCPC; "Press Releases, 1948–1949," NCAC Records, box 5.

105 "NCAC News Release," August 8, 1948, NCAC Records, box 5, folder labeled "Publicity/Press Releases—1946–1948," SCPC; "New Evidence of the Militarism of America," February 13, 1949, NCAC Records, box 5, folder labeled "Publicity/Press Releases—1946–1948"; "Foes of Draft Warn of Militarization," *New York Times*, February 14, 1949.

106 "News Release," January 25, 1959, NCAC Records, box 5, folder labeled "Press Releases—1950–1959."

antidraft groups affiliated with the NCAC were based on a single college campus.[107]

The NCAC turned out a never-ending stream of mimeographed (and sometimes printed) flyers, brochures, and pamphlets with titles like *Now Is the Time for Action: UMT Hearings Announced*; *Redouble Your Efforts*; and *There Is Still Time: Act Now before the Draft Comes Up for a Vote*. Figures 6.4 and 6.5 are examples. These publications were designed to aid the work of the local affiliates. Many NCAC flyers featured pleas to write or visit members of Congress, with headings like: "Will You Write Your Own Congressman?" "Visits to Congressman Imperative!" and "Telephone Today!" One two-sided mimeographed sheet was headed *No Draft Is Needed: Summary of Fifteen Arguments against Draft Extension*. A mimeographed sheet entitled *What You Can Do to Oppose Conscription* urged readers to be informed, write letters to the editors of newspapers and magazines, and write to members of Congress. The NCAC office in Washington sent out frequent updates on the progress of specific UMT and Selective Service System bills as the Senate and House held hearings and both houses scheduled floor votes. Some of these updates included remarkably detailed analyses of congressional processes, along with sophisticated head counts showing how various senators and representatives were expected to vote.[108]

The NCAC performed a unique and valuable function for the broader anticonscription movement with this work, for no other organization had so extensive a grassroots network or produced such frequent and detailed materials for local groups. The War Resisters League, Fellowship of Reconciliation, and Women's International League for Peace and Freedom had neither the national office staff nor the grassroots network that the NCAC had (and WRL, at least, was too radical to undertake extensive efforts aimed at Congress). NSBRO was heavily geared toward serving young men who were members of the peace churches who had issues pertaining to CO status and with finding suitable alternative civilian service. The CCCO worked with resisters and, later, with COs and had a network of draft counselors, but it had nothing like the NCAC's network of groups and individuals that would lobby Congress.

---

107 Multiple flyers and brochures, NCAC Records, box 9, folders labeled "Colorado Council Against Conscription," "Committee to Oppose Conscription," and "Greater Philadelphia Council Against Conscription." For an obituary of Gene Sharp, see "Gene Sharp, Global Guru of Nonviolent Resistance, Dies and 90," *New York Times*, Feb. 2, 2018.

108 Multiple flyers in NCAC Records, box 5a, ser. C, folder labeled "Calls to Action/Issue Statements 1945–1949."

Peacetime Conscription —the Way to War!

Powerful forces are now at work to saddle America with a permanent program of militarism - conscription of all 18-year-old boys, and the largest peacetime military budget in history!

Such a program cannot fail to arouse fear and suspicion in other nations. It would doom the United Nations to failure. It would intensify the international arms race that has already begun, and would inevitably lead to World War III which may well be the end of modern civilization.

Peacetime conscription (universal military training) is the very core of the campaign to militarize America. But the facts of history show that conscription defeats its own ends. Here is the record:

1. Conscription undermines national health!

   The frequency of venereal disease, mental disease, and suicide in the Army, even in peacetime, is several times greater than among the civilian population.

2. Conscription militarizes education!

   The Army's latest compulsary training plan would force almost every college to accept an Army unit in order to get male students.

3. Conscription threatens national security!

   It spurs other nations to arm against us, and lessens the chances of success in the settlement of disputes by peaceful means.

4. Conscription menaces democracy!

   No army can be democratic. Youth are regimented and indoctinated with the ideal of force and violence.

5. Conscription is wasteful!

   Conscription would cost more each year than America spends on its entire system of public education!

YOUR HELP IS NEEDED in this fight to save Democracy in America!

Here is what you can do:

1. Write letters to your representatives in the government asking them to vote against peacetime conscription. Here are some names and addresses

The Honorable Arthur Vandenberg Senate Office Building Washington, D.C. My dear Senator Vandenberg: May I urge you to do all in your power to oppose peacetime military conscription for the following reasons:	The Honorable Homer Ferguson Senate Office Building Washinton, D.C. My dear Senator Ferguson:
The Honorable Chan Gurney, Chairman of the Senate Armed Forces Committee, Washington, D.C. My dear Senator Gurney:	The Honorable Jesse Wolcott House of Representatives Office Building Washington, D.C. My dear Mr. Wolcott: The Honorable Walter G.Andrews Chairman of the House Armed Services Committee, Washington.

2. Send for a free copy of "Ask the Man Who Owned One" from National Council Against Conscription, 1013 - 18th St., N.W. Washington 6, D.C.

3. Get more copies of this leaflet for distribution among your friends!

FIGURE 6.4 Flyers by the National Council Against Conscription urging Americans to voice their opposition to conscription during the Cold War
COURTESY OF THE SWARTHMORE COLLEGE PEACE COLLECTION ARCHIVES

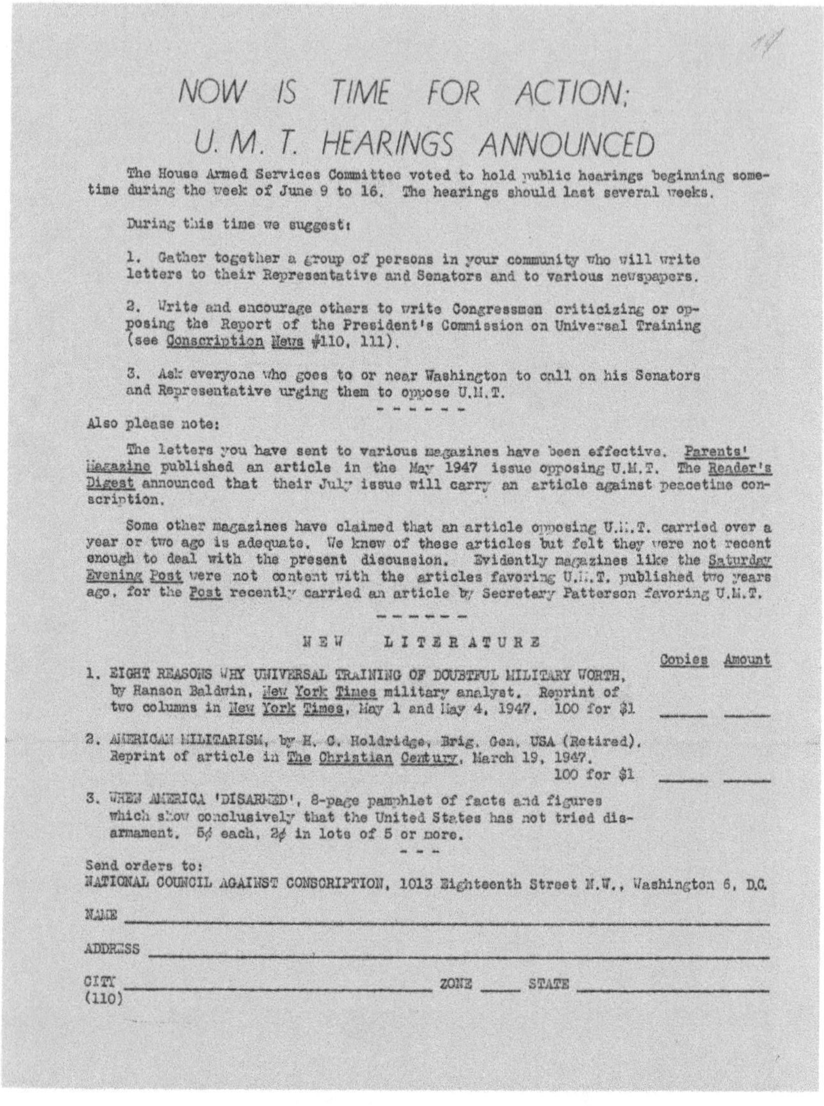

NOW IS TIME FOR ACTION;

U. M. T. HEARINGS ANNOUNCED

The House Armed Services Committee voted to hold public hearings beginning sometime during the week of June 9 to 16. The hearings should last several weeks.

During this time we suggest:

1. Gather together a group of persons in your community who will write letters to their Representative and Senators and to various newspapers.

2. Write and encourage others to write Congressmen criticizing or opposing the Report of the President's Commission on Universal Training (see Conscription News #110, 111).

3. Ask everyone who goes to or near Washington to call on his Senators and Representative urging them to oppose U.M.T.

\- - - - - -

Also please note:

The letters you have sent to various magazines have been effective. Parents' Magazine published an article in the May 1947 issue opposing U.M.T. The Reader's Digest announced that their July issue will carry an article against peacetime conscription.

Some other magazines have claimed that an article opposing U.M.T. carried over a year or two ago is adequate. We knew of these articles but felt they were not recent enough to deal with the present discussion. Evidently magazines like the Saturday Evening Post were not content with the articles favoring U.M.T. published two years ago, for the Post recently carried an article by Secretary Patterson favoring U.M.T.

\- - - - - -

N E W L I T E R A T U R E

	Copies	Amount
1. EIGHT REASONS WHY UNIVERSAL TRAINING OF DOUBTFUL MILITARY WORTH, by Hanson Baldwin, New York Times military analyst. Reprint of two columns in New York Times, May 1 and May 4, 1947. 100 for $1	____	____
2. AMERICAN MILITARISM, by H. C. Holdridge, Brig. Gen. USA (Retired). Reprint of article in The Christian Century, March 19, 1947. 100 for $1	____	____
3. WHEN AMERICA 'DISARMED', 8-page pamphlet of facts and figures which show conclusively that the United States has not tried disarmament. 5¢ each, 2¢ in lots of 5 or more.		

\- - -

Send orders to:
NATIONAL COUNCIL AGAINST CONSCRIPTION, 1013 Eighteenth Street N.W., Washington 6, D.C.

NAME ____________________

ADDRESS ____________________

CITY ____________________ ZONE ____ STATE ____________

(110)

FIGURE 6.5 Flyers by the National Council Against Conscription urging Americans to voice their opposition to conscription during the Cold War
COURTESY OF THE SWARTHMORE COLLEGE PEACE COLLECTION ARCHIVES

9 The 1948 Statute and Segregation in the Military

A. Philip Randolph was president of the Sleeping Car Porters Union (a member of the American Federation of Labor), but when he testified in the Senate about the Selective Service bill on March 31, 1948, he did so on behalf of a different organization (he worked with several), the Committee Against

Jimcrow in Military Service and Training. Randolph promised that if segregation were not banned in conscription, there would be mass civil disobedience: "Passage now of a Jim Crow draft may only result in a mass civil disobedience movement along the lines of the magnificent struggles of the people of India against British imperialism.... In resorting to the principles and direct-action techniques of Gandhi ... Negroes will be serving a higher law."[109]

Senator Wayne Morse (R-OR), who two decades later would become known as a vocal opponent of the Vietnam War, was so outraged that he told Randolph that in wartime such statements would constitute treason.[110] The *New York Times* ran the story the next morning on the front page under the headline "Congress Told UMT Racial Bars Would Unleash Civil Disobedience," but the *Times* seemed to treat the exchange between Randolph and Morse as merely sensational, nothing more. Nevertheless, the *Times* quoted no fewer than four paragraphs of Randolph's testimony, including the following: "I personally pledge myself to openly counsel, aid and abet youth, both white and Negro, to quarantine any Jim Crow conscription system, whether it bear the label of UMT or Selective Service.... From coast to coast in my travels I shall call upon Negro veterans to join this civil rights movement and to recruit their younger brothers in an organized refusal to register and be drafted."[111]

Like the *Times*, the Senate seemed unaware of what it was dealing with. Perhaps in an effort to tamp down the controversy, the Senate committee requested testimony from Truman Gibson, the only Black member of Truman's Commission on UMT, who predictably gave unqualified support to UMT. But if procuring Gibson's testimony had been meant to counter the controversy Randolph's testimony had sparked, the effort failed. Two days later, Representative Adam Clayton Powell, Jr. (D-NY), wrote to the Senate Armed Services Committee denouncing Gibson as an Uncle Tom, and saying, "There aren't enough jails in America to hold the Negro people who will refuse to bear arms in a Jim Crow army."[112]

---

109 C. P. Trussell, "Congress Told UMT Racial Bars Would Unleash Civil Disobedience," *New York Times*, April 1, 1948.

110 Morse and Ernest Gruening (D-AL) were the only senators to vote against the Gulf of Tonkin Resolution on August 7, 1964. No one in the House voted against the resolution. In 1948 Morse was a Republican, although he had become a Democrat by 1964.

111 Trussell, "Congress Told UMT Racial Bars"; *Senate Hearings on Universal Military Training*, statement of A. Philip Randolph, national treasurer of the Committee Against Jimcrow, March 31, 1948, 685–94.

112 *Senate Hearings on Universal Military Training*, statement by Truman Gibson, March 30, 1948, 642–47; Adam Clayton Powell, Jr., to Senate Armed Services Committee, April 1, 1948, 964.

What the *New York Times* (and perhaps Congress) missed is that Randolph's testimony was just one part of a carefully planned, well-orchestrated, national campaign. The previous autumn, on October 10, 1947, a group of civil rights activists meeting at Randolph's office in Manhattan formed the Committee Against Jimcrow in Military Service and Training.[113] Grant Reynolds, a former army chaplain and later attorney, was named chairman; Randolph was named treasurer. A two-person legal committee was established, consisting of Charles Hamilton Houston, who preceded Thurgood Marshall as the general counsel of the NAACP, and the illustrious civil rights lawyer Belford Lawson, Jr.[114] On November 23, 1947, the Committee Against Jimcrow went public with a press release announcing that it had a number of prominent members, including the world heavyweight boxing champion, Joe Louis.[115]

On March 22, 1948, the Committee Against Jimcrow met with Truman at the White House to urge the president to "[insist] upon anti-segregation and civil rights amendments in any Universal Military Training program and any revival

---

113 "Negroes Threatened by U.M.T. Segregation, Not for Publication," published by the Committee Against Jimcrow in Military Service and Training, NCAC Records, box 7, folder labeled "Race Relations and UMT: Involvement with the Committee Against Jimcrow in Military Service and Training, 1945–1947," SCPC. Randolph called the October 10 meeting in response to an August 14, 1947, letter from Bill Worthy asking Randolph to do so. Worthy told Randolph that he (Worthy) had met with John Swomley of NCAC about the pending conscription legislation, and "both of us felt the serious threat of a new form of *government-sponsored* segregation touching a whole generation of youth merits dramatic action. I agreed to contact you to ascertain if, around the middle or end of October, you would be willing to call a meeting in New York of perhaps a dozen key Negro leaders to plan action." Worthy to Swomley, August 14, 1947, NCAC Records, box 7, "Race Relations and UMT" folder, emphasis in the original. A preparatory meeting for the October 10 meeting was held in New York on September 27 at A. Philip Randolph's office, attended by Randolph, Bayard Rustin, Bill Worthy, and others.

114 See, for example, the Supreme Court case *New Negro Alliance v. Sanitary Grocery*, 303 U.S. 552 (1938), which established the right to boycott and picket retail stores that did not hire Blacks and laid the legal groundwork for the national "Don't Buy Where You Can't Work" campaign. Lawson was lead counsel for the (successful) New Negro Alliance and argued the case before the Supreme Court. Thurgood Marshall, then a staff lawyer for the NAACP, *assisted* Lawson in *New Negro Alliance*. In *Henderson v. United States*, 339 U.S. 816 (1950), which struck down racial segregation in the dining cars of interstate trains and was one of the forerunners of *Brown v. Board of Education*, Lawson was once again the lead attorney for the winning side. Amicus briefs were filed in the case by, among others, Arthur Goldberg on behalf of the Congress of Industrial Organizations and Marshall on behalf of the NAACP; both Goldberg and Marshall went on to serve as Supreme Court justices.

115 "Joe Louis among Negro Leaders on Newly Formed Committee Against Jimcrow Training Bill," November 23, 1947, NCAC Records, box 7, "Race Relations and UMT" folder.

of Selective Service." The delegation included Reynolds, Randolph, Houston, and Bill Worthy, who was then the action director of the Congress of Racial Equality. "In regard particularly to a revival of Selective Service, Negro veterans are well qualified to insist upon broad, unequivocal anti-segregation and civil rights safeguards for prospective draftees," the delegation declared.[116]

When Randolph gave his Senate testimony on March 31, newspaper editors should have realized that there was nothing casual or accidental in his references to civil disobedience; the testimony was preceded by a press release from the Committee Against Jimcrow outlining exactly what Randolph was going to say. The press release was titled "Negro Labor Leader Projects Civil Disobedience Program Under Jimcrow Draft," and it ended by quoting Randolph as saying: "I can only repeat that this time Negroes will not take a jimcrow [*sic*] draft lying down. The conscience of the world will be shaken as by nothing else when thousands upon thousands of us second-class Americans choose imprisonment in preference to permanent military slavery."[117]

While all this was going on, the Committee Against Jimcrow was getting Civil Disobedience Pledge cards signed by young men who promised that they would not cooperate with any draft unless the military were fully desegregated. Their printed material said, in part: "Civil disobedience is not new. It is not a strange thing to America. It is as old as the Boston tea party; as much a part of the Negro's struggle for freedom as the underground railroad—a direct violation of the fugitive slave law. A jimcrow draft would violate the American ideals of equality and justice to such a degree that it is the duty of the people to maintain those ideals in spite of the law by refusing to register under a jimcrow draft." The attached pledge card stated: "I hereby pledge ... not to register or be drafted for jimcrow military service or training." The pledge forms had spaces for signature, name, address, telephone number, and date and were to be returned to Randolph.[118]

116 Press release of the Committee Against Jimcrow in Military Service and Training, "Negro Leaders Urge 5-Point Program on President Truman," March 22, 1948, NCAC Records, box 7, "Race Relations and UMT" folder, SCPC.

117 "Negro Labor Leader Projects Civil Disobedience Program under Jimcrow Draft," March 31, 1948, press release by the Committee Against Jimcrow, NCAC Records, box 7, "Race Relations and UMT" folder.

118 Bayard Rustin, Bill Sutherland, Wilfred Lynn, and Arthur Wright, *Questions on Civil Disobedience*, p. 2, published by the Action Committee c/o Committee Against Jimcrow in Military Training and Service, NCAC Records, box 7, "Race Relations and UMT" folder, SCPC. The manual also says at the top, "Bulletin 11, June 17, 1948."

Meanwhile, two Black World War II draft resisters, Bayard Rustin and Bill Sutherland, together with Wilfred Lynn and Arthur Wright, produced and circulated a lengthy training manual on civil disobedience entitled *Questions on Civil Disobedience*. Written in a question-and-answer format, the manual addressed these questions, among others: "Is civil disobedience un-American?," "Is it right to resist a law which has been passed after considerable discussion in a democratic state?," and "Is non-violence as practiced by Gandhi applicable to the struggle for Negro Freedom here?" The 1948 draft resistance effort organized by the Committee Against Jimcrow was a direct precursor to the civil rights movement's civil disobedience campaigns that marked the next fifteen years and today are more widely known.[119]

In Congress, hearings on the Selective Service on the House side were held in April 1948. On April 19, Reynolds, chairman of the Committee Against Jimcrow, testified. He politely but firmly repeated Randolph's admonition of the previous month concerning civil disobedience. This time, however, there could be no mistaking the import of his statements: Reynolds presented the first batch of signed resistance pledge cards to the committee.[120]

∵

On June 17, 1948, several people associated with the Committee Against Jimcrow—including Randolph, Reynolds, Rustin, Muste, and Sutherland—met at St. Philip's Parish House in Manhattan to discuss formation of a new, parallel entity that would be closely aligned with the existing committee. The idea was to establish separate committees to handle two different kinds of work. One would lobby Congress and the president to abolish segregation in the military; the other would organize civil disobedience (draft resistance) so long as segregation in the military prevailed. At the meeting, Rustin, Sutherland, Worthy, and Arthur Wright, all of whom were doing staff work for the Committee Against Jimcrow, were asked to "draw up [specific] recommendations for subsequent meetings."[121]

---

119 Rustin et al., *Questions on Civil Disobedience*.

120 "House Committee Hears Civil Disobedience Appeals from Reynolds and Randolph," April 19, 1948, press release by the Committee Against Jimcrow in Military Training and Service, CDGA collection League for Non-Violent Civil Disobedience Against Segregation, SCPC. Despite the title of this press release, Randolph himself was unavailable to attend the House hearings, so a statement reiterating his Senate testimony was read into the record by a colleague, Alfred Black. *House Hearings on H.R. 6274 and H.R. 6401*, statements from Randolph and Reynolds, April 19, 1948, 6420–22 and 6422–32.

121 Minutes of a Meeting to Explore the Possibility of Establishing a Committee to Give Support to Those Who Refuse to Register and Serve in Jimcrow Armed Forces, June 17,

On June 23, they presented their proposals to the full group.[122] The minutes of that meeting were taken by Rustin. The group agreed on a name for the new entity: League for Nonviolent Civil Disobedience Against Military Segregation. The staff were to be Rustin and Houser, executive cosecretaries; Sutherland, field secretary; Worthy, public relations; and Wright, research and fieldwork. The national chairmen were Randolph and Donald Harrington of the New Jersey Council Against Conscription. The group was deliberate in naming two executive cosecretaries (one Black, one white) and two national chairmen (one Black, one white).[123]

The group's minutes made clear that the (old) Committee Against Jimcrow and the (new) League for Nonviolent Civil Disobedience "are to work so closely together in program and action ... to be *two aspects of a single movement*."[124] The same people were involved in both groups, and they recognized "that officers, staff, and volunteer workers of both groups will, according to circumstances of time and place, emphasize one, the other, or both aspects."[125] Figures 6.6 and 6.7 show two of the flyers produced by the new organization.

On June 28, the League for Nonviolent Civil Disobedience issued a press release that listed Rustin as contact person and quoted Randolph's explanation that the new organization was supporting draft resistance but not draft evasion. Specifically, the league urged men to "openly refuse to register" and openly "refuse to be inducted." It did *not* encourage men to "quietly ignore registration" or to "feign illness, fake dependents and resort to other subterfuges." "The strength of our movement lies in truth and in those men who openly will inform the government that they are violating an unjust law," the release quoted Randolph as saying.[126]

On July 26, 1948, Truman issued his Executive Order 9981, which is widely viewed today as having desegregated the armed forces. In fact, Truman's executive order did nothing of the sort; it merely said, "It is hereby declared to be the policy of the President that there shall be equality of treatment and opportunity for all persons in the armed forces without regard to race, color, religion or

1948, CDGA Records, folder labeled "League for Nonviolent Civil Disobedience Against Military Segregation," SCPC.

122 Report of Steering Committee, June 21, 1948, CDGA Records, folder labeled "League for Nonviolent Civil Disobedience Against Military Segregation," SCPC.

123 Minutes of a Meeting Establishing a Committee to Form a Civil Disobedience Movement to Give Support to Those Who Refuse to Register in Segregated Armed Services, June 23, 1948, CDGA Records, folder labeled "League for Nonviolent Civil Disobedience Against Military Segregation," SCPC.

124 Ibid., emphasis in the original.

125 Report of Steering Committee, June 21, 1948.

126 "For Immediate Release," June 28, 1948, CDGA Records, folder labeled "League for Nonviolent Civil Disobedience Against Military Segregation," SCPC

FIGURE 6.6 Flyers from the League for Nonviolent Civil Disobedience Against Military Segregation urging men to refuse registration and military service unless the military was desegregated
COURTESY OF THE SWARTHMORE COLLEGE PEACE COLLECTION ARCHIVES

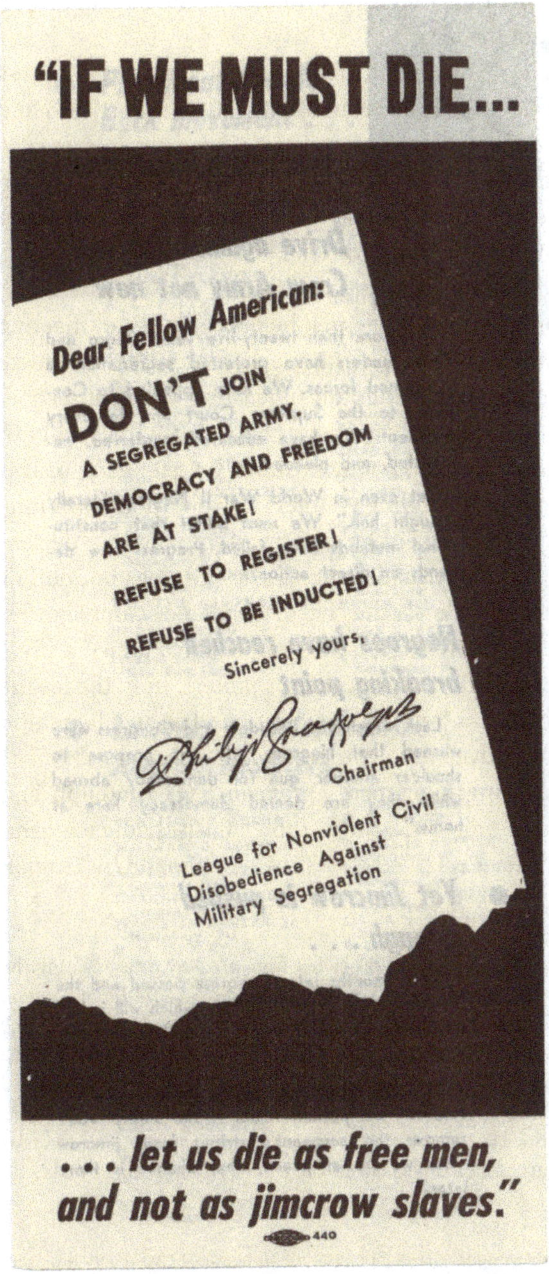

FIGURE 6.7 Flyers from the League for Nonviolent Civil Disobedience Against Military Segregation urging men to refuse registration and military service unless the military was desegregated
COURTESY OF THE SWARTHMORE COLLEGE PEACE COLLECTION ARCHIVES

national origin."[127] At the time of the executive order, "separate but equal" still constituted full equality under the law.[128]

The next day the League for Nonviolent Civil Disobedience issued a press release denouncing the presidential order. Randolph "called upon Negro and white youth who love freedom to refuse to register." The press release also noted that "Rustin said that the executive order is merely the continuation of that destructive 'separate but equal' pattern which has kept the American Negro is semi-slavery since the end of the Civil War."[129]

Nevertheless, it was clear that the league needed further internal discussion of the import of Truman's order, and to that end it held an emergency meeting in Chicago on August 2. Randolph, Muste, Sutherland, and others attended; Rustin once again took the minutes. Randolph urged that the civil disobedience campaign be continued, and his view was unanimously backed both by those present and by those who had been consulted by telephone or had sent in memoranda or telegrams, including chapters in New York, Baltimore, Chicago, Indianapolis, Denver, and Los Angeles. The group voted to "declare that the President's order … is totally inadequate because it does not outlaw segregation" and "to continue relentlessly the civil disobedience campaign."[130]

On August 6, the League for Nonviolent Civil Disobedience issued a press release announcing its determination to continue its protest. In it, the league denounced the presidential order as "totally inadequate" and quoted Randolph as saying that he has "no alternative but to counsel civil disobedience."[131]

But then something surprising happened. On August 13, Randolph and Reynolds reversed course and announced that the civil disobedience campaign should be called off. They explained that statements made by Truman

---

127 For a discussion of how far short Truman's order fell from actually desegregating the armed forces, see Rawn James, Jr., *The Double V: How Wars, Protest, and Harry Truman Desegregated America's Military* (New York: Bloomsbury Press, 2013), 128–30. For the more conventional view, see Deneen L. Brown, "How Harry S. Truman Went from Being a Racist to Desegregating the Military," *Washington Post*, July 27, 2018. See also the National Archives website, which describes Exec. Order No. 9981, 13 Fed. Reg. 4313 (July 26, 1948), as the "Desegregation of the Armed Forces," https://www.archives.gov/milestone-documents/executive-order-9981, last accessed April 28, 2022. The text of Truman's order makes no mention or reference to desegregation.

128 Plessy v. Ferguson, 163 U.S. 537 (1896).

129 "For Immediate Release," July 27, 1948, CDGA, "League for Nonviolent Civil Disobedience" folder, SCPC.

130 Minutes of the National Coordinating Committee Meeting, August 2, 1948, CDGA, "League for Nonviolent Civil Disobedience" folder, SCPC.

131 "For Release, Friday, August 6, 1948," CDGA, "League for Nonviolent Civil Disobedience" folder, SCPC.

after his order had been issued made clear that the order was aimed at the eventual elimination of segregation in the military, and, at any rate, the order had undermined public support for the proposed civil disobedience campaign. The sudden turnaround was shocking to many of the staff and most active members of the League for Nonviolent Civil Disobedience; on August 18, many of them signed a long statement rejecting the conclusion of Randolph and Reynolds and vowing to continue the civil disobedience campaign. Signers of the statement included Houser, Muste, Rustin, and Sutherland.[132] However, Randolph and Reynolds had effectively made further efforts by the League for Nonviolent Civil Disobedience impossible. Shortly afterward, Randolph and Muste jointly signed a last fund appeal for the now-defunct organization asking for contributions to retire the debts of the entity and allow it to die an honorable death.[133]

10 Legal Cases Pertaining to the Conscription Statute

The 1948 conscription statute provided that the Justice Department would review the cases of men administratively appealing decisions of local draft boards to deny their CO applications. Often, the Justice Department used the FBI to do background investigations of the men involved in these cases, so that when the cases returned to the Selective Service Appeal Board for decision, the board would have both an FBI file and a written Justice Department recommendation pertaining to the registrant. Selective Service procedure did not allow the registrant to review these documents when the board heard his appeal. This left the registrant in the impossible position of not knowing what information might be in those materials, whether the information was accurate, and what assertions he might need to address or refute in order to succeed in his appeal. This led to a significant amount of litigation.

In 1953, the US Supreme Court addressed the controversy. In *United States v. Nugent*, the court ruled that the existing Selective Service procedure denying registrants access to the documents that would influence (or decide) the outcome of his case was perfectly legal. The court held that so long as the registrant was free to present any materials or evidence that he wished, the

132 This untitled one-page document attached to an untitled three-page document signed by nineteen people states, “On the following pages you will find our complete statement,” August 18, 1948, CDGA, “League for Nonviolent Civil Disobedience” folder, SCPC.

133 A copy of the appeal is attached to A. J. Muste to A. Philip Randolph, n.d., CDGA, “League for Nonviolent Civil Disobedience” folder, SCPC.

procedure was constitutional. Justices Frankfurter, Douglas, and Black wrote an impassioned dissent, arguing that basic fairness, simple justice, and due process required that these materials be available to registrants.[134]

Two years later, in 1955, in a pair of companion cases handed down the same day (and written by the same justice), the Supreme Court reversed itself. This time the former dissenters formed the majority. In *Gonzales v. United States*, the Supreme Court held that the full copy of the Justice Department recommendation must be provided to registrants. In *Simmons v. United States*, the court additionally held that a fair summary of the FBI file must also be provided to registrants before their appeal hearing.[135]

Gonzales and *Simmons* were decided on March 14, 1955. Later that year the Selective Service statute was up for renewal in Congress. Hershey was so outraged by these two decisions that on June 10, 1955, he made a surprise appearance before the Senate Armed Services Committee to urge Congress use the pending legislation to undo these Supreme Court decisions. Congress ignored Hershey's entreaty.[136]

Between the world wars, a pair of US Supreme Court cases (*Schwimmer* and *Macintosh*) had held that it was permissible for Congress to pass laws barring aliens who would be COs from becoming naturalized United States citizens (see chapter 4). Accordingly, the oath for citizenship at the time included a statement that you were willing to bear arms for the United States (and people who would not swear to that were ineligible for naturalization). Naturalization is a privilege, not a right, the court reasoned; thus, Congress can set whatever requirements it wishes for naturalization. As I have shown, these two cases in the naturalization context later became the basis of the legal rule that there is no inherent right to be a CO in the conscription context.[137]

In 1946, the US Supreme Court overruled *Schwimmer* and *Macintosh*. In *Girouard v. United States*, the Supreme Court said, "Refusal to bear arms is not necessarily a sign of disloyalty or a lack of attachment to our institutions. One

134 United States v. Nugent, 346 U.S. 1 (1953).

135 Gonzales v. United States, 348 U.S. 407 (1955); Simmons v. United States, 348 U.S. 397 (1955).

136 "Hershey CO Proposal Fails to Pass Senate Committee," *NSBRO Reporter* 12, no. 7 (July 1955): 1.

137 United States v. Schwimmer, 279 U.S. 644 (1929); United States v. Macintosh, 283 U.S. 605 (1931).

may serve his country faithfully and devotedly, though his religious scruples make it impossible for him to shoulder a rifle."[138]

11 Conclusion

During the Cold War, both strands of the American zeitgeist were fairly reflected in public attitudes toward conscription. On the one hand, this was the period of the Cold War, and the United States saw the first sustained peacetime draft in our history. And even after the peacetime draft was in full operation, the political pressure to go beyond merely *selective service* to implement universal military training was significant.

On the other hand, the public revulsion against war and conscription was widespread. While the 1950s are often characterized as a period of complacent conformity and conservatism, that was certainly not an accurate reflection of the public attitude toward conscription during this period. Scores of small, medium, and large grassroots organizations opposing the peacetime draft sprang up in states, cities, and towns across the country, and World War II was barely over when the NCAC was born to coordinate the work of those homegrown antidraft committees.

For the first two years that peacetime conscription was on the books, the Selective Service System almost ceased to exist. The majority of draft boards had no full-time staff, months went by when not one person in the country was drafted, and many men casually ignored their obligations under the conscription statute. Many Americans were delighted to be rid (or nearly rid) of the draft. It took the start of the Korean War to rejuvenate the US draft apparatus.

The number of men who performed alternative civilian service as COs during this period (14,727) was larger than the number men who served in CPS camps during World War II (12,000), even though the number of newly registering men during this period was significantly lower than the number of World War II era registrants (even excluding men older than forty-five during World War II who were never actually drafted). And the number of men who considered themselves to be COs and who filled out CO application forms during this period was vastly higher than the actual number of COs, since men would be given any other classification for which they were eligible before a draft board would even consider a CO claim.

138 Girouard v. United States, 328 U.S. 1 (1946). This time, the court split 6–3, with Stone, Reed, and Frankfurter dissenting.

Before the 1948 conscription statute was enacted, Break With Conscription saw one thousand people committing felony civil disobedience against the draft simultaneously in thirty-seven states across the country. Break With Conscription is little known today, but it was probably the largest coordinated example of public nonviolent civil disobedience in the United States up to that time. During the Cold War, hundreds of open conscription resisters went to prison (some more than once), and tens of thousands of men quietly refused to cooperate with the Selective Service despite the risk of serious penalties.

There was also significant cooperation and cross-fertilization between the peace and civil rights movements during this period, a situation that benefited both movements. While Rustin and Houser were planning the February 12, 1947, Break With Conscription events, they were also planning the first Freedom Ride, which was held that April. That first Freedom Ride, called the Journey of Reconciliation, was sponsored by the Fellowship of Reconciliation and the Congress of Racial Equality, and many participants were World War II resisters recently released from prison: Wally Nelson, George Houser, Bayard Rustin, Jim Peck, and Igal Roodenko. Rustin was on the staff of FOR when he served as the principal organizer of the August 28, 1963, March on Washington at which Martin Luther King, Jr., delivered his "I Have a Dream" speech, and when Rustin left FOR's staff, he became executive secretary of the War Resisters League. The only person to participate in both the first and the second Freedom Rides (in May 1961) was Jim Peck, who had been one of the leaders of the Danbury prison strike that first succeeded in desegregating a US prison. During this period, Peck worked half time for WRL and half time for CORE.

CHAPTER 7

The Vietnam War Era

On August 7, 1964, Congress passed the Gulf of Tonkin Resolution, which successive presidents used as a blank check for their military actions in Indochina. For purposes of this chapter, the Vietnam era is defined as August 1964 through January 1973, when the Paris Peace Agreement was signed.[1]

On July 28, 1965, President Lyndon B. Johnson announced that he was ordering an additional 50,000 troops to Vietnam; this increased the number of US soldiers in Vietnam to 125,000, up from 23,000 at the beginning of the year. Johnson also announced a doubling of monthly draft calls. For comparison, the average monthly call for inductions during the first six months of 1961 had been 2,167 men; during the first six months of 1966, that figure was 27,230. In May 1961, the call for inductions was zero; in May 1966, that figure was 40,600. Inductions continued to soar during the rest of the Vietnam era, as shown in Table 7.1.[2]

During the Vietnam era, the number of local draft boards grew from 4,050 boards (with 16,275 members) on June 30, 1964, to just more than 4,100 boards (with 24,499 members) on June 30, 1972.[3] In 1967, more than 95 percent of local board members were older than forty, and more than a fifth were older than

1 For general background and context on antidraft activity during the Vietnam era, see Charles DeBenedetti and Charles Chatfield, *An American Ordeal: The Antiwar Movement of the Vietnam Era* (Syracuse, NY: Syracuse University Press, 1990); Nancy Zaroulis and Gerald Sullivan, *Who Spoke Up? American Protest against the War in Vietnam, 1963–1975* (New York: Holt, Rinehart & Winston, 1984); Alice Lynd, ed., *We Won't Go: Personal Accounts of War Objectors* (Boston: Beacon Press, 1968); Michael Ferber and Staughton Lynd, *The Resistance* (Boston: Beacon Press, 1971); and Daniel Berrigan et al., *Delivered into Resistance* (New Haven, CT: Advocate Press, 1969).

2 John D. Pomfret, "Johnson Orders 50,000 More Men to Vietnam and Doubles Draft; Again Urges U.N. to Seek Peace," *New York Times*, July 29, 1965.

3 US Selective Service System, *Semi-Annual Report of the Director of Selective Service for the Period January 1 to June 30, 1972 to the Congress of the United States pursuant to the Military Selective Service Act of 1967* (Washington, DC: Government Printing Office, August 15, 1973), 40, 61. In mid-1967, with the enactment of amendments to the draft law, the annual reports that the Selective Service had provided to Congress since 1950 (see chap. 6) became semiannual reports, cited here as, for example, *Semi-Annual Report*, January–June 1972. (Although this report has a publishing date, not all do.)

 | DOI:10.1163/9789004546684_009

TABLE 7.1 Inductions by year, Vietnam era

1964	150,808
1965	103,328
1966	343,481
1967	298,559
1968	341,404
1969	262,646
1970	203,707
1971	153,631
1972	25,273

SOURCE: US SELECTIVE SERVICE SYSTEM, *SEMI-ANNUAL REPORT OF THE DIRECTOR OF SELECTIVE SERVICE FOR THE PERIOD JULY 1–DECEMBER 31, 1973 TO THE CONGRESS OF THE UNITED STATES PURSUANT TO THE MILITARY SELECTIVE SERVICE ACT OF 1967* (WASHINGTON, DC: GOVERNMENT PRINTING OFFICE, N.D.), APP. 14, 32.

seventy. All were male. Two-thirds were veterans, including more than a fifth who were World War I veterans.[4]

In 1967, 213 draft board members were Black, about 1.3 percent of the total, at a time when the US population was 11 percent Black.[5] In late 1965, 22.8% of the enlisted men in combat units in Vietnam were African Americans, and in 1966 22.4% of all Army troops killed in action were Black. In 1966, the former Confederate states of Alabama, Arkansas, Louisiana, and Mississippi had no Black draft board members.[6] In roughly the same period, the populations of these states were 26.2 percent, 18.3 percent, 29.8 percent, and 36.7 percent Black, respectively.[7] Of Georgia's 509 draft board members serving on 164 local boards, exactly one was Black, at a time when 25.8 percent of the population of

4 National Advisory Commission on Selective Service, *In Pursuit of Equity: Who Serves When Not All Serve?* (Washington, DC: Government Printing Office, 1967), 73, table 1.1; 5; and 74, table 1.3 (hereafter, Marshall Commission report). The National Advisory Commission is often referred to as the Marshall Commission, after its chair, Burke Marshall, who had served as head of the Justice Department's Civil Rights Division from 1961 to 1964.

5 Marshall Commission report, 75, table 1.6; US Census Bureau and Bureau of Labor Statistics Staff, *Recent Trends in Social and Economic Conditions of Negroes, July 1968* (Washington, DC: Government Printing Office, 1968).

6 Marshall Commission report, 80–81, table 1.8.

7 These Census Bureau figures are for 1970. Campbell Gibson and Kay Jung, *Population Division: Historical Census Statistics on Population Totals by Race, 1790 to 1900*, Working Paper no. 56 (Washington, DC: US Census Bureau, September 2002), tables 15, 18, 33, and 39, respectively.

Georgia was Black.[8] The lack of racial diversity on draft boards was not unique to the South; in 1966, twenty-two states had no Black draft board members, including New Jersey, Rhode Island, and Vermont. From 1957 through 1966, Jack Helms, an Imperial Wizard of the Ku Klux Klan, was the chairman of the largest draft board in Louisiana, Local Board 42 in New Orleans.[9] Thus, during this period the KKK was better represented on Louisiana draft boards than Blacks were.

From the reinstitution of the draft in 1948 until the beginning of the Vietnam era, the Selective Service met or exceeded Department of Defense (DOD) calls for inductions every year. This changed during the Vietnam era. According to official Selective Service documents, during fiscal years 1968 to 1970, the Selective Service failed to meet the calls for induction provided by DOD.[10]

However, examination of Selective Service reports from that era reveals that the Selective Service consistently tried to hide the extent of its shortfalls. For example, for January through June 1968, DOD calls were for 212,200 men, and the Selective Service reported that it furnished 206,206 inductees. However, the Selective Service inappropriately counted 21,178 *volunteers* as inductees, so the correct number of involuntary inductions was 185,028. (Note that in Selective Service documents an inductee is involuntarily conscripted, as distinct from an enlistee, who volunteers.) For July through December 1968, DOD called for 86,800 men. The Selective Service claimed that it exceeded that call by inducting 90,200 men; however, 16,550 of those were volunteers. Thus, while the Selective Service claimed to be exceeding DOD calls, it was actually falling significantly short. For January through June 1969, DOD calls were for 180,100 men; the Selective Service inducted only 172,446 men. The shortfall was worse than it appears, because 13,780 of that number were actually volunteers. For July through December 1969, DOD called for 109,800 men; the Selective Service falsely claimed that it had exceeded the call by inducting 111,140 men, but 11,632 of them were volunteers.[11] Once again, the Selective Service failed to meet its quota. These shortfalls would prove important when considered in their wider context.

8 Marshall Commission report, 26 and 80, table 1.8; *Semi-Annual Report*, July–December 1967, app. 4, 50; Gibson and Jung, *Population Division*, table 25.

9 James E. Westheiter, *Fighting on Two Fronts: African Americans and the Vietnam War* (New York: New York University Press, 1997), 25.

10 *Semi-Annual Report*, July–December 1973, app. 14, 32. For a contrary view, see George Q. Flynn, *The Draft, 1940–1973* (Lawrence: University Press of Kansas, 1993), 184, which asserts that during the Vietnam era "at no time was there a shortage of draftees."

11 *Semi-Annual Report*, January–June 1968, 6, 17; July–December 1968, 12–13; January–June 1969, 14; and July–December 1969, 22–23.

As draft calls increased—along with opposition to the draft—the Selective Service was forced to lower the intellectual standards for induction twice: once in October 1966 and again in January 1967. After the second change, men who scored as low as the tenth percentile on the Armed Forces Qualifying Test were considered suitable for induction.[12]

Throughout this period, the Selective Service valorized its decentralized structure and the autonomous and independent nature of its forty-one hundred local draft boards. In one of its reports to Congress, the Selective Service boasted, "The boards represent a truly decentralized method of operation.... The procedure is democratic and typically American. The unpaid board members are neighbors of the registrants. Their decisions are based on knowledge of the local conditions."[13]

The downside to this decentralization was that every draft board was, in effect, a fief unto itself, and there was little consistency in how different draft boards treated men in the same circumstances. Lawrence Baskir and William Strauss were senior officials on President Gerald Ford's postwar Presidential Clemency Board (Baskir was general counsel and Strauss was chief executive officer). After the war ended, they published a book, *Chance and Circumstance: The Draft, the War, and the Vietnam Generation*, in which they analyze the chaotic, wildly inconsistent ways in which different draft boards operated. They observed, "Left to their own devices, the four thousand draft boards developed four thousand different policies."[14]

On July 2, 1966, well before the expiration of the draft law in mid-1967, Johnson created the blue ribbon National Advisory Commission on Selective Service, often referred to as the Marshall Commission after its chairman, Burke Marshall. The Marshall Commission was charged with evaluating the current functioning and fairness of the Selective Service and making recommendations for legislative improvements. The Marshall Commission submitted its book-length findings in February 1967, in advance of extensive debate in

---

12 "Acceptability Standards Are Revised," *Selective Service* 16, no. 10 (October 1966): 1; "Mental Standards Lowered Again," *Selective Service* 17, no. 2 (January 1967): 4. *Selective Service* was the monthly house organ of the Selective Service System during this period and is different from the formal annual (and, later, semiannual) reports that the Selective Service made to Congress. In October 1969 the name of the monthly changed to *Selective Service News*, and the publication stopped including volume and issue numbers.

13 *Semi-Annual Report*, July–December 1967, 6; James W. Davis, Jr., and Kenneth M. Dolbeare, *Little Groups of Neighbors: The Selective Service System* (Chicago: Markham, 1968).

14 Lawrence M. Baskir and William A. Strauss, *Chance and Circumstance: The Draft, the War, and the Vietnam Generation* (New York: Alfred A. Knopf, 1978), xvii, 24.

Congress about the Selective Service later that year. The Marshall Commission report was called "the most comprehensive study of conscription undertaken in this period" by two scholars who produced their own book-length analysis of the Selective Service the following year. A key problem, the Marshall Commission found, was "the philosophy of local autonomy around which the present system is organized."[15] The commission identified three distinct ways in which this philosophy of local autonomy played out.

- Policies were wildly inconsistent from state to state and from draft board to draft board. For example, some states (e.g., Alabama) required fifteen semester hours of classes for a registrant to be eligible for a student deferment, some states required twelve semester hours (e.g., Louisiana), and some states left the definition of "full-time student" to the individual college (e.g., Oklahoma). Some draft boards reclassified men when the reason for a previous deferment no longer existed; some boards did not. And 27 percent of the registrants surveyed had been illegally reclassified I-A even when the reason for their former deferment still existed.[16]
- State or national Selective Service headquarters often sent incoherent policy directives to local boards. The Marshall Commission referenced draft boards that received two different sets of "directly conflicting" instruction on the same topic.[17] The Local Board Memorandum No. 107, pertaining to COs, dated July 6, 1970, provides an especially piquant example of this (described in the discussion of conscientious objectors later in this chapter).
- Arguably most embarrassing was that local boards often simply ignored the law, the regulations, and instructions from the national and state Selective Service offices. For example, the Marshall Commission found that about half the men classified as II-A (occupational deferment) simply did not qualify for any occupational deferment.[18]

This report was not produced by the pacifist War Resisters League but by a presidential commission that included, among others, former secretary of defense Thomas Gates and the former Marine Corps commandant General David Shoup, a World War II recipient of the Medal of Honor. The Marshall

15 Davis and Dolbeare, *Little Groups of Neighbors*, 24; Marshall Commission report, 83.
16 Marshall Commission report, 169, 26.
17 Ibid., 27.
18 Ibid.

Commission made many recommendations for improving the Selective Service. These recommendations, and Congress's response, are discussed later in this chapter; for now, it suffices to say that none of the commission's major recommendations was adopted.

Hershey retired as director of the Selective Service in February 1970; after a brief interregnum with an acting director, Curtis Tarr became director on April 16, 1970. As a result of Congress's failure to reform the Selective Service in 1967 (or at any other time), most of the problems identified by the Marshall Commission in 1967 remained. In early 1972, Tarr reported to the president and to Congress on the cesspool of illegal activities, inconsistency of operations, and general incompetence at the Selective Service that he had inherited from Hershey.

As Tarr informed Nixon, when Tarr took over, "each local board could [and did] operate in a unique fashion. It was the policy of the National Headquarters to impose a minimum number of restraints on local initiative." This led to frequent, blatantly illegal actions by local boards. Unfortunately, Tarr continued, "the Federal courts had begun to take a much more critical view of the operation of the system. Many decisions condemned bluntly the discrepancies and differing practices in selective service operations."[19]

Tarr told Nixon that "immediately ... it became evident that standards, procedures, and constraints varied widely from one State to another, and indeed from one local board to another in the same State." He told Congress that the Selective Service had been "a loosely coordinated collection of semi-autonomous local boards." Thus, five years after the Marshall Commission report, each local draft board remained a separate fief. Tarr told Congress in mid-1972 that he had belatedly "launched a comprehensive program to inform people serving the System about their responsibilities" under the law and existing regulations.[20]

"But even more discouraging was our inability to train" staff, Tarr said. Incompetent staff caused widespread "procedural errors which led to erroneous inductions and prosecutions, both of which were unfortunate ... and grossly unfair to registrants." For the first time in Selective Service history, Tarr instituted "an important procedural change ... now we make certain that a registrant has committed a violation ... before any action is taken in the judicial process."[21]

19 Tarr to Nixon, April 30, 1972, reprinted in *Semi-Annual Report*, January–June 1972, 31–33.

20 Ibid.; "February 28, 1972 Statement of Curtis W. Tarr before the Judiciary Committee of the United States Senate," reprinted in *Semi-Annual Report*, January–June 1972, 38–47.

21 Ibid., 40–41.

These words were not written by the Central Committee for Conscientious Objectors but by the director of the Selective Service reporting in his official capacity. In 1972, after eight years of war in Vietnam and less than a year before the Paris Peace Agreement, the Selective Service was—*for the first time*—instituting a system to ensure that a man had actually violated the draft law before he was criminally prosecuted. Before these changes were made toward the end of the war, federal prosecutors had found it impossible to prosecute 106,000 of the criminal cases referred to it by the Selective Service because of serious procedural violations committed by the agency that would have made it impossible to obtain a conviction.[22]

The mess that Tarr inherited from Hershey at the Selective Service was not a mistake or an accident. For decades, Hershey had touted the decentralization and autonomy of local draft boards as the great virtue of his system. In 1953, Hershey had written that decentralization was the single most important factor contributing to the system's success—that "classifications are made in the communities by local board members who live in those communities."[23] In 1951, a congressional committee challenged Hershey about autonomous draft boards' reaching inconsistent results, and Hershey defended the system by saying that the alternative was dictatorship: "We will never have absolute uniformity ... so long as we have liberty. When that is gone, you can have uniformity."[24]

1 Early Demonstrations and Preliminary Legal Skirmishes

On October 15, 1965, antiwar protesters held coordinated demonstrations in cities across the United States. The next day, there were coordinated

22 Baskir and Strauss, *Chance and Circumstance*, 12, 69.

23 Lewis B. Hershey, "The Liberties We Cherish Are OUR Liberties," *Selective Service* 3, no. 7 (July 1953): 1. In his biography of Hershey, George Q. Flynn makes this point several times. For example, Flynn writes, "Thousands of individual volunteers, operating in their own communities, decided which of their neighbors would go to war. Hershey envisioned the local boards as little schools of how democracy functioned during a national emergency. Finally, and most pragmatically, this decentralized approach would fragment problems and deflect opposition to the draft." Flynn, *Lewis B. Hershey: Mr. Selective Service* (Chapel Hill: University of North Carolina Press, 1985), 77.

24 "No, They Don't Have That Kind of Liberty in Some Places Abroad We Might Name," *Selective Service* 1, no. 3 (October 1951).

demonstrations against US policy in Vietnam in cities around the world, including London, Rome, Brussels, Copenhagen, and Stockholm.[25]

In Ann Arbor, Michigan, on October 15, thirty-nine people (all but one were students at the University of Michigan) were arrested while conducting a nonviolent sit-in at the local draft board. Hershey was outraged and ordered that the arrested students have their student deferments revoked and that the students be reclassified I-A.[26] (The reclassifications were promptly challenged in court and were declared illegal in *Wolff v. Selective Service.*) In its December 27, 1965, issue, the *Nation* ran an article entitled "Punishment by Conscription: General Hershey's Big Stick," in which the magazine criticized this "punitive use" of the draft law.[27]

Hershey penned a sharp rejoinder, which the *Nation* published in its January 3, 1966, issue. (Hershey was so pleased with his response that he reprinted it in *Selective Service*, the agency's house organ, and wrote about it in his annual report to Congress at the end of the fiscal year.) The gist of Hershey's reply was that it is "the greatest misunderstanding … that this procedure is something new in Selective Service, designed for this particular occasion." On the contrary, said Hershey, it had *for decades* been Selective Service policy to put Selective Service delinquents at the front of the line in the Order of Call, the regulation that designates in what order men who are classified I-A will be called.[28]

Hershey's obfuscation was so egregious that it fairly qualifies as a lie. It was true that the Selective Service's Order of Call since the 1950s had put delinquents first, even before volunteers.[29] However, both the statute and the regulations defined a delinquent as one who "shall knowingly fail or neglect or refuse to perform any duty required of him" under the draft law.[30] Thus, a delinquent was one who refused to register or refused to fill in a classification

25 Douglas Robinson, "Policy in Vietnam Scored in Rallies throughout U.S.," *New York Times*, October 16, 1965, l, and "10,000 in City Parade Join World Protest on Vietnam," *New York Times*, October 17, 1965.

26 Flynn, *Draft, 1940–1973*, 184.

27 Wolff v. Selective Service, 372 F.2d 817 (2d Cir. 1970); Carl Cohen, "Punishment by Conscription: General Hershey's Big Stick," *Nation*, December 27, 1965, 520–22.

28 "Director Offers Comments on Michigan Student Incident," *Selective Service* 16, no. 3 (February 1966): 4; US Selective Service System, *Annual Report of the Director of Selective Service for the Fiscal Year 1966 to the Congress of the United States Pursuant to the Universal Military Training and Service Act as Amended* (Washington, DC: Government Printing Office, 1967), 43.

29 Exec. Order 10659, 21 Fed. Reg. 1079 (Feb. 15, 1956); Exec. Order 11241, 30 Fed. Reg. 11129 (Aug. 26, 1965).

30 Section 12(a) of the statute, as quoted in Wolff v. Selective Service Local Board, 372 F.2d 820, 820–21 n.1 (2d Cir. 1967). Tracking the statute, the Selective Service wrote a regulation that defined a delinquent as someone who "has failed to perform any duty or duties

questionnaire or refused to submit to a preinduction physical when called. The Ann Arbor sit-in was the first time Selective Service had ever punitively reclassified a registrant as I-A for simply protesting the war, not for failing to perform an obligation under the statute.[31] (To be sure, the Michigan students had engaged in civil disobedience. For that, the appropriate remedy would have been arrest and prosecution for violation of whatever local ordinance, such as trespass, they might have violated, not punitive draft reclassification.) Nevertheless, in late 1965, Hershey's action remained a one-off expression of the general's pique. It was illegal, but it was not yet official government policy.

On October 21, 1967, hundreds of thousands of antiwar demonstrators marched in Washington, DC. That night, about fifty thousand of them besieged the Pentagon and stayed overnight. Many burned draft cards overnight, and hundreds of demonstrators were arrested the next day. Johnson was outraged and called Hershey to demand that something be done about draft card burners.[32] In response, five days later Hershey sent both a memorandum and a letter to every draft board in the country. The memorandum was rather more narrowly drawn and provided that when a draft resister destroyed or returned a draft card, the resister should be reclassified I-A, treated as delinquent, and immediately drafted. The accompanying letter, however, was broader and provided that any "illegal activity" (such as the Michigan sit-in) should have the same results. As George Q. Flynn wrote of the Hershey letter, "The Ann Arbor procedure became official policy." The policy became public when an article about it appeared in the *New York Times* on November 9, with a sidebar that included the full text of both the memorandum and the letter.[33]

Not surprisingly, Hershey's action provoked an avalanche of lawsuits. The National Student Association brought a legal challenge, as did many individuals.[34] Attorney General Ramsey Clark was skeptical that Hershey's ukase was

---

required of him under the selective service law." 32 C.F.R. § 1642.4, cited in Gutknecht v. United States, 396 U.S. 295, 297 (1970).

31 Flynn, *Draft, 1940–1973*, 184, 214–16.

32 Joseph A. Loftus, "Guards Repulse War Protesters at the Pentagon," *New York Times*, October 22, 1967, 1; Ben A. Franklin, "War Protesters Defying Deadline Seized in Capital," *New York Times*, October 22, 1967; Flynn, *Draft, 1940–1973*, 215; Michael S. Foley, *Confronting the War Machine: Draft Resistance during the Vietnam War* (Chapel Hill: University of North Carolina Press, 2003), 153.

33 Flynn, *Draft, 1940–1973*, 215; Neil Sheehan, "Hershey Talked with White House," *New York Times*, November 9, 1967.

34 National Student Association v. Hershey, 412 F.2d 1103 (D.C. Cir. 1969); Gilchrist v. United States, 430 F.2d 631 (10th Cir. 1970); Bucher v. Selective Service, 421 F.2d 24 (3d Cir. 1970).

legal. The White House aide Joseph Califano tried to mediate the spat between Hershey and Clark, and on December 8, 1967, the administration released a joint statement in the names of the two men, announcing that a new unit of the Justice Department had been created to prosecute draft violators. "It has long been the law that a registrant who violates any duty" may be declared delinquent and immediately drafted, the joint statement said.[35]

Hershey clearly believed that he had prevailed over Clark. Hershey held what he described as "a lengthy press conference" to explain that the reclassification of protesters was "in no way a new position assumed by the System. It was merely a reiteration" of long-standing policy. Hershey also insisted that, despite the many pending lawsuits, the courts were helpless to intervene: "There is no question that this is entirely an administrative procedure. It can in no manner be considered judicial."[36]

The contents of the December 8 joint statement led Flynn, in his book *The Draft: 1940–1973*, to take the view that the Clark-Hershey squabble ended in something of a draw; the historian Michael S. Foley believes that Hershey prevailed.[37] But the joint statement does not tell the whole story.

The first case challenging the new policy to reach the Supreme Court was the case of James Oestereich, a divinity student who had been reclassified I-A pursuant to the Hershey letter after he had returned his draft card. There was nothing remarkable in the Supreme Court's terse decision declaring both the Hershey Memorandum and Selective Service policy "blatantly lawless" (and it was blatantly obvious that Hershey's boast that his policy was beyond the reach of the courts was hollow).[38]

What was remarkable—and was missed by Foley and underappreciated by Flynn—was that the Department of Justice, whose legal function includes defending governmental policy in the courts, declined to do so in this case.[39] This is an extraordinarily rare occurrence; for example, in February 2011, when the Obama administration announced that it would not defend the Defense

---

35 National Student Association v. Hershey, 412 F.2d 1103 (D.C. Cir. 1969); Gilchrist v. United States, 430 F.2d 631 (10th Cir. 1970); Bucher v. Selective Service, 421 F.2d 24 (3d Cir. 1970); "Joint Statement on the Draft issued by Ramsey Clark and Lewis Hershey," reprinted in "Joint Statement on the Draft," *New York Times*, December 10, 1967.

36 *Semi-Annual Report*, July–December 1967, 21–23; Lewis B. Hershey, "Enforcement of the Law," *Selective Service* 17, no. 12 (December 1967): 1–2.

37 Flynn, *Draft, 1940–1973*, 154; Foley, *Confronting the War Machine*, 154.

38 Oestereich v. Selective Service, 393 U.S. 233 (1968). For a comic account of James Oestereich's antics when he refused induction in Boston on February 26, 1968, see Foley, *Confronting the War Machine*, 248–49.

39 Flynn acknowledges the role of Solicitor General Griswold in *Lewis B. Hershey* but treats the matter mostly as a tiff between Hershey and Griswold that left Hershey indignant (266).

of Marriage Act (outlawing same-sex marriages) in the Supreme Court, the administration's announcement caused an enormous stir both in the mainstream press and in law reviews.[40] In the government's brief to the court in the *Oestereich* case, US Solicitor General Erwin Griswold (the former longtime dean of Harvard Law School) made three astonishing acknowledgments:

- Hershey's letter invited local draft boards to act illegally.
- There was, point-blank, no authority in the draft law that would permit punitive reclassifications.
- Even if the statute allowed for punitive reclassification, which it did not, such a provision would have been clearly unconstitutional.

For these reasons, the *government* told the court that Jim Oestereich was right and the Selective Service was wrong.[41] During oral argument of the case, Griswold agreed with Melvin Wulf, the ACLU attorney representing Oestereich, that the Selective Service had acted illegally. Referring to "General Hershey's letter and memorandum," Griswold told the justices there were "multiple problems ... there was no indictment by a grand jury, no right to counsel, no trial by a jury and indeed no trial." Griswold went on: "I find nothing in the statute which authorizes a draft board under any circumstances" to act as it did.[42]

Also remarkable here is how long the Selective Service's illegal conduct continued. Hershey's action after the Ann Arbor sit-in was illegal, as the court had made abundantly clear in *Wolff* (and in other cases). Thus, Hershey knew he was acting illegally when he issued his memorandum and letter in October 1967. The Supreme Court's decision in *Oestereich* made that clear (again) in unambiguous language. However, the Selective Service *still* did not comply with the law. Despite the Supreme Court's ruling that Hershey's action had been "blatantly lawless," the Selective Service continued its illegal policy. When the ACLU's Melvin Wulf again argued the (same) issue before the court in 1970, in *Breen v. Selective Service*, the Nixon administration actually defended the Selective Service. The Supreme Court was not amused. "We are ... unable to

40 Charlie Savage and Sheryl Gay Stolberg, "In Shift, U.S. Says Marriage Act Blocks Gay Rights," *New York Times*, February 24, 2011; Kathleen Tipler, "Obama Administration's Non-Defense of DOMA and Executive Duty to Represent," *Maryland Law Review* 73, no. 1 (January 2013): 287–311.

41 Erwin Griswold et al., Principal Brief of Respondent in Oestereich v. Selective Service System Local Board 11, No. 68-46, 1968, U.S. Supreme Court Briefs Lexis 3 (September 1968).

42 October 24, 1968, Transcript of Oral Argument before the Supreme Court in Oestereich v. United States, 393 U.S. 233 (1968) (No. 68-46). The quotations used in text appear on pp. 23, 24, and 28 of the transcript.

distinguish this case from *Oestereich*," the court said.[43] In effect, the justices said, "We meant what we said last time."

2 The 1967 Conscription Statute

In February 1967, the Marshall Commission presented its report to the president. As noted earlier, it found serious problems with the Selective Service, and it proposed a raft of reforms. The 1963 extension of the draft law was set to expire on June 30; on March 6, 1967, Johnson sent a message to Congress recommending a set of changes based upon the recommendations of the Marshall Commission. Johnson recommended:

- Tightening eligibility for all deferments, eliminating deferments for most graduate students, and starting a national "public discussion" about the advisability of eliminating undergraduate deferments.
- Replacing the current order of call with a "random system of selection"—that is, a lottery.
- Addressing the wildly inconsistent practices of local draft boards that the Marshall Commission had criticized by radically restructuring the Selective Service by dividing the country into eight regions, reducing the approximately 4,100 existing draft boards to 300 to 500, and requiring those boards to apply a single set of consistent national standards.
- Banning the process whereby men could enlist in the reserves after they had received an induction order. This had long been a favored way for white middle-class men to ensure that, even if they received induction orders, they could evade service in Vietnam.[44]

Congress was not impressed with Johnson's recommendations. With regard to undergraduate student deferments—which had always been permissive—Congress changed the law to make the II-S classification mandatory for all undergraduates. Congress also explicitly made draft selection by lottery illegal. As for restructuring the Selective Service, Congress simply ignored the president's proposal. This ensured that the chaotically inconsistent status quo, in which each local draft board acted in its own idiosyncratic (often illegal) way, would continue. Finally, Congress required enlistments in both the reserves

43 Breen v. Selective Service, 396 U.S. 460 (1970).

44 Max Frankel, "Johnson Plans Draft by Lottery, with 19-Year-Olds First; Would Cut Deferments Sharply," *New York Times*, March 7, 1967 (the full text of the president's message to Congress appears on p. 32).

and the National Guard be retained for those seeking to enlist after receiving induction orders.[45]

Many additional proposed changes to conscription did not get past congressional debate. Senator Ernest Gruening (D-AL), one of only two senators to have voted against the Gulf of Tonkin Resolution, proposed an amendment whereby conscripts could not be posted to Southeast Asia without their consent. Donald Rumsfeld, then a Republican representative from Illinois, and Senator Mark Hatfield (R-OR) proposed abolishing conscription and replacing it with an all-volunteer army.[46] A representative of the United Church of Christ proposed a change to the CO provisions that would allow selective objection to a particular war.[47]

The 1967 statute did make a change in the applicable standard for COs, which I discuss later in this chapter.

Congress tried to address the enormous amount of litigation engendered by the draft. This included both criminal cases, in which the government prosecuted people for refusing to register, refusing induction, burning draft cards, and the like, and civil cases, such as *Wolff*, in which men (or organizations) sued the Selective Service, complaining of illegal procedures (such as the infamous Hershey memorandum in *Wolff*).

On the criminal side, Congress required that courts give precedence on their dockets to draft violations and that if the Department of Justice failed to handle such cases "as expeditiously as possible," it explain to Congress "in writing the reasons for its failure to do so." This provision did not work (see the discussion under Prosecutions later in this chapter).

On the civil side, Congress sought to divest the federal courts of jurisdiction to hear cases brought by individuals against the Selective Service. Under Section 10(b)(3) of the 1967 law, courts would have no jurisdiction over Selective Service matters except in criminal cases brought by the government against draft violators. Congress's legislative history of the statute made clear that this provision was written into the bill to make future lawsuits like *Wolff* impossible.[48] It did not work. *Oestereich*, for example, was a case that this new

45 An Act To amend the Universal Military Training and Services Act, Pub. L. No. 90-40, 81 Stat. 100 (1967). See generally, Davis and Dolbeare, *Little Groups of Neighbors*, 26–28.

46 *Amending and Extending the Draft Law and Related Authorities: Hearings on S.1432, April 12, 13, 14, 17,18, and 19, Before the Committee on Armed Services United States*, 90th Cong., 1st sess. (1967), 164 (hereafter, *1967 Senate Hearings*). On abolishing the draft, see *1967 Senate Hearings*, statements of Donald Rumsfeld, 425–44, and Mark Hatfield, 481–90.

47 *1967 Senate Hearings*, statement of United Church of Christ, 598–602.

48 A Bill To Amend the Universal Military Training and Service Act of 1967, S. Doc. No. 1432, at 30–31 (1967), as cited in Fein v. Selective Service System, 405 U.S. 365, 372 n.4 (1972). See

provision would have barred. As the Supreme Court noted in *Oestereich*, and in other cases, allowing such a provision to stand would be unconstitutional.[49] Section 10(b)(3) had no real-world effect. In fiscal year 1970, 447 new civil cases were filed against the Selective Service in federal courts; in FY1971, 695 such new civil cases were filed.[50] (However, while the courts widely ignored the provision, the Supreme Court never formally struck it down as unconstitutional).[51]

3 Nixon, the Lottery, and the End of the Draft

On October 17, 1968, Richard Nixon, then running for president, proposed ending the draft.[52] On May 13, 1969, after his election, Nixon asked Congress for legislative changes to the draft law, including but not limited to establishment of a lottery to determine the order in which men would be inducted. Congress acted promptly to establish a lottery but waited until 1971 to address Nixon's other proposals. On November 26, 1969, Nixon signed into law a one-sentence amendment that repealed the ban on a Selective Service lottery that Congress had enacted in 1967. On the same date, he signed Executive Order 11497, which created a national draft lottery and reduced the time that men classified I-A would be vulnerable to receiving an induction order from seven years (ages 19 to 26) to one year (at age 19). This latter provision was viewed as a way of

also 90 Cong. Rec. 15426 (1967), in which the legislative intent was also cited, as stated in *Fein*, 405 U.S. 365.

49 Fein, 405 U.S. at 373; Oestereich, 393 U.S. at 237–38.

50 US Courts, *Annual Report of the Director of the Administrative Office of the U.S. Courts* (Washington, DC: Government Printing Office, 1974), 128, table 19; Amy J. Rutenberg, *Rough Draft: Cold War Military Power and the Origins of Vietnam-Era Draft Resistance* (Ithaca, NY: Cornell University Press, 2019). Rutenberg incorrectly states, "Because of the way Congress had constructed the laws governing Selective Service, the courts did not have jurisdiction" (180). It is true that Congress tried to so construct the draft law, but the effort failed. For a contemporaneous, contrary view asserting that section 10(b)(3) had some efficacy, see Robert M. O'Neil, "Review of Selective Service Reclassifications," *George Washington Law Review* 37, no. 3 (March 1969): 536–63.

51 Charles Donahue, Jr., "The Supreme Court vs. Section 10(b)(3) of the Selective Service Act: A Study in Ducking Constitutional Issues," *UCLA Law Review* 17, no. 5 (June 1970): 908–74.

52 Nixon Agnew Campaign Committee, press release on all-volunteer force, October 17, 1968, https://www.nixonfoundation.org/wp-content/uploads/2012/01/Candidate-Nixon-Statement-on-an-All-Volunteer-Force.pdf, last accessed December 6, 2022. For an extensive compilation, edited by a former director of the Selective Service, of internal governmental documents relevant to the Nixon administration's successful effort to end conscription, see Bernard Rostker, *I Want You! The Evolution of the All-Volunteer Force* (Santa Monica, CA: Rand, 2006).

reducing the uncertainty for draft-eligible men who had previously spent a seven-year period of exposure to being drafted.[53]

On December 1, 1969, the first draft lottery since World War II was held to select random birth dates for order of induction for men aged nineteen through twenty-five. The new lottery went into effect on January 1, 1970.[54]

The lottery was a complete fiasco. As Tarr, the Selective Service director, acknowledged in his report to Congress, it failed for three major reasons. Because men were constantly flowing into and out of the I-A pool (because of changes in personal circumstances), draft boards found themselves drafting men with higher lottery numbers before calling men with lower numbers (who had newly entered the I-A pool). This vitiated the reason for having a lottery.

To address this unexpected problem, the Selective Service decided to extend the (newly created) one year of eligibility for some men with lower lottery numbers who had not been drafted. This was done to eliminate the perceived unfairness of drafting men with higher lottery numbers first, but it also undermined the much-touted new policy of reducing the period within which men would be liable to conscription.

Finally, as soon as it became clear that men with lottery numbers higher than 195 would not be called in 1970, large numbers of students with higher numbers voluntarily relinquished their student deferments to game the system. By thus "exposing themselves" to the draft during a period when they would not actually be drafted, these students exhausted their one-year period of exposure and effectively immunized themselves from being drafted in the future. Moreover (as Tarr noted with considerable chagrin), as soon as the enormous nationwide network of draft counselors became aware of this trick, they alerted men with high lottery numbers that it would be to their advantage to voluntarily relinquish their student deferments.[55]

∴

The last major change in the draft law before conscription ended was signed into law on September 28, 1971.[56] The statute had several major provisions.

53 An Act to Amend the Military Selective Service Act of 1967 to authorize modifications of the system of selecting persons for induction into the Armed Forces under this Act, Pub. L. No. 91-124, 83 Stat. 220 (1969); *Semi-Annual Report*, July–December 1969, 7.

54 *Semi-Annual Report*, July–December 1969, 7.

55 Ibid., July–December 1970, 1–3.

56 An Act to Amend the Selective Service Act of 1967, Pub. L. No. 92-129, 85 Stat. 348 (1971) (hereafter, 1971 Amendment to Selective Service Act).

It extended the authority to draft for two years, which was in keeping with Nixon's announced hope and intention to eliminate the draft by mid-1973.

It rescinded the portion of the 1967 amendments that had removed presidential discretion with regard to undergraduate deferments; this allowed the Nixon administration to proceed with its stated desire to phase out student deferments. In his testimony to Congress, Tarr had said, "The major inequity remaining in Selective Service comes from granting deferments to [students]. Doing so disadvantages those who, for one reason or another, do not go to a college or university." Tarr was correct about the inequity of the student deferment, which disproportionately benefited white middle-class registrants. However, although the Nixon administration touted this change as an example of its move toward greater fairness with the draft, it *retained* student deferments for all students who then held them.[57]

The statute also eliminated the draft-board-by-draft-board quotas that were clearly not working and replaced them with a uniform national call. In keeping with the newly instituted lottery, the uniform national call would be by lottery number.[58]

4 Evaders

During the Vietnam era, many men thought it would be inconvenient or irksome to serve two years in the army; others believed that, with a shooting war on, being drafted could be dangerous. Men who were also not keen to risk prison and all that that entailed had many ways to avoid or evade the draft. Some were legal; some skated close to the line of illegality; some were clearly illegal. In the actual situation, men who committed illegal acts faced varying degrees of risk if they were caught.

The law required men to present themselves to their local draft board for registration on, or within five days after, their eighteenth birthday. For those inclined to evade the system, probably the easiest way to do so was to "forget" or otherwise simply fail to register. In 1976, Ramsey Clark, the former attorney general, put the number of such clandestine nonregistrants during the Vietnam era at nearly 2 million.[59] Not registering was clearly illegal. Nevertheless, these

57 Ibid.; "February 23, 1971 Testimony of Curtis Tarr before House Armed Services Committee," reprinted in Curtis W. Tarr, "Undergraduate Student Deferments," *Selective Service News* (December 1971): 3 in special supplement; 1971 Amendment to Selective Service Act; "1971 Changes in Draft Policy," special issue, *Selective Service News* (January–February 1972): pt. 2, 1.

58 1971 Amendment to Selective Service Act; "1971 Changes in Draft Policy," 3.

59 Richard D. Lyons, "Full Amnesty Apparently Has a Long Way to Go," *New York Times*, January 25, 1976.

quiet (nonpublic) nonregistrants faced almost zero risk—for two reasons. Selective Service generally had no organized way to track down or locate quiet nonregistrants. Also, Selective Service always said, "The mission of the System is to put men into the Armed Forces, not to have them imprisoned."[60] Thus, on those rare occasions when quiet nonregistrants were caught, they were given the opportunity to register late rather than be prosecuted.

Another way to evade service was to arrange to have a physical or psychological problem that got you classified IV-F. Some men asked friendly physicians or psychiatrists to write to the draft board about a physical or mental problem (say, bone spurs in the feet) that would be disqualifying. Acknowledging that you were gay was a surefire out. (In 1970, James McClain was criminally prosecuted for the crime of sending obscene material through the mail when he mailed his draft board explicit photographic evidence that he was gay. A draft board clerk had to take five days off to recover from the shock of receiving the evidence.) Other men sought (sometimes creative) ways to flunk the preinduction physical examination. From 1956 to 1973, the percentage of men who failed the preinduction physical rose from less than 36 percent to 57.9 percent. And even men who passed a preinduction physical could try to fail the induction physical. During the first six months of 1970, 15.8 percent to 21.4 percent of the men who passed preinduction physicals still managed to flunk the induction physical each month.[61]

And there was another, excellent way to avoid both receiving an induction order and going to prison. For most of the war, the policy of the Justice Department was not to prosecute men for refusing to show up for their preinduction physical; instead, US attorneys would wait to seek an indictment for the same man when he later refused induction. However, the policy of the Selective Service was not to send induction notices to men who had not passed their preinduction physicals (and passing required attending). In this way, men who simply ignored a notice to attend a preinduction physical were often permanently lost to the system. Untold thousands of men escaped the draft this way before the Selective Service belatedly changed this policy on June 16, 1970, with the issuance of Local Board Memorandum 106.[62]

A favorite way to evade the draft was by "pyramiding" legitimate (or semilegitimate) deferments sequentially, one after another. For most of the war, men were liable to receive induction orders only until their twenty-sixth

60 *Semi-Annual Report*, July–December 1969, 18.

61 Baskir and Strauss, *Chance and Circumstance*, 43–44; "Draft Board Shocked," *WIN Magazine* 6, no. 15 (September 15, 1970): 28; *Semi-Annual Report*, July–December 1973, app. 11, 31, and January–June 1970, 4.

62 Ibid., 5, 38.

birthday. Many men were eligible for student deferments while they were undergraduates, and, for part of the war, deferments were also available for graduate school. After college, a man could file a new application to be a CO. If turned down, he could go through multiple levels of appeal: an appearance before his local draft board, an appeal to the state appeals board, a presidential appeal. These appeals could easily take a year or two, and Selective Service could not issue an induction order while an appeal was pending. Moreover, as early as the 1950s, courts required that a local draft board reopen a man's case every time he submitted new information in support of a requested classification (including CO).[63] After years of appeals were over, it was usually a simple matter to file some new information with a draft board, have one's case reopened, and start the multiyear process of appeals all over again. In 1949, state boards considered a mere 6,920 appeals; in 1969, that number had shot up to 165,064 (or 61.8 percent of the total draft call of 1969, which was 266,900 men).[64]

Men for whom none of these things worked could (and did) flee to Canada. The US-Canada extradition treaty provided that each country would extradite to the other country people who were accused of acts that were crimes in the extraditing country. Because Canada had no draft law, running afoul of conscription was not an offense in Canada. Estimates of how many men fled to Canada during the Vietnam era vary. John Hagan estimates that 50,000 Americans of draft age fled to Canada in "the largest northward exodus since the American revolution." Flynn puts the figure at 60,000 to 100,000; Baskir and Strauss say that 100,000 "was close to the truth."[65]

As noted earlier, 106,000 men escaped conscription because their draft boards had made procedural or clerical errors somewhere along the way. One factor that explains this stunningly high figure is that courts required that when a man was turned down for a classification that he had sought, the draft board must state the reason for the denial in writing.[66] The requirement seems simple enough, but harried, overworked draft board clerks often did not do this.

63 Selective Service regulations stated that reopening a man's case when new information was submitted was permissive. 32 C.F.R. § 1625.13. However, courts consistently interpreted the process as required: *Mulloy v. United States*, 398 U.S. 410, 415 n.3 (1970), which cites multiple cases, including *United States v. Burlich*, 257 F. Supp. 906, 911 (S.D.N.Y. 1966); *United States v. Vincelli*, 215 F.2d 210 (2d Cir. 1954), also supports this proposition.

64 *Semi-Annual Report*, July–December 1970, 43; *Semi-Annual Report*, January–June 1972, 52.

65 John Hagan, *Northern Passage: American Vietnam War Resisters in Canada* (Cambridge, MA: Harvard University Press, 2001), 35; Flynn, *Draft, 1940–1973*, 180; Baskir and Strauss, *Chance and Circumstance*, 180.

66 United States v. Jones, 468 F.2d 454 (3d Cir. 1972); United States *ex rel.* Morton v. McBee, 310 F. Supp. 328 (N.D. Ill. 1972).

Exacerbating matters for the Selective Service was the nationwide network of thousands draft counselors associated with antiwar and antidraft organizations that provided free draft counseling to hundreds of thousands of men. These counselors ensured that men were aware of highly technical aspects of the draft, such as the availability of time-consuming appeals, the obligation of draft boards to reopen cases upon receipt of new information, and the requirement that classification denials be explained in writing.[67]

5 Conscientious Objectors

During the Vietnam era, the number of men seeking I-O status (CO available for alternative civilian service) soared (see Table 7.2). The Selective Service

TABLE 7.2 Number of conscientious objectors by year, Vietnam era

Year	I-Os (COs awaiting civilian placement)	I-Ws (working COs)	Total I-Os and I-Ws
1963	7,893	2,147	10,040
1964	10,549	2,322	12,881
1965	10,423	3,182	13,605
1966	9,345	5,547	14,892
1967	11,041	6,367	17,408
1968	13,341	6,402	19,743
1969	15,855	8,743	24,598
1970	28,188	10,221	38,409
1971	36,713	12,677	49,390
1972	9,920	9,758	19,678

SOURCE: US SELECTIVE SERVICE SYSTEM, *SEMI-ANNUAL REPORT OF THE DIRECTOR OF SELECTIVE SERVICE FOR THE PERIOD JULY 1–DECEMBER 31, 1973 TO THE CONGRESS OF THE UNITED STATES PURSUANT TO THE MILITARY SELECTIVE SERVICE ACT OF 1967* (WASHINGTON, DC: GOVERNMENT PRINTING OFFICE, N.D.), APP. 13, 32.

Note: Selective Service used the I-W classification to denote both men currently performing alternative civilian service and those who had completed their alternative civilian service (see Table 6.3). The numbers of men classified I-W in this table reflect only those who were currently performing alternative civilian service.

67 Baskir and Strauss, *Chance and Circumstance*, 37–48.

combined the figures for men classified I-A (immediately eligible for combatant duty) and I-A-O (available for noncombatant military service); thus, no separate count exists for men classified I-A-O.

Recall that in 1948 Congress had sought to narrow the definition of CO by requiring that COs not merely come their position through religion but must also believe in God. That language remained unchanged in successive amendments to the statute. In the 1960s, the Central Committee for Conscientious Objectors brought a court challenge to this language that eventually made its way to the Supreme Court (see chapter 6).

On March 8, 1965, the Supreme Court decided *United States v. Seeger*. The three men who claimed to be COs in that case (Seeger, Jakobson, and Peter) had challenged the cited language as violating the freedom of religion guarantees in the First Amendment. This argument was similar to Walter Nelles's argument for the National Civil Liberties Bureau in his Supreme Court amicus brief in the *Arver* case during World War I. The argument had been plied repeatedly since 1918 but had never been successful—until *Seeger*. In *Seeger*, the court held that, despite the clear language of the statute, COs did not have to base their objection on belief in a Supreme Being: "We think it clear that the beliefs which prompted [Seeger's] objection occupy the same place in his life as the belief in a deity holds in the lives of" others. Although the court recognized the First Amendment issue in *Seeger*, its holding sidestepped it. Instead the court grounded its ruling in statutory interpretation, saying that Congress had not really meant to exclude men such as Seeger from CO status. In *Seeger*, the statutory language excluding political, sociological, or philosophical beliefs remained untouched.[68]

Nevertheless, the court's ruling in *Seeger* worked a major expansion of the definition of what constituted a CO, and not everyone was happy about that. When Congress was considering the 1967 amendments to the draft law, the retired general Mark W. Clark testified, falsely, that the court's holding in *Seeger* meant that every man could decide for himself whether he is a CO: "There is danger in permitting the individual to be the judge in his own case. If the current interpretation of the draft law as made by the Supreme Court in the *Seeger* case is permitted to stand unchanged, it will, for practical purposes, permit individual registrants the right to determine for themselves whether or not they wish to bear arms in support of their country." If Clark actually believed this—and it appeared that he did, because he repeated it three times in his testimony—then he was a fool, because *Seeger* said nothing of the sort.[69]

68 United States v. Seeger, 380 U.S. 163 (1965).

69 *1967 Senate Hearings*, statement of Mark W. Clark, April 12, 1967, 21, 35, 36.

If Clark did not believe what he was saying, he was a liar. In any event, Congress understood, correctly, that the 1948 language had been meant to restrict, not expand, the definition of who was a CO. In an effort to undo the Supreme Court's holding in *Seeger*, Congress deleted the Supreme Being language in the 1967 amendments but left intact the language prohibiting merely political, sociological, or philosophical beliefs.

Congress's efforts to restrict who could be a CO failed. Three years after the 1967 amendments, the Supreme Court, in *Welsh v. United States*, expanded the earlier *Seeger* decision. Where *Seeger* had said that religious beliefs did not have to include a belief in God, *Welsh* held that a CO's beliefs did not even have to be religious in the classical sense. Strong ethical beliefs that hold the same place in a secular man's life as religion holds for a religious man qualify. Once again, the court sidestepped the constitutional issue and rooted its holding in statutory interpretation.[70]

The *Welsh* decision badly shook up the Selective Service. The agency questioned "whether [after *Welsh*] the System can continue to function." The problem was that the "Selective Service will certainly become involved in more litigation. Many more men will probably appeal their cases as a legal means of delay, particularly when such a delay offers ... the possibility of avoiding service." So unhappy was the Selective Service that more than two years after the court's ruling, Tarr admitted to Congress that the system had not yet revised its Form 150 for COs to account for the fact that COs need no longer be religious.[71]

Within days of the court's ruling, however, the Selective Service issued Local Board Memorandum No. 107 to explain the *Welsh* ruling. The memorandum was incoherent. It stressed that a "personal moral code" did *not* qualify a man to be a CO under the law but that "solely moral or ethical beliefs" did qualify.[72]

Bill Raley, a law professor, argues that the combined effects of *Seeger* and especially *Welsh* effectively killed the draft. Raley points out that, in the first month after *Welsh*, there were 100,000 new CO applications, that *Welsh* "provided draft law violators with a silver bullet defense," and that, after *Welsh*, the number of CO classifications granted actually exceeded the number of men inducted. Raley's assessment of the far-reaching consequences of these Supreme Court decisions is fair, but his conclusion that these decisions "killed

---

70 Welsh v. United States, 398 U.S. 333 (1970).

71 *Semi-Annual Report*, January–June 1970, 8, and January–June 1972, 45.

72 "National Headquarters July 6, 1970 Local Board Memorandum 107," which was reprinted in the *Semi-Annual Report*, January–June 1970, 39–40. The same direct contradiction was repeated several times throughout the document.

the draft" (Raley's words) is overblown, for it ignores multiple other factors at work.[73]

6 Draft Resistance

Perhaps the most important factor that Raley's analysis ignores is the role of draft resisters and the national network that the resisters created, which they called the Resistance. An important distinction must be observed here. Draft *resisters* were not draft *evaders*. Resisters engaged in total noncooperation with the draft; their World War I counterparts were known as absolutists. If a resister came to his position of noncooperation before his eighteenth birthday, he would publicly refuse to register; if he came to his position after he had already registered, he would send his draft card back to the government (often to his own draft board, sometimes to another government official) or publicly destroy his draft card; both were symbolic ways of severing ties with the Selective Service.[74] If called for induction, resisters would publicly refuse induction.

A flyer by the New York Resistance chapter put it this way: "The Resistance is a community of young people who refuse to cooperate in any way with the Selective Service System. We accept no deferments, carry no draft cards, and refuse to submit to induction. In doing so, we have directly challenged the illegitimate war-making power our government has taken for itself. As a result, we of the Resistance face prison—not as an end, but as a consequence of our acts." That is, members of the Resistance were the opposite of draft *dodgers*, who cooperated with the Selective Service and zealously (and not always legally) sought exemptions or deferments. A flyer by the Twin Cities Resistance explained, "We are against draft-dodging, because that's the easy way out.... Our responsibility is ... not to run away from it with deferments or exemptions

73 Bill Raley, "How Conscientious Objectors Killed the Draft: The Collapse of the Selective Service during the Vietnam War," *Cleveland State Law Review* 68, no. 2 (2020): 151. Raley also argues that the Selective Service statute's CO provision clearly violated the establishment clause of the First Amendment by discriminating in favor of some religious beliefs and against others (although he does not cite Walter Nelles as the progenitor of that argument). Raley surmises that the Supreme Court may have been reluctant to eliminate the CO provision by judicial fiat and therefore elected to expand the CO provision to address and cure the constitutional defect.

74 Elmer, *Felon for Peace*, 54–56.

or emigrations, but to stand up to the System: to confront the war-makers head-on: to resist."[75]

An interesting precursor to draft resistance during the Vietnam era was a series of (mostly student) "We Won't Go" statements published starting in April 1964. In that month, a group of 87 young people issued a statement self-consciously modeled on a similar statement that French students had used during the war in Algeria.[76] On December 29, 1966, 100 student leaders—student council presidents and college newspaper editors—released a similar statement. This one made the front page of the *New York Times*, which noted that "as elected student leaders, the students represent a far more moderate university group than the members of the student New Left."[77] On April 17, 1967, the Harvard *Crimson* printed a statement signed by 86 Harvard students that said, "[We] declare our intention to refuse military service while the United States is fighting in Vietnam."[78] In May 1967, 250 students from twenty-five medical schools published a similar statement.[79]

Of course, as others have noted, for people holding safe student deferments, announcing "We Won't Go" was a mostly vacuous statement.[80] Specifically, such a statement leaves open the possibility of avoiding the draft by cooperating with it and evading an induction order by retaining the student deferment, feigning homosexuality, faking bone spurs in a foot, or emigrating to Canada. Nevertheless, these We Won't Go statements were often a sort of halfway house to actual resistance. For example, the Chicago Area Draft Resistance evolved from an earlier We Won't Go entity, as did the Boston Draft Resistance Group (after the list of signers of the Harvard statement had grown from the original 86 to 330).[81]

75 One-page flyer titled "The Resistance," n.d., CDGA Resistance box, folder labeled "The Resistance"; two-page flyer, "The Fourth of December," n.d., CDGA Resistance box, folder labeled "1968 Copies of Anti-Draft Statements."

76 Lawrence O'Kane, "87 Youths Sign an Ad Declaring They Won't Fight in Vietnam," *New York Times*, April 26, 1964; Ferber and Lynd, *Resistance*, 59.

77 Terence Smith, "Student Leaders Warn President on Doubts on War," *New York Times*, December 30 1967, 1; Ferber and Lynd, *Resistance*, 60. Ferber and Lynd refer to the *New York Times* article but mistakenly report that it appeared on December 29, 1966.

78 "The Boston Draft Resistance Group," n.d., Collected Document Group–A (CDGA), Boston Draft Resistance Group box, folder labeled "Position Papers, Form Letters—1967," Swarthmore College Peace Collection (SCPC), Swarthmore College, Swarthmore, PA.

79 Ferber and Lynd, *Resistance*, 64.

80 Foley, *Confronting the War Machine*, 77; Ferber and Lynd, *Resistance*, 47–65, 81.

81 Ferber and Lynd, *Resistance*, 62; "Boston Draft Resistance Group," and Paul Hamburg and Bill Hunt, "Boston Draft Resistance Group," n.d., both in the same CDGA box and folder cited in n101.

In mid-1965, in the wake of some sporadic draft card burnings and antidraft protests, Congress unintentionally gave the draft resistance movement a huge boost. On August 20, a photo appeared in *Life* magazine of Chris Kearns burning his draft card. In response, Congress quickly passed a one-sentence amendment to the draft law making burning a draft card its own separate offense, a felony punishable by five years in prison and/or a fine of as much as $10,000. Johnson signed the statute into law August 30.[82] The new law provided a powerful way for would-be draft resisters who had previously registered (and thus acquired draft cards) to sever their ties with the Selective Service in a dramatic public way that often garnered considerable publicity. It also provided a way for the wives, girlfriends, or colleagues of would-be resisters to associate themselves with illegal antidraft activity by burning the draft card of the new resister. In this way, the new law was similar to the World War II statute that required men as old as sixty-five to register: both laws created opportunities for people who were not themselves draft eligible to violate the draft law directly. During the Vietnam era, many people took advantage of this opportunity.

The first person to publicly burn a draft card after the new law was passed was David Miller of the Catholic Worker movement, who was described at the time by Jim Forest, also of the Catholic Worker, as "a tall, Nordic athlete who could grace any recruiting poster."[83] Miller burned his draft card in New York on October 15, 1965, at one of the nationwide demonstrations on that date. This was big news. The next day, Miller's action was on the front page of the *New York Times*, *St. Louis Post-Dispatch*, *San Francisco Examiner*, *Los Angeles Times*, *Chicago Tribune*, *Philadelphia Daily News*, and other newspapers.[84] Photographs of Miller, in suit and tie while burning his draft card, appeared in the *New York Daily News*, *St. Petersburg* (FL) *Times*, *Tucson* (AZ) *Citizen*, and other newspapers.[85]

82 "Doubling the Draft: When They Tag You I-A," *Life*, August 20, 1965, 30; An Act To amend the Universal Military Training and Service Act of 1951, as amended, Pub. L. No. 89-152, 79 Stat. 586 (1965).

83 Douglas Robinson, "Policy in Vietnam Scored in Rallies throughout the U.S.," *New York Times*, October 16, 1965; James H. Forest, "In Time of War," in Berrigan et al., *Delivered into Resistance*, 6.

84 Robinson, "Policy in Vietnam Scored in Rallies"; "Thousands Join Street Rallies Denouncing U.S. Viet Nam Policy," *St. Louis Post-Dispatch*, October 16, 1965; "The Antis ... and the Anti-Antis," *San Francisco Examiner*, October 16, 1965; "Campus Groups Protest Viet Policy across U.S.," *Los Angeles Times*, October 16, 1965; "Pickets Tussle in Loop over U.S. Viet Policy," *Chicago Tribune*, October 16, 1965; "Cops Braced Coast-to-Coast for More Anti-War Protests," *Philadelphia Daily News*, October 16, 1965.

85 Henry Lee, "Pacifist Burns Draft Card in Viet Protest," *New York Daily News*, October 16, 1965; "Viet Nam Protests Staged across U.S.," *St. Petersburg* (FL) *Times*, October 16, 1965; "Demonstrator Burns Draft Card," *Tucson* (AZ) *Citizen*, October 16, 1965.

On November 6, 1965, five men, including Tom Cornell of the Catholic Peace Fellowship and David McReynolds of the War Resisters League, burned their draft cards at a rally in New York, while A. J. Muste looked on.[86] If anything, this was even bigger news. Articles, with large photographs of the draft card burning, appeared on the front pages of the *New York Times*, *Philadelphia Inquirer*, *Los Angeles Times*, *St. Louis Post-Dispatch*, and other newspapers. Articles and photographs appeared on the inside pages of the *San Francisco Examiner*, *Des Moines Register*, *Jackson* (TN) *Sun*, *Indianapolis Star*, and other papers.[87]

On March 31, 1966, David Reed, David O'Brien, David Benson, and John Phillips burned their draft cards in Boston at a demonstration organized by the New England Committee for Nonviolent Action.[88]

Nearly all these men were prosecuted in due course, and several challenged the constitutionality of the law making burning a draft card (one's own or someone else's) illegal. The first such case to reach the US Supreme Court was David O'Brien's; he was represented by the ACLU. His argument was that burning a draft card should be protected under the First Amendment as symbolic speech. The court rejected the argument and upheld the conviction (and the statute) with only Justice William O. Douglas dissenting. In recent decades, the *O'Brien* decision has been the subject of much well-deserved criticism by the legal academy. The constitutional scholar Michael Dorf has called *O'Brien* "the worst of all possible worlds"; James McGoldrick has said *O'Brien* should be "buried with finality"; and Jed Rubenfeld has opined that O'Brien "is deeply, demonstrably flawed."[89]

∴

The first large-scale draft card burning of the Vietnam era occurred in New York on April 15, 1967, the same day as a massive, legal antiwar demonstration. As

86 Douglas Robinson, "5 Draft Card Burners Doused at Rally," *New York* Times, November 7, 1965; Elmer, *Felon for Peace*, 60.

87 Robinson, "5 Draft Card Burners Doused at Rally"; "5 Pacifists in N.Y. Burn Draft Cards," *Philadelphia Inquirer*, November 7, 1965, 1; "5 Pacifists Burn Draft Cards as FBI Watches," *Los Angeles Times*, November 7, 1965, 1; "Draft Cards Are Burned by Five at Rally," *St. Louis Post-Dispatch*, November 7, 1965, 1; "5 Pacifists Burn Cards," *San Francisco Examiner*, November 7, 1965, 5; "5 Burn Draft Cards, Doused," *Des Moines Register*, November 7, 1965, 4; "Five Men Burn Draft Cards," *Jackson* (TN) *Sun*, November 7, 1965, 5; "Five Burn Draft Cards," *Indianapolis Star*, November 7, 1965.

88 Foley, *Confronting the War Machine*, 205.

89 United States v. O'Brien, 391 U.S. 367 (1968); Michael C. Dorf, "Incidental Burdens on Fundamental Rights," *Harvard Law Review* 109, no. 6 (April 1996): 1204; James M. McGoldrick, "*United States v. O'Brien* Revisited: Of Burning Things, Waving Things, and G-Strings," *University of Memphis Law Review* 36, no. 4 (Summer 2006): 905; Jed Rubenfeld, "The First Amendment's Purpose," *Stanford Law Review* 53, no. 4 (April 2001): 771.

late as the evening before the card-burning event, organizers had no idea how many men would participate; on the evening before the burning, 57 men said they would take part the next day. However, 175 draft cards were burned, putting Bruce Dancis, one of the lead organizers of the event, into a position similar to the one that George Houser had been in after the Break With Conscription demonstrations in February 1947: delighted that they had had far more participation than they had expected.[90]

Also during the spring of 1967, a small group on the West Coast, including Stanford University student council president David Harris (who later was briefly married to the folk singer Joan Baez), Lennie Heller, and Dennis Sweeny, conceived of a new organization that would be composed of those who explicitly refused any cooperation—registration, carrying draft cards, induction—with the Selective Service.[91] Harris, Heller, and Sweeny came up with the name for their new creation, the Resistance (with a capital *R*), deliberately invoking the French Resistance to the Nazis during World War II.[92] The initial plan of the Resistance organizers was to conduct a large-scale, nationwide draft card burning on October 16, 1967.[93] The idea of the Resistance caught on, and by January 1968 the Resistance had "regional offices" in Berkeley, Boston, Chicago, Los Angeles, and Washington, DC, with thirty smaller chapters throughout the country.[94]

To be sure, the Resistance was never a unitary organization; although at its height there were chapters in thirty to fifty cities, the Resistance was a largely anarchic confederation of draft noncooperators who differed from each other in style and tone and often did not even know about each other's existence.[95] A couple of examples will illustrate this point. While Harris and Sweeny were organizing on the West Coast what they believed would be the first large draft card burning or turn-in of the Vietnam era on October 16, 1967, Dancis was organizing what actually was the first large draft card burning of the Vietnam

90 Bruce Dancis, *Resister: A Story of Protest and Prison during the Vietnam War* (Ithaca, NY: Cornell University Press, 2014), 88–102; Eric Prokosch, "Draft Resistance," *WIN Magazine* 3, no. 8 (April 28, 1967): 6.

91 David Harris (1946–2023) wrote three partially overlapping memoirs. His first, *Goliath* (New York: Avon Books, 1970), consisted of his philosophical musings. His second, *Dreams Die Hard: Three Men's Journey through the Sixties* (San Francisco: Mercury House, 1982), described the intertwined histories of the resistance organizer Dennis Sweeny, the politician Allard Lowenstein, and Harris. Harris's third, *Our War: What We Did in Vietnam and What It Did to Us* (New York: Random House, 1996), described his reflections on the Vietnam War.

92 Ferber and Lynd, *Resistance*, 89–91; Foley, *Confronting the War Machine*, 77–78.

93 Harris, *Dreams Die Hard*, 181.

94 "Press Release, January 31, 1968," folder labeled "Leaflet, pamphlets, press releases," and a five-page attendance list for the resistance conference, folder labeled "Correspondence," both in the CDGA box labeled "The Resistance" (hereafter, CDGA Resistance box), SCPC.

95 Elmer, *Felon for Peace*, 54–55.

era on April 15, 1967, on the East Coast. Neither group knew about the other.[96] The West Coast Harris-Sweeny group has been described as "a unique blend of cowboys, Nietzsche, drugs, Jung [and] motorcycles,." while other resisters consistently appeared at public events in ties and jackets.[97]

A second major, unintended government gift to the Resistance was the revelation of General Lewis Hershey's memorandum entitled "Channeling." Although first circulated to draft boards in 1965, news of the memorandum first appeared in January 1967 on the front page of the house organ of Students for a Democratic Society, *New Left Notes*. Hershey's memorandum had been contained in a Selective Service orientation kit. In the memorandum, Hershey explained that only one function of the Selective Service was to obtain some men for induction into the army; an equally important, parallel function was, through deferments and exemptions, to channel all men into areas that the government deemed to be in the national interest. This "is the American or indirect way of achieving what is done by direction in foreign countries where choice is not allowed," Hershey said. Referring to a draft registrant, Hershey said channeling "engenders a sense of fear, uncertainty, and dissatisfaction"; this is a good thing, according to Hershey, because it "motivates him."[98]

Not many people read *New Left Notes,* but the story was picked up widely. To take but one example, *Ramparts* magazine reprinted the original text of the memorandum in its entirety, alongside an article about the memorandum, which said, in part, "It is a classic totalitarian document, its language cloaked in reasonableness, which makes it clear that the government has assumed the right to control and direct the life choices of young Americans by the manipulative use of the draft and its exemptions." Much later the historian Michael Foley, referring to the publication (and contents) of the memorandum, wrote, "These were stunning revelations."[99]

Soon, chapters of the Resistance all over the country printed and distributed copies of the channeling memorandum. Often the document came with

96 Dancis, *Resister*, 102; Lenny Heller and Steve Hamilton, "Plan for Implementation, Spring 1967" (undated, but on p. 3 someone scrawled, "Very early plan"), CDGA Resistance box, folder labeled "Leaflets, Pamphlets, Releases"; Fred Rosen, "Draft Resistance," *WIN Magazine* 3, no. 12 (June 30, 1967): 4; Richard Boardman, "Resistance Is, April 30, 1968 [Not for Publication]," CDGA, Draft Resistance/Evasion/Desertion box, folder labeled "Draft Resistance/Evasion/Desertion," SCPC.

97 Ferber and Lynd, *Resistance*, 78; Dancis, *Resister*, 100.

98 Peter Henig, "Selective Service System: On the Manpower Channelers," *New Left Notes* 2, no. 3 (January 20, 1967): 1; US Selective Service System, "Channeling," a document dated 1965 that was included in the service's orientation kit for draft boards nationwide, in CDGA Resistance box, folder labeled "The Resistance," SCPC.

99 "The Redress of Their Grievances: An Editorial Essay," *Ramparts* 6, no. 5 (December 1967): 28; Foley, *Confronting the War Machine*, 62.

a notation like the one on the brochure printed by ten Resistance offices in California: "This is an unedited, official Selective Service document."[100] It may be ironic that the single document most widely printed and distributed by the Resistance was penned by the director of the Selective Service, but members of the Resistance knew what they were doing. Hershey's "Channeling" memorandum was a great recruitment tool for the Resistance.

After January 1967, the personal statements of many draft resisters cited and quoted "Channeling." In particular, they cited the memorandum to explain why they were unwilling to apply for CO status or accept a student deferment: channeling men who did not go into the army was a central purpose of the Selective Service, and these resisters wanted no part of that.[101]

Not surprisingly, in the wake of the April 15 draft card burning in Central Park's Sheep Meadow, a federal grand jury was convened to investigate the demonstration. By law, the proper role of a grand jury is limited to investigating events after the fact (and deciding whether those events ought to give rise to a criminal prosecution). However, by autumn, federal authorities had decided to use this grand jury improperly in an attempt to intimidate and deter people from participating in the upcoming national draft card turn-in planned for October 16. On October 5, federal prosecutors met with FBI agents in New York. According to a previously secret memorandum to FBI Director J. Edgar Hoover, obtained through a Freedom of Information Act request, "It was decided at this conference that subpoenas will be served by U.S. Marshals during the week of 10/9/67, with a view in mind that this action would act as a possible deterrent to those individuals who anticipated participating in the [draft card turn-in] demonstration scheduled for the week of 10/16/67." The memorandum added, "This matter will be followed very closely."[102]

After April 15, the next big draft card burning/turn-in was the October 16 event planned by the Resistance on the West Coast. In the wake of the April 15 success, Resistance organizing grew rapidly all over the country. On October 16,

100 See multiple documents in the CDGA Resistance box, in particular, the first of several folders labeled "The Resistance" and the third of several folders labeled "Leaflets, Pamphlets, Press Releases," as well as a folder labeled "Philadelphia Resistance," SCPC.

101 See multiple documents in the CDGA Resistance box, in particular, the folder labeled "1968 Copies of Anti-Draft Statements," the first of several folders labeled "The Resistance," the third of several folders labeled "Leaflets, Pamphlets, Press Releases," and folder labeled "Philadelphia Resistance."

102 Airtel Memorandum from SAC [special agent in charge], New York, to Director, FBI, October 5, 1967, Subject: "Card Burning Demonstration, NYC, April 15, 1967," obtained by the author through a Freedom of Information Act request.

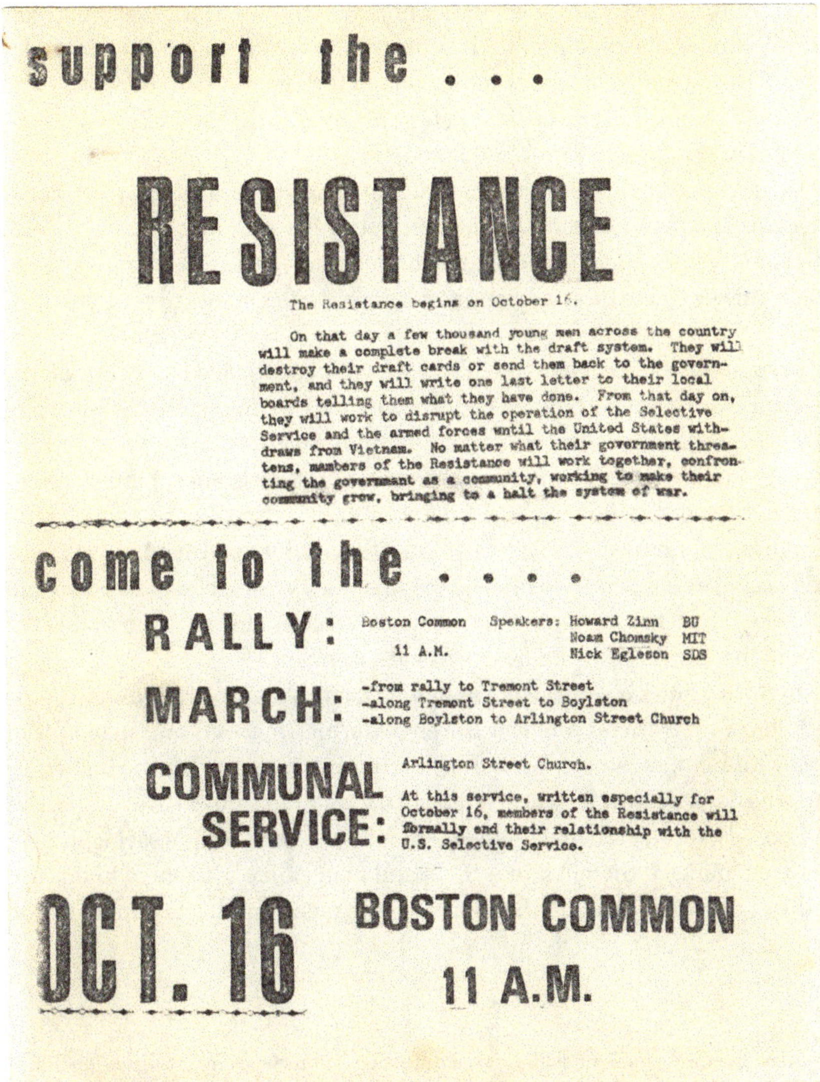

FIGURE 7.1 Publicity flyer for the October 16, 1967, national draft card turn-in organized by the Resistance
COURTESY OF THE SWARTHMORE COLLEGE PEACE COLLECTION ARCHIVES

twelve hundred to fourteen hundred men burned their draft cards (or returned them to the government), and demonstrations and protests were held in thirty cities nationwide (see the flyer for the Boston rally in Figure 7.1).[103] The draft

103 Unsigned flyer by the New York Resistance, "The Resistance/Dec. 4," CDGA Resistance box, folder labeled "The Resistance"; New York Resistance, "Press Release," n.d., CDGA Resistance

card turn-in/burning coincided with a broader Stop the Draft Week that saw tens of thousands of antidraft demonstrators—with hundreds of arrests for civil disobedience—gathering in many cities across the country.[104] The October 16 draft card turn-in also led to some interesting (and unusual) repercussions, which I discuss later in this chapter.

The next nationally coordinated draft card turn-in organized by the Resistance was less than two months later, on December 4, 1967; six hundred more draft cards were burned or returned to the government. The next one was April 3, 1968; this time one thousand resisters in thirty cities across the country took part.[105]

Throughout its existence, members of the decentralized network of the Resistance came up with many creative methods to dramatize their actions. Steve Suffet ate his draft card in late 1966. In May 1967, instead of turning himself in to start a three-month sentence for draft resistance, John Stephens remained at work at the New England Committee for Nonviolent Action's farm in Voluntown, Connecticut. When US marshals came to arrest him, his wife, Candy, handcuffed herself to him. By the time they were separated, CNVA staffers had completely surrounded the marshals' vehicles and sat down so that the car couldn't move. It took hours to extricate Stephens.[106]

On April 3, 1968, a young woman named Dena Shoshana Rubin tore up the draft card of her friend Michael Dover (with his approval) and mailed it to the government, together with her own letter explaining what she had done. The same day, two New Jersey men returned their draft cards to their respective draft boards; fourteen people signed letters to the draft boards saying in part, "I have encouraged them, I honor them and I am at one with them in this act. Further, I advocate that all other draft-eligible men take this or similar action

box, the third of several folders labeled "Leaflets, Pamphlets, Press Releases"; page 2 of four-page brochure by the Philadelphia Resistance, "What Is Resistance? Strategy/Tactics /Purpose," n.d., CDGA, Combined box, folder labeled "Philadelphia Resistance," SCPC.

104 Wallace Turner, "Antiwar Demonstrations Held outside Draft Boards across U.S./119 Arrested on Coast," *New York Times*, October 17, 1967; "The Oakland Seven," n.d., brochure about Stop the Draft Week demonstrations in Oakland, California, CDGA, Collective box, folder labeled "The Oakland Seven," SCPC. The *Times* article describes events in Oakland, Chicago, Boston, Cincinnati, Philadelphia, Oregon, Ithaca, and Washington, DC.

105 Philadelphia Resistance, "What Is Resistance?" 2; two-page New York Resistance press release dated March 29, 1968, in CDGA Resistance box, the third of several folders marked "Leaflets, Pamphlets, Press Releases," SCPC.

106 Steve Suffet, "Why I Ate My Draft Card," *WIN Magazine* 3, no. 3, no. 2 (January 27, 1967): 3; Steve Trimm, "American Tragedy," *WIN Magazine* 3, no. 12 (June 30, 1967): 6.

as soon as possible."[107] (These letters were remarkably similar in content and tone to the letter that had gotten Larry Gara prosecuted in 1948.)

On May 6, when C. J. Hinke, a high school student, refused to register in Newark, New Jersey, dozens of his supporters accompanied him to the Federal Building to insist that they be arrested along with him. On June 13, when Don Baty was scheduled to be arraigned for induction refusal, he took sanctuary in Washington Square Methodist Church in New York and was joined by hundreds of supporters who later tried nonviolently to block his arrest by US marshals.[108] On June 17, several women, including Irene Johnson and Jill Bosky, burned draft cards of male friends in Washington, DC, in a demonstration sponsored by the CNVA.[109] On July 2, when a seminarian refused induction at the Whitehall Street Induction Center in New York, he chained himself to his wife as well as to three ministers and a rabbi.[110]

In April 1969, when Bob Eaton, a Quaker, was indicted for his draft resistance, he took sanctuary in a Friends Meeting House, where he chained himself to fourteen friends; it took "a crowd of FBI agents, burly Philadelphia policemen and Highway Patrolmen armed with everything from trained police dogs to heavy cable cutters" to arrest Eaton. On September 5, when David Rosen burned his draft card at his draft board in Manhasset, New York, he first tore it into several pieces, one of which was burned by his friend Pam Fadem.[111]

In February 1971, John Luginbill, walked into his draft board in Portland, Indiana, asked to see his own file (which, by law, Selective Service was obligated to produce on request), walked out with the file, and destroyed it. On April 9, 1971, a man took his draft file from his Toledo, Ohio, draft board and destroyed the file.[112]

---

107 Dena Shoshana Rubin to Whom It May Concern, April 3, 1968; Michael J. Dover to "Whoever May Receive This," n.d., both in CDGA Resistance box, folder labeled "Copies of Antidraft Statements," SCPC; "To the Authorities of the Selective Service System," n.d., CDGA Resistance box, folder labeled "1968 Copies of Antidraft Statements," SCPC.

108 New York Resistance, "For Release April 30," CDGA Resistance box, third of several folders labeled "Leaflets, Pamphlets, Press Releases," SCPC; "Sanctuary for Draft Resister Don Baty," undated flyer, CDGA Resistance box, the third of several folders labeled "Leaflets, Pamphlets, Press Releases."

109 Marian Mollin, *Radical Pacifism in Modern America: Egalitarianism and Protest* (Philadelphia: University of Pennsylvania Press, 2006), 104.

110 "Clergy and Laity Concerned about Vietnam Press Release," July 2, 1968, CDGA Resistance box, the third of several folders labeled, "Leaflets, Pamphlets, Press Releases."

111 Ofelia Alayeto, "Bob Eaton Busted on Alternative Service," *WIN Magazine* 5, no. 11 (June 1, 1969): 25; Elmer, *Felon for Peace*, 55.

112 John Luginbill, "Celebrate Life," undated flyer, CDGA, Peacemakers Collected Records, box 2, folder labeled "Efforts: War/Draft Resistance"; John Luginbill, "Celebrate Life,"

7 The Call to Resist Illegitimate Authority and the Spock-Coffin Indictment

On October 2, 1967, "A Call to Resist Illegitimate Authority" was released at a press conference in New York. The Call to Resist was conceived by Paul Lauter, peace education secretary for the American Friends Service Committee; Noam Chomsky, the linguist, anarchist, and public intellectual; and others as a vehicle for people not eligible for the draft (by virtue of age or sex) to directly support acts of resistance by young men. The Call to Resist listed several ways of resisting illegitimate authority, including members of the military refusing immoral orders to serve in Vietnam and civilians refusing induction or otherwise resisting the draft. It then said: "We believe that each of these forms of resistance against illegitimate authority is courageous and justified.... We will continue to lend our support to those who undertake resistance to this war." The statement continued by enumerating specific ways that the signers would aid resisters. The 320 prominent signers included clergy, professors, and others. The Reverend William Sloane Coffin, Jr., chaplain of Yale, chaired the October 2 press conference and announced that the signers supported the national draft card turn-in later that month. Also participating in the press conference were the pediatrician Benjamin Spock; Mitchell Goodman, the writer and antiwar activist; and Marcus Raskin, codirector of the Institute for Policy Studies, all of whom were signers of the Call to Resist.[113]

On Monday, October 16, the national draft card turn-in was held. The event in Boston, at Arlington Street Church, was the second-largest turn-in in the country that day, with 281 draft cards burned or returned. Michael Ferber, one of the organizers, delivered a speech encouraging draft resistance. Coffin both spoke and received the draft cards turned in by resisters. Other speakers included the historian Howard Zinn.[114]

On Friday, October 20, four days after the national draft card turn-in and on the eve of the massive antiwar demonstration in Washington, a delegation from

---

Peacemaker 24, no. 3 (February 27, 1971); untitled, one-page typed sheet with an attached, undated clipping from an unidentified newspaper about this incident, in CDGA, Peacemakers Collected Records, box 2, folder labeled "Efforts: War/Draft Resistance," SCPC. The typed sheet says that the event occurred on April 9, 1971, but the sheet itself is undated.

113 [Untitled Article], *Resist*, no. 1 (November 1967): 1; Martin Gansberg, "War Foes Are Promised Churches as Sanctuary," *New York Times*, October 3, 1967. A copy of the January 5, 1968, indictment of Coffin, Ferber, and others, the Spock-Coffin Indictment No. 68-1-F (1968), 5, is in the author's files.

114 "Trial of the Boston Five," *Resist*, no. 12 (June 10, 1968): 3; Foley, *Confronting the War Machine*, 90–109; Spock-Coffin Indictment, No. 68-1-F (1968), 5.

the Call to Resist, including Coffin, Goodman, Raskin, and Spock, appeared at the Department of Justice in Washington. They carried a briefcase holding 992 of the draft cards that had been returned nationwide the previous week, including 214 of the cards that Coffin had collected in Boston. The delegation was there to return the resisters' cards to the government; this was their way of showing direct complicity with the action of the young resisters. Before the group entered the building, Coffin stood on the steps and announced to the press, "We hereby publicly counsel these men to continue their refusal to serve in the armed forces ... and we pledge ourselves to aid and abet them in all the ways we can. This means that if they are now arrested for failing to comply with the law ... we too must be arrested, for in the sight of the law, we are as guilty as they are." Goodman said, "We will encourage them and aid and abet and counsel them in every way we know how." The group hoped to present the returned draft cards to Attorney General Ramsey Clark, but they were shunted off to an assistant deputy attorney general.[115]

On January 4, 1968, Spock, Coffin, Goodman, Raskin, and Ferber were indicted in Boston on a single count of conspiring to aid and abet draft resistance. Unlike the 139-page indictment of Eugene Debs in 1918, the Spock-Coffin indictment was a mere ten (double-spaced) pages. It focused on three events: the October 2 press conference in New York, the October 16 draft card turn-ins at Arlington Street Church and elsewhere, and the October 20 visit to the Justice Department to turn in the draft cards collected four days earlier. The indictment included a list of eleven so-called overt acts that the defendants had committed in furtherance of their conspiracy, a section mostly written in plain English. The defendants had done precisely what the government claimed they had.[116]

Nationwide support for the defendants was immediate and strong. On January 6, 1968, in St. Louis, 91 people sent a telegram to Clark saying that they had aided and abetted the draft resisters and would continue to do so. On January 14, at a meeting at New York's Town Hall, 1,000 people signed a scroll saying that they were aiding and abetting draft resisters. An editorial in the January 22, 1968, issue of the *Nation* opined, unpersuasively, that "the defendants can argue that what they did was nothing more, essentially, than a

---

115 E. W. Kenworthy, "Thousands Reach Capital to Protest Vietnam War," *New York Times*, October 21, 1967; "Trial of the Boston Five," *Resist*, no. 12 (June 10, 1968): 3; Tom Wells, *War Within: America's Battle over Vietnam* (New York: Henry Holt, 1994), 194–95; Spock-Coffin indictment, 5–6; United States v. Spock, 416 F.2d 165 (1st Cir. 1969); Foley, *Confronting the War Machine*, 107–9, 225–31.

116 Fred P. Graham, "Spock and Coffin Indicted for Activity against Draft," *New York Times*, January 6, 1968; "Trial of the Boston Five," 3; Spock-Coffin indictment.

legitimate extension of the right of free speech." But Michael Foley is exactly right when he says of the Spock-Coffin defendants, "They were asking for it." Foley is also correct that many other resistance organizers could just as well have been indicted for conspiracy to aid and abet draft resistance and that it was a blunder for the government to prosecute well-known figures such as Spock and Coffin instead of lesser-known organizers.[117]

On January 28, when the defendants were arraigned in Boston, 1,000 people attended a demonstration of support at which twenty-nine draft cards were turned in to the aiders and abettors Dave Dellinger; Philip Berrigan, the antiwar priest from Baltimore; and the Harvard chaplain Richard Mumma, who destroyed the cards. One of the cards had belonged to Patch Dellinger, Dave Dellinger's son, who had turned it in. The same day, 4,000 people attended a support meeting in New York where Spock and Ferber spoke; one man burned his induction order at the meeting. A few days later in New York, the writer, anarchist, and public intellectual Paul Goodman; the editor, film critic, and World War II draft opponent Dwight Macdonald; and Dave McReynolds, a leader in the pacifist movement, appeared in front of television cameras to tear up the draft cards of two resisters; they mailed the torn cards, together with their confession, to the attorney general. By November 1967, 1,500 people in forty-two states had signed the original Call to Resist Illegitimate Authority that had helped Spock and Coffin get indicted, and by April 1969, 5,000 people had signed it.[118]

The five men were tried in federal court in Boston that spring. On June 13, 1968, the jury acquitted Raskin and convicted the remaining four defendants.[119]

The four convicted defendants appealed, relying heavily on the First Amendment argument suggested by the editorial in the *Nation*. The group's appellate lawyers were an all-star team that included the leading radical attorney, Leonard Boudin; James St. Clair, who later represented Nixon in the impeachment proceedings; and the former US Supreme Court justice Arthur Goldberg. The

---

117 "The Spock Indictment," *Nation* 206, no. 4 (January 22, 1968): 99; Foley, *Confronting the War Machine*, 230–31.

118 Paul Lauter, "The Indictments: Response," *Resist*, no. 4 (January 19, 1968): 1; Denise Levertov, "Town Hall Evening: A Report," *Resist*, no. 4 (January 19, 1968): 3; Alan Graubard, "The Arraignments: Boston, Jan. 28 and 29," *Resist*, no. 5 (February 2, 1968): 1; John H. Fenton, "Spock, Coffin and Three Others Arraigned on Charges of Conspiracy to Aid Draft Resisters," *New York Times*, January 30, 1968; "4,000 Here Cheer Spock and Ferber; War Policy Decried," *New York Times*, January 30, 1968; Mitchell Goodman, "An Exchange on Resistance," *New York Review of Books* 12, no. 7 (April 10, 1969): 40.

119 John B. Fenton, "Dr. Spock Guilty with 3 Other Men in Antidraft Plot," *New York Times*, June 14, 1968.

appellate court rejected the argument that actively urging and encouraging men to resist the draft—and actually helping them to do that by destroying their draft cards (or returning them to the government)—was mere speech. Speech is speech, and action is action, the court said. The appellate court cited some of the legal precedents discussed in previous chapters—*Schenck* (World War I) and *Gara* (Cold War)—for the proposition that the government can make it illegal to counsel draft evasion, resistance, or avoidance. The appeals court, however, reversed the convictions of Spock and Ferber, saying that there was insufficient evidence to support their convictions. The appeals court also found technical problems with the convictions of Coffin and Goodman, vacated those convictions, and sent the case back to the trial court, directing that Coffin and Goodman be retried without repetition of the procedural errors. The federal government declined to reprosecute Coffin and Goodman. Thus, in the end, none of the five defendants stood convicted.[120]

8 The Draft Board Raids

On October 27, 1967, a few days after the siege of the Pentagon, four men walked into a Baltimore draft board and poured blood over the files as an antiwar and antidraft protest. When they were done, they patiently waited for the FBI to arrive to arrest them.[121]

In May 1968, after the four men had been tried and convicted in federal court but before they had been sentenced for the Baltimore raid, two of those men, Philip Berrigan and Tom Lewis, joined with seven others (including Philip's brother, Daniel Berrigan, who also was a priest) in a new act of civil disobedience.[122] The group, which came to be known as the Catonsville Nine, entered Local Board 33 in Catonsville, Maryland, a Baltimore suburb. They scooped 378 draft files of men classified I-A into wire trash baskets, took the files outside, and burned them in the parking lot with napalm they had made themselves

120 United States v. Spock, 416 F.2d 165 (1st Cir. 1969); John H. Fenton, "U.S. Court Upsets Spock Conviction in Fight on Draft," *New York Times*, June 12, 1969.

121 "Blood Poured on Baltimore Draft Files; Priest, Minister, 2 Others Seized," *New York Times*, October 28, 1967.

122 Philip Berrigan (1923–2002), a World War II veteran and longtime peace activist, was the author of several books, including *A Punishment for Peace* (New York: Macmillan, 1963) and *Prison Journals of a Priest Revolutionary* (New York: Ballantine Books, 1971). For a double biography of Daniel and Philip Berrigan, see Murray Polner and Jim O'Grady, *Disarmed and Dangerous: The Radical Lives and Times of Daniel and Philip Berrigan* (New York: Basic Books, 1997).

from a recipe they had found in a Green Beret manual. Once again, the raiders waited patiently for arrest.[123]

During the coming months and years, similar protests multiplied. As they did, they also changed format in three salient ways. Each of the changes was designed to make the draft board raids more effective at stopping conscription.

First, draft board raiders began attacking multiple draft boards simultaneously. Sometimes these different draft boards were all in one large office complex. For example, on October 31, 1969, a group of draft board raiders destroyed files at the Indianapolis draft board complex that housed forty-four draft boards.[124] On the night of May 24–25, 1969, the Chicago 15 destroyed files at a Chicago draft board complex that housed thirty-four draft boards.[125] In other cases, raiders simultaneously hit draft boards at multiple locations. For example, on the night of November 7–8, 1969, the Boston Eight raided six draft boards in Boston at four locations throughout the city. On May 19, 1970, eleven people who called their group We the People destroyed files at three draft boards in Philadelphia (all at one location) and at another draft board in Lansdowne, Pennsylvania.[126]

In addition, instead of attacking draft boards in broad daylight during office hours, as the protesters in Baltimore and Catonsville had done, protesters started breaking into draft boards on nights and weekends as a way of increasing the time that they would have to destroy files. In many or most of these cases, the raiders later appeared publicly at a demonstration or press conference and distributed statements or press releases explaining what they had done. For example, on August 2, 1969, the New York Eight destroyed files at five draft boards in the Bronx, and on August 15, the group destroyed files at six draft boards in Jamaica, Queens. They surfaced at a press conference in New York on August 21.[127] On July 1, 1969, a group of feminist women calling themselves Women Against Daddy Warbucks destroyed files at a complex that

123 "9 Seize and Burn 600 Draft Files," *New York Times*, May 18, 1968; Elmer, *Felon for Peace*, 67–69.

124 "Notes Are Left at Draft Board," *Indianapolis News*, November 1, 1969; "County Draft Boards to Open Tomorrow Despite Vandalism," *Indianapolis Star*, November 2, 1969; "Beaver 55 Guilty," *Resist Newsletter*, no. 43 (June 18, 1970): 3.

125 Elmer, *Felon for Peace*, 69; "Mark Twain Column," "'Oh, Happy Day' at the Chicago Draft Board," *WIN Magazine* 5, no. 15 (September 1, 1969): 9.

126 "Six Draft Offices in Boston Invaded," *New York Times*, November 9, 1969; Elmer, *Felon for Peace*, 65–93; "We the People," n.d., CDGA, Philadelphia Resistance records box, folder marked "Philadelphia Resistance," SCPC.

127 "Eight Admit Vandalizing Offices of Two Draft Boards Here; 3 Priests in Group," *New York Times*, August 22, 1969.

housed thirteen draft boards in Manhattan; on July 3, the women appeared at a press conference in Rockefeller Center and tossed around confetti made from shredded draft files.[128] The Boston Eight surfaced publicly at a press conference a week after their draft board raids.[129]

Finally, draft board raiders developed a detailed and sophisticated understanding of the cross-reference system that draft boards used for their files. This included chronological ledgers (Selective Service Form 102), alphabetical file cards, and minutes of local board meetings. Often but not always, draft board raiders were meticulous and thorough in their destruction of not only I-A files but of all or much of this cross-reference system. When this occurred, it rendered draft boards inoperable for considerable periods of time.[130]

Estimates of the number of such draft board raids vary widely. Charles Meconis puts the number at several dozen. Baskir and Strauss say that there were "hundreds of similar [to Catonsville] raids on draft boards." Hundreds of separate raids is almost certainly too high a figure. Marian Mollin says, correctly, that Meconis's figures "are the most well documented and … the most reliable."[131] However, Meconis's list misses many cases of draft file destruction, perhaps in part as a result of the ambiguity of what constitutes a raid "similar to Catonsville." In any event, omissions from Meconis's list include, but are not limited to, the following: the June 10, 1968, bombing of the Berkeley, California, draft boards; destruction by arson of all the files of the Xenia, Ohio, draft board in mid-1968; destruction of the files of four draft boards at four locations in

---

128 Will Lissner, "6 War Protesters Seized on 5th Ave.," *New York Times*, July 4, 1969. The authorities botched the arrests of the women when they appeared at Rockefeller Center the day after their clandestine raid on the draft boards. Two women who were not involved in the action, Linda Forest and Barbara Webster, were arrested along with four of the five who were members of Women Against Daddy Warbucks. For unknown reasons the fifth woman, Maggie Geddes, was not arrested until later. This left Geddes free to travel to Canada the next day to deliver a gift of confetti made from shredded draft files that she presented to a delegation of women from the National Liberation Front who also were visiting Canada at the time. Geddes delivered the confetti with a letter of solidarity that said, in part, "In accepting these files, please share with us the hope that such things as bombers, guns, conscription and oppression will cease to exist for both of our peoples." Mollie Flanagan, "Women's Day at the Draft Board," *WIN Magazine* 5, no. 14 (August 6, 1969): 38, and untitled addendum by Maggie Geddes.

129 Elmer, *Felon for Peace*, 65–93.

130 Ibid., 89–90; "We the People" statement; Mollie Flanagan, "Women's Day at the Draft Board," *WIN Magazine*, 5, no. 14 (August 6, 1969): 38; "Beaver 55 Strikes Again," *WIN Magazine* 5, no. 22 (December 15, 1969): 18.

131 Charles A. Meconis, *With Clumsy Grace: The American Catholic Left, 1961–1975* (New York: Seabury Press 1979), 153–66; Baskir and Strauss, *Chance and Circumstance*, 66–67; Mollin, *Radical Pacifism*, 240n45.

northern Ohio in raids between October 1969 and January 1970; an August 20, 1970, arson at the draft board in Reisterstown, Maryland, that destroyed files; and a second raid, by a different group, of the Providence, Rhode Island, draft board complex, this one on Christmas Eve 1970.[132] Because some single raids involved multiple, even dozens, of draft boards (forty-four boards in Indianapolis, thirty-four boards in Chicago), it is clear that hundreds of draft boards were affected.

The Selective Service steadfastly maintained that these raids did not actually slow or stop conscription.[133] But this was not true. A secret FBI memorandum dated November 10, 1970, published after the FBI office in Media, Pennsylvania, was burglarized, showed that the Lansdowne draft board, which had been raided six months earlier (May 19) by We the People, was still unable to carry out even routine functions.[134] As historians have noted, when draft board raids rendered certain draft boards inoperable, the Selective Service attempted to shift unmet quotas to other draft boards in the state. However, because those other draft boards were not meeting their own quotas, this attempted redistribution often failed.[135] During the war both evasion and open resistance frequently caused draft boards to fail to meet their quotas. In this context, it is not surprising that boards were unable to help make up the deficits of other boards after their files had been destroyed.

The draft board raids slowed or stopped conscription in an additional way. Selective Service regulations required that men classified I-A be drafted in a specific order, and one of the most successful ways that men were able to beat charges of refusing induction was to show that they had been called out of order. Thus, even the cruder draft board raids—those that failed to damage or to completely destroy the cross-reference system—could prevent draft boards

132 "Berkeley Blast Damages Selective Service Office," *New York Times*, June 11, 1968; William Worthy, "Anti-War Sabotage," *Resist*, no. 15 (August 27, 1968): 1; "Ohio Draft Board Rip-Off," *WIN Magazine* 6, no. 4 (March 1, 1970): 19; "Maryland Draft Board Fire," *WIN Magazine* 6, no. 15 (September 15, 1970): 22; "Vandals Burn Records at Fields Point," *Providence (RI) Journal*, December 25, 1970; "Statement of the Providence Three," an anonymous flyer distributed immediately after the conclusion of a lecture given by the lawyer Bill Kunstler and the author at Brown University's Sayles Hall on January 5, 1971. A copy of the flyer is in the author's files.

133 *Semi-Annual Report*, January–June 1970, 1; "Six Draft Offices in Boston Invaded," *New York Times*, November 9, 1969. The *Semi-Annual Report* refers to Delaware and Rhode Island raids; the *Times* story quotes Colonel Paul Feeney on the Boston raid.

134 "11/10/70 Memorandum to SAC, Indianapolis, From SAC Philadelphia," *WIN Magazine* 8, nos. 4–5 (March 1972): 80.

135 Elmer, *Felon for Peace*, 124–25.

from drafting even men whose files had not been damaged because the boards feared botching the required order of induction.[136]

These draft board raids sparked several controversies, including within the peace and antidraft movements. Some pacifists believed that destruction of property was not nonviolent and that the secrecy involved in planning nocturnal raids violated the truth-and-openness ethic that pacifists hold dear. On the other hand, most participants in these actions considered themselves to be pacifists. WRL, FOR, CNVA, and the Catholic Peace Fellowship all had pacifist staff and members who both *believed* that the raids were nonviolent and who participated in them. These groups also had staff and members who deplored such actions.

This debate played out in part in peace movement publications. The November–December 1969 issue of *WRL News* had three articles by three WRL staffers, all under the headline "The Draft Board Raids Controversy: Con and Pro." Two took the con position, including Jim Peck, who wrote, "Raiding, looting, and destruction cannot command respect." The third article supported the raids.[137] The September 1, 1970, issue of *WIN Magazine*, published jointly by the War Resisters League and the New York Workshop in Nonviolence, had four articles about draft file destruction. One was by Maggie Geddes, one of the Women Against Daddy Warbucks, who had just destroyed files in Manhattan. A second was written by a member of the Chicago 15 about their destruction of files at thirty-four Chicago draft boards, explaining their effort in some detail. A piece by Bradford Lyttle said, "From the classical pacifist and nonviolent viewpoint, looting of draft boards and the destruction of draft files … cannot be condoned." The fourth article, by Jim Peck, likened the raids to vigilantism.[138] Participants in the August 1969 Peacemakers Orientation Program in Nonviolence discussed the issue in some detail, and the December 6, 1969, issue of the *Peacemaker* included an article entitled "Thinking about Destruction of Draft Files," which consisted of many comments, both pro and con.

Draft board raids have been widely referred to as Catholic actions and their participants as members of the Catholic Left. The subtitle of Charles A. Meconis's book on the draft board raids is *The American Catholic Left, 1961–1975*. In discussing draft board raids, Mollin writes about the "Catholic resistance

---

136 Baskir and Strauss, *Chance and Circumstance*, 72–73.

137 Igal Roodenko, Jim Peck, and Jerry Elmer, "The Draft Board Raids Controversy: Con and Pro," *WRL News*, no. 156 (November–December 1969): 3–4.

138 Maggie Geddes, "Draft File Rip Off," 5; "Twain Column," "'Oh, Happy Day,'" 9; Bradford Lyttle, "Ends-Means," 5; Jim Peck, "The Emperor's New Clothes," 8, all in *WIN Magazine* 5, no. 15 (September 1, 1969).

actions between Catonsville and the 1973 withdrawal from Vietnam." It is certainly true that there was disproportionately heavy representation of Catholic priests and nuns (and of former priests and nuns) in the draft board raids. In their public statement explaining their action, the Catonsville Nine said, "We are Catholic Christians who take the gospel of our faith seriously." Five of the fifteen paragraphs of that statement quoted a recent papal encyclical.[139] Today it would be improbable (to say the least) for activists to explain their radical civil disobedience by quoting at length an official church teaching. But the characterization of the draft board raids as Catholic actions was overstated. The raid on the Providence, Rhode Island, draft board was carried out by a pair of Jewish teenagers. Two Quaker-leaning women organized the raid in Rochester, New York; they made a deliberate decision to exclude people associated with the Catholic Left.[140] The We the People group that raided the Philadelphia and Lansdowne draft boards was composed of young people—students, dropouts, and recent graduates aged nineteen to twenty-three—with no identifiable religious affiliations (or motivations) in their publicity materials. Even in the October 1967 Baltimore Four blood pouring—widely seen as a quintessentially Catholic Left action—half the members were not Catholic: James Mengel was a United Church of Christ minister, and David Eberhardt was a self-identified Episcopalian who had grown up as a Presbyterian.[141]

Another controversy concerned the adulation of many activists for the Berrigans. Mollin writes that "venerated as priests, the Berrigans cast themselves as modern prophets and apostles carrying out the work of God and Christ himself."[142] Mollin is correct about this. And while it is not surprising that the Berrigans, steeped as they were in Catholicism, would hold such a view of themselves, what is astonishing is how widespread the view was among others. In a single published essay, one writer says all of the following about the Berrigans: they are "like Saint Paul"; "Christ returned"; "like Moses receiving the commandments"; "like Old Testament prophets"; "elegant prophet"; and "dapper saint." In a single paragraph, the usually more tempered Harvard theologian Harvey Cox likened the Berrigans to Moses, Jeremiah, and other "prophets and

---

139 Meconis, *With Clumsy Grace*; Mollin, *Radical Pacifism*, 175; Jerry Elmer, "Remembering My Friend, Dan Berrigan," *RI Future*, May 2, 2016, rifuture.org, last accessed May 6, 2022.

140 Elmer, *Felon for Peace*, 116.

141 "We the People" statement; "Blood Poured on Baltimore Draft Files; Priest, Minister, 2 Others Seized," *New York Times*, October 28, 1967; David Eberhardt, email to author, December 11, 2021.

142 Mollin states that "these attitudes arose without question largely because they so closely reflected the patriarchal and hierarchical underpinnings of the Catholic Church." Mollin, *Radical Pacifism*, 172.

seers." Another popular book described Dan Berrigan as "a living saint" and Phil Berrigan as being akin "to Old Testament prophets like Amos or Hosea or to John Brown, the great abolitionist."[143] Many in the broader antidraft movement found such hagiography distasteful or repellent.

9 Prosecutions

Not surprisingly, the number of criminal prosecutions for Selective Service violations increased dramatically during the Vietnam era, as shown in Table 7.3. In fiscal year 1961, the number of Selective Service prosecutions initiated was 251; by FY1972, that number had increased to more than 5,100.

In 1971 and 1972, both the Selective Service and the Department of Justice were overwhelmed with Selective Service cases. On June 30, 1972, the 5,424

TABLE 7.3 Number of new selective service prosecutions by fiscal year, Vietnam era

Fiscal year	Cases
1965	380
1966	663
1967	1,335
1968	1,826
1969	3,305
1970	3,725
1971	4,539
1972	5,142
1973	3,043
1974	1,008
1975	274

SOURCES: FOR FY1965–67: US COURTS, [1968] *ANNUAL REPORT OF THE DIRECTOR OF THE ADMINISTRATIVE OFFICE OF THE UNITED STATES COURTS*, 130, UNNUMBERED TABLE; FOR FY1967–71: US COURTS' ANNUAL REPORT FOR 1971, 146, TABLE 23; FOR FY1972–75, US COURTS' ANNUAL REPORT FOR 1975, 194, TABLE 17.

143 Francine du Plessix Gray, *Divine Disobedience* (New York: Alfred A. Knopf 1970), 65, 72, 117, 113, 137; Harvey G. Cox, Jr., "Tongues of Flame: The Trial of the Catonsville Nine," in Stephen Halpert and Tom Murrray, eds., *Witness of the Berrigans* (New York: Doubleday, 1972), 21; Shawn Francis Peters, *The Catonsville Nine: A Story of Faith and Resistance in the Vietnam Era* (New York: Oxford University Press, 2012), 324, 323.

pending Selective Service cases made up more than 21 percent of all cases pending in federal courts.[144]

Worse still, the federal government had a backlog of more than thirty thousand Selective Service criminal cases awaiting indictment or other disposition with no ability to handle or process such an enormous number. These were cases of men who had refused induction, refused preinduction physicals, failed to report for alternative service, or had filed false information. The Selective Service was forced to call in Army, Navy, and Air Force Reserve lawyers to work on the backlog. Unsurprisingly, half the cases in the backlog proved impossible to prosecute because the Selective Service had acted illegally or draft boards had made careless mistakes. But being unable to prosecute only 50 percent of referred cases is actually quite good compared with the Selective Service's overall performance. During the Vietnam era, the Department of Justice considered fewer than 20 percent of cases referred by the Selective Service to be worth bringing to trial.[145]

Given the percentage of cases that the DOJ declined to prosecute at all (that is, the weakest cases), its conviction rate was stunningly low. The conviction rate in 1969 was 51.6 percent; for each of the three years from 1970 to 1972, the conviction rate hovered between 33.4 percent and 36.2 percent; and in 1973, the conviction rate dropped to 27.9 percent.[146] This created a vicious cycle in which federal prosecutors became reluctant to prosecute draft cases. As Tarr said in mid-1971, "Unfortunate experience with Selective Service cases naturally dulled the ardor of a U.S. Attorney to prosecute violations of the Selective Service law."[147]

10 Legal Cases Pertaining to the Conscription Statute

In *Gillette v. United States*, the Supreme Court upheld the CO provision in the draft law that required COs to oppose all wars; Gillette had argued that

144 *Semi-Annual Report*, July–December 1972, 47. A pending case is one in which an indictment has been returned—that is, a criminal prosecution has been commenced—but the case has not yet been concluded.

145 *Semi-Annual Report*, July–December 1972, 48, 44, 48–49. The Justice Department prosecuted 25,240 cases while declining to pursue 106,000 because of obvious defects.

146 Chart F, "Disposition of Defendants Charged with Violations of Selective Service Acts," attached to "White House Press Conference of Laurence H. Silberman, Deputy Attorney General, and Byron V. Pepitone, Director, Selective Service, Sept. 16, 1974," box 8, folder labeled "Memoranda—Justice Dept.," Charles E. Goodell Papers, Gerald R. Ford Presidential Library, Ann Arbor, MI.

147 *Semi-Annual Report*, January–June 1971, 9.

objectors to a particular war ought to be recognized. Relying on *Seeger* and *Welsh*, Gillette made a First Amendment argument that the requirement to oppose all wars violated his religious freedom; this was a somewhat modified version of Walter Nelles's position. The Supreme Court rejected the First Amendment argument, saying that the requirement in the law was "neutral and secular." Douglas dissented and, using Nelles's reasoning, would have expanded CO status to selective objectors.[148]

In *Holmes v. United States*, a man who had refused induction claimed that Congress had no constitutional authority to conscript in peacetime, that is, in the absence of a formal declaration of war. He was convicted and sentenced. On appeal, the Supreme Court denied certiorari, that is, it declined to even hear the case and left the lower court decision (conviction and sentencing) in place. Justices Douglas and Potter Stewart dissented, saying, "The question whether there can be conscription when there has not been a declaration of war has never been decided by the Court. It is an important question. It is a recurring question."[149]

In *Mitchell v. United States*, the draft resister David Mitchell appealed his conviction using a Nuremberg defense—that the United States was committing crimes against humanity and the Nuremberg principles provided that citizens have an affirmative duty not to participate in those crimes. Once again, the Supreme Court denied certiorari; once again, Douglas dissented. In his dissent, Douglas quoted Supreme Court Justice Robert H. Jackson, who had been a prosecutor at Nuremberg: "If certain acts in violation of treaties are crimes, they are crimes whether the United States does them or whether Germany does them, and we are not prepared to lay down a rule of criminal conduct against others which we would not be willing to have invoked against us."[150]

Toussie v. United States presented the question of when the statute of limitations runs out on the crime of refusing to register. The draft law required a man to register on his eighteenth birthday or within five days thereafter. If the obligation to register ends five days after a man's eighteenth birthday, then the statute of limitations (which is five years) runs out when a man is twenty-three years and five days old. Conversely, if the obligation to register is a continuing one, a man who doesn't register can be prosecuted after he turns twenty-three years and five days of age. Robert Toussie was prosecuted eight years after he had not registered. Until *Toussie*, it had been widely understood that the obligation to register was continuing. However, in 1970 the Supreme

148 Gillette v. United States, 401 U.S. 437 (1970).

149 Holmes v. United States, 391 U.S. 936, 949 (1968).

150 Mitchell v. United States, 386 U.S. 972, 973 (1967).

Court held otherwise and set Toussie free. Congress disagreed and wrote into the 1971 draft law that the obligation to register was a continuing one.[151]

Gutknecht v. United States was a companion case to *Oestereich* (though the cases were technically not consolidated on the court's docket). The cases were argued at the same time by the same lawyers (on both sides) and raised similar—but not identical—issues. Dave Gutknecht was a draft resister who had turned in his draft card on October 16, 1967, in the first national draft card turn-in. On December 20, he was punitively accelerated in the order of induction in the immediate aftermath of the Hershey memorandum of October 26, 1967. In *Oestereich*, the technical question was whether the Selective Service could punitively reclassify a man as I-A for returning a draft card (and the court ruled in Oestereich's favor). *Gutknecht* raised the related issue of whether the Selective Service could punitively accelerate the induction of a man who was already I-A when he returned his draft card. The court ruled in Gutknecht's favor.[152]

In the wake of the *Oestereich/Gutknecht/Breen* trilogy of cases, a great many resisters who had burned or returned their draft cards, and had refused induction after their orders for induction had been accelerated, appealed their convictions. (*Breen* was a 1970 Supreme Court case—discussed earlier in this chapter—that, like *Oestereich* and *Gutknecht*, challenged Hershey's instruction to draft boards regarding punitive treatment of draft opponents.) For example, Joe Gilchrist, a Cornell student who had returned his draft card on April 10, 1968, was among them. On August 21, 1970, the Court of Appeals—citing *Oestereich*, *Gutknecht*, and *Breen*—threw out Gilchrist's conviction.[153] Gilchrist still ended up in prison. Sixteen days after his conviction was reversed, on September 6, 1970, Gilchrist was arrested for destroying draft files at four local draft boards in Rochester, New York.

11 Conclusion

Many factors help to explain why the Vietnam era draft was so unsuccessful. To start, the Selective Service faced three broad categories of problems.

---

151 Toussie v. United States, 397 U.S. 112 (1970); An Act to Amend the Military Selective Service Act of 1967, 92 Pub. L. No. 92-129, 81 Stat. 100, September 28, 1971.

152 Gutknecht v. United States, 396 U.S. 295 (1970). For a critical view of the Supreme Court's reasoning in the *Oestereich/Gutknecht/Breen* trilogy of cases, see "The Supreme Court, 1969 Term: Delinquency Status and Punitive Reclassification," *Harvard Law Review* 84, no. 1 (November 1970): 222–29.

153 Gilchrist v. United States, 430 F.2d 631 (10th Cir. 1970).

In the first category are the government's stupid and/or illegal actions that created a system more fragile and more susceptible to successful court challenges than it needed to be. None of these things automatically inure in conscription. It was not necessary to the conscription apparatus that forty-one hundred local draft boards were not merely permitted but actually encouraged to operate independently of one another and independently of the applicable statutes and regulations. Hershey insisted that this be done. Nor was it necessary to hire local board staff that were incompetent and leave them untrained. It was not necessary for Hershey to issue written instructions to draft boards to reclassify protesters I-A after the courts had ruled that this was illegal. Hershey insisted on doing all these things anyway. In the last instance, this led to the stunning result that the government's own lawyer told the Supreme Court that Hershey's instructions were illegal and wrong. These were all unforced errors.

In the second category were the small, separate actions taken by literally millions of men to evade service without openly confronting the Selective Service. Sometimes but not always, these moves were motivated in whole or in part by opposition to the war, but almost always they were also motivated by at least some self-interest in avoiding the inconvenience, injury, or death that could result from military service in time of war. Sometimes but not always, these actions skated close to illegality (such as a man's exaggerating a minor physical infirmity in order to be classified IV-F) or were clearly illegal (such as a man's inventing nonexistent bone spurs in order to be classified IV-F).

These two factors, taken together, created a system that was rickety.

Into this context comes the third category of problems for the Selective Service: the thousands of men and women who chose to confront the system openly and illegally with nonviolent civil disobedience. At the height of the war, more than five thousand people faced criminal prosecutions for antidraft civil disobedience in a single year. One-sixth of the prison population was composed of violators of Selective Service law. The Resistance—disorganized though it was—conducted events at which hundreds or thousands of men, in dozens of small and large cities nationally, burned or returned draft cards on the same day. Tens of thousands of other people openly aided and abetted these actions, often pressing the government with signed statements, to prosecute them too. Small groups of people burglarized hundreds of draft boards all over the country, carefully and methodically destroying primary files and cross-reference systems, making the draft boards inoperable—and then holding press conferences to announce their handiwork, thereby inviting the government to prosecute them.

In his "Letter from Birmingham Jail," Martin Luther King, Jr., wrote that civil disobedience has the greatest chance of success when it is part of a broader context or movement. That was certainly the case with opposition to the Vietnam era draft. The Selective Service System was already fragile, both from self-inflicted wounds and from widespread evasion. After the Tet Offensive in February 1968, public sentiment also turned against the war. This was the broader context in which the Resistance worked, and to the extent that the Resistance was effective, the broader context contributed to that success.[154]

∴

The resisters posed a special problem for the government: whether to prosecute them. Many individual draft resisters during the Vietnam era—men who had publicly refused to register, burned draft cards, refused induction—were never prosecuted. Interestingly, and perhaps surprisingly, many of the people who publicly destroyed draft files were also not prosecuted. For example, among other cases, the men and women who claimed responsibility for destroying files at numerous draft boards in New York (August 1969), Boston (November 1969), Philadelphia (February 1970), and Delaware (June 1970) were never indicted or prosecuted.

This lack of prosecution was certainly not for lack of information from the FBI about draft resistance activities. Director J. Edgar Hoover sent a memorandum to all FBI offices in which he directed FBI agents to pay special attention to the Resistance:

> IDENTITIES OF INDIVIDUALS SURRENDERING SELECTION SERVICE CARDS AND THE IDENTITY OF THE INDIVIDUALS RECEIVING THEM ALONG WITH ANY STATEMENTS MADE SHOULD BE NOTED AND APPROPRIATE NOTES MADE FOR EVIDENTIARY PURPOSES. CLOSE COVERAGE SHOULD BE GIVEN TO ANY SPEECHES GIVEN AT RALLIES, PARTICULARLY TO EXHORTATIONS OF OVERT REFUSALS TO COMPLY WITH THE SELECTIVE SERVICE ACT.[155]

An example of this "close coverage" can be seen in the FBI's response to the June 17, 1968, Women's Draft Card Burning in Washington, DC. Three pacifist organizations sponsored the demonstration: the New England Committee for Nonviolent Action, WRL, and New York Workshop in Nonviolence. Eleven

154 Martin Luther King, Jr., "Letter from Birmingham Jail," in *The Radical King: Martin Luther King, Jr.*, ed. Cornel West (Boston: Beacon Press, 2015), 129.

155 The undated memorandum from FBI Director J. Edgar Hoover to "All SACs" addressed "Anti-draft Activities, Counseling, Aiding and Abetting, SSA [Selective Service Act]" and was reprinted in *WIN Magazine* 8, nos. 4–5 (March 1972): 74. The capitalization, punctuation, and underscoring used here appeared in the original.

women publicly burned the draft cards of their husbands or male friends. In an article in the CNVA newsletter, Citti Allsup said that she participated in the event in order "to be a woman with other women who would be taking the same risks as male draft resisters."[156]

FBI files obtained through a Freedom of Information Act request, and not previously published, reveal the extent of the surveillance of the event. In advance of the demonstration, Hoover personally approved the use of video surveillance, saying, "It is imperative that this matter be given attention and that the Bureau be currently advised of all information as it is developed." Three FBI agents posing as reporters took video footage of the event; another agent took 16 black-and-white photographs; another took 32 photographs; another took 37 photographs; yet another agent took 64 photographs; and numerous agents interviewed twenty of the participants in various parts of the country about the event from June 1968 until March 12, 1969. One FBI report about the event noted that the participants in the demonstration were "totally overt." Another said that "sound and silent movie footage[,] combined with Bureau Agent eye witnesses, more than adequately establishes that each of the subjects interviewed on tape did[,] of their own volition, assume an active role in the card burning activity."[157] Yet not one of the women draft card burners was prosecuted.

To be clear: nothing whatsoever was improper about FBI surveillance of the women's draft card burning. The participants were violating the law and were doing it deliberately and publicly for reasons that seemed good to them. The FBI's job is to investigate violations of federal law.

The question here is what are we to make of the highly selective prosecutions of draft resisters during the Vietnam era?

It is almost certain that the (highly) selective prosecutions of draft resisters were deliberate, not inadvertent, because of the FBI's thorough surveillance of the Resistance and resisters. With the authorities having devoted enormous personnel resources to gathering information, it is improbable that decisions not to use the information (for prosecution in some cases) were made casually.

Yet it may be impossible to surmise what complex set of reasons might have led to the highly selective prosecutions. Tom Wells, in his masterful *The War Within*, posits, based on a later interview with Ramsey Clark, that the former attorney general was a closet dove who was reluctant to prosecute resisters. But

---

156 Citti Allsup, "Sisters Say Yes," *Direct Action for a Nonviolent World*, no. 90 (June–July 1968): 1; "CNVA Press Release," n.d., obtained from the FBI through the FOIA, in the author's files.

157 The FBI files I obtained include an undated Airtel from Hoover to SACs in New Haven and Washington and five separate memoranda dated June 21, 1968; June 24, 1968; March 21, 1969; June 18, 1968; and September 11, 1968. Copies of all these documents remain in my files.

even if Clark's comments to Wells were not a post hoc rationalization, this could be only a partial explanation, for in January 1969 Clark was replaced as attorney general by the law-and-order John N. Mitchell, yet the selective prosecutions continued. Baskir and Strauss argue that the problem was the sheer caseload: "The criminal justice system became swamped to the point where Selective Service prosecutions became almost impossible."[158] This, too, is unlikely to be the entire explanation. If the problem were solely caseload, the solution would be to bring on additional staff, just as the Selective Service did in the early 1970s.

But even if it is impossible to know reasons, it is possible to posit a consequence of the highly selective prosecutions. Early Resistance organizers, in postwar reminisces, disagree about whether the Resistance managed to fill the prisons. Bruce Dancis, who organized the New York draft card burning on April 15, 1967, believes that the Resistance did not accomplish that. David Harris, who organized the draft card turn-in on October 16, 1967, believes the opposite. "We *did* fill up the prisons. Christ, I was in the prisons—let me tell you: they were full."[159] But perhaps both Dancis and Harris are addressing the wrong question. The correct question may be not whether the resisters filled the prisons but what the consequences were of the government's (apparent) policy of selective prosecution.

Draft resistance posed a classic dilemma for the government. With regard to prosecuting resisters, the government was damned if it did and damned if it didn't. On the one hand, the government could prosecute and imprison thousands of resisters, including the country's most prominent pediatrician, Yale's chaplain, and lots of priests and nuns. But such prosecutions would create waves of good publicity for the resisters and help them to achieve their goals of filling the prisons and bringing pressure on the government to stop the war and the draft. On the other hand, the government could decline to prosecute the resisters, thereby in effect announcing to the world that you can burn your draft card, refuse induction, and burglarize draft boards with (at least relative) impunity. This latter course could only have the effect of encouraging such activity.

In the event, the government, by pursuing a course of selective prosecution, had the worst of both worlds. The government prosecuted and imprisoned thousands of resisters, which did create enormous political pressure against the draft and the war. Yet the government did not prosecute thousands of others, thereby encouraging further resistance. However, the government may not have had a better option, because however many of the thousands of resisters it prosecuted, that level of prosecutions would inevitably redound to the detriment of the government.

158 Wells, *War Within*, 131–33, 234–37; Baskir and Strauss, *Chance and Circumstance*, 68.

159 Dancis, *Resister*, 330–31; Wells, *War Within*, 269 (emphasis in original).

CHAPTER 8

Conclusion

Writing in the *Nation* in March 1968, in the immediate aftermath of the Spock-Coffin indictment, Jean Carper, author of a book about the Vietnam era draft, wrote, "For the first time in its history our government faces a draft rebellion not only by the young but by the old, not only men but women, not only by individuals alone but by groups." In a similar vein, the first sentence of Sherry Gershon Gottlieb's 1991 book on the Vietnam era draft was, "The Vietnam era was the first time since the Civil War that draft evasion was widespread."[1] This is the conventional view and it is widespread. But it is also wrong. The depths of American hostility to conscription—starting before the Civil War and continuing through the Vietnam era—have been significantly underrecognized and underappreciated.

Today, the War of 1812 is thoroughly forgotten; indeed, the subtitle of Donald R. Hickey's book on that war was *A Forgotten Conflict*. Also forgotten, along with the entire war, is that President James Madison asked Congress to enact conscription during that war but was turned down because conscription was un-American. No one today much remembers Daniel Webster's anticonscription speech on the floor of the House of Representatives in 1814, but when conscription was proposed in 1917 and 1940, draft opponents—members and witnesses alike—made many speeches in Congress citing and quoting Webster.[2]

Gottlieb is at least correct that today the massive opposition and resistance to conscription during the Civil War is widely recognized. Some Americans today know about the antidraft riots in New York City (although many probably

1 Jean Carper, "The Real Crime of Dr. Spock," *Nation* 206, no. 11 (March 11, 1968): 327; Jean Carper, *Bitter Greetings: The Scandal of the Military Draft* (New York: Grossman, 1967); Sherry Gershon Gottlieb, *Hell No We Won't Go: Resisting the Draft during the Vietnam War* (New York: Viking, 1991), ix.

2 For speeches made by senators about World War I, see 65 Cong. Rec. S1364–65 (April 27, 1917), statement of Robert M. La Follette, R-WI, and 65 Cong. Rec. S1327 (April 27, 1917), statement of Thomas W. Hardwick, D-GA. For speeches by House members about World War I, see 65 Cong. Rec., H.R. 1188 (April 26, 1917), statement of Robert Crosser, D-OH, and 65 Cong. Rec., H.R. 1188 (April 26, 1917), statement of Denver S. Church, D-CA. For World War II era House debates see the statements of Norman Thomas and Catherine Fitzgibbon during the hearings before the Committee on Military Affairs, House of Representatives, on H.R. 10132, 76th Cong., 3d sess. (1940).

 | DOI:10.1163/9789004546684_010

do not realize that those remain the worst riots in US history, even today). Yet, even in the case of the Civil War, the full dimensions of the opposition to conscription have been significantly understated.

For three-quarters of a century, successive historians have reported the shocking statistic that more than 20 percent of the men drafted in the North during the Civil War refused to go.[3] But even these figures vastly undercount draft refusal in the North. By recognizing only men who refused induction orders, scholars have missed the substantially larger numbers and percentages—larger by half—of men who illegally evaded draft registration in the first place: the skulkers and skedaddlers. An additional 32.4 percent never enrolled, including more than half the eligible men in Delaware, Maine, and New Hampshire. Although scholars have commented on the high level of illegal draft refusal during the Civil War, they have seriously understated what the level actually was. In fact, 53.2 percent of men refused the draft in the North, not the 20.8 percent that successive generations of scholars have cited.

Draft opposition during the Civil War was not only widespread in the Union, affecting every state, but also violent. Thirty-eight enrollment officers were murdered doing their jobs, sometimes ambushed, and sixty more were wounded. Their files were stolen and destroyed; their offices were burned to the ground. It sometimes took whole companies of cavalry to quell antidraft protests and rioting.

Statistical records for the Confederacy were less detailed and complete, but we know that draft evasion was so widespread that in many places it was the rule, not the exception. In the Confederacy, there were not just a few fugitive bands of draft refusers; these bands were ubiquitous, heavily armed, and well organized. Most notably, many lived among the people and survived as well and as long as they did because they were popular and received widespread support.

Antidraft sentiment during World War I is also underappreciated. In fact, depending on whose figures are used, either hundreds of thousands or millions of men quietly refused to register during the World War I draft (which lasted but eighteen months).[4] This, even though nonregistration was a felony, and draft

---

3 Eugene C. Murdock, *One Million Men: The Civil War Draft in the North* (Madison: State Historical Society of Wisconsin, 1971), 350–56; James W. Geary, *We Need Men: The Union Draft in the Civil War* (DeKalb: Northern Illinois University Press, 1991), 83, tables 2 and 3; and Peter Levine, "Draft Evasion in the North during the Civil War, 1863–1865," *Journal of American History* 67, no. 4 (March 1981): 816.

4 PMG *3d*, 12, 53; Christopher Capozzola, *Uncle Sam Wants YOU: World War I and the Making of the Modern American Citizen* (New York: Oxford University Press, 2008), 30; and John Whiteclay Chambers II, *To Raise an Army: The Draft Comes to Modern America* (New York: Free Press, 1987), 211.

evaders were actively pursued, not only by federal agents but by thousands of American Protective League vigilantes. Boatloads of well-off men fled the country to avoid the draft.[5] On the eve of draft registration, fifteen thousand people packed Madison Square Garden in New York for an antidraft rally, but antidraft organizing went on all over the country, in large cities like Chicago, Detroit, and San Francisco and in small places like Somerville, Massachusetts, and Lawrence and Olathe, Kansas.[6] Anticonscription and antiregistration leaflets came flying off printing presses in Seattle, Philadelphia, West Virginia, Michigan, and elsewhere.[7]

For days leading up to the June 5, 1917, registration day, the front-page stories in the *New York Tribune, Cincinnati Enquirer, San Francisco Chronicle, Los Angeles Times*, and other newspapers were not about government-sponsored registration but about antiregistration agitation and organizing.[8] Students organized chapters of the Anti-Militarism League, urging nonregistration, at Oberlin, Yale, Harvard, Columbia, City University of New York, and other campuses.[9] And not all the anticonscription activity came from Ivy League students like Owen Cattell, Charles Phillips, and Eleanor Parker at Columbia or from Yiddish-speaking anarchists like Emma Goldman and Alexander Berkman. A heavily armed group of fifty draft opponents organized in Snyder and Abilene, Texas; it took the US Cavalry to put down armed conscription opponents in Tennessee.[10] The Green Corn Rebellion involved five hundred mostly illiterate Black and white farmers in rural Oklahoma.[11] Especially active antidraft organizing in Kansas led to front-page stories in the *Kansas City* (MO) *Times, Kansas City* (MO) *Star, Topeka* (KS) *Daily Capital, Lawrence* (KS) *Daily Journal World,* and (*Lawrence,* KS) *Daily Gazette.* The headline in the *Topeka Daily Capital* was typical: "AntiDraft Plot Here Results in Arrest of 5." These articles were not about registration but about opposition to it.[12]

5 "Won't Miss Draft on Foreign Soil," *New York Times*, May 31, 1917.

6 "New Police Arms Awe Socialists," *New York Times*, June 1, 1917; "Borders Watched as Wilson Warns Draft Runaways," *New-York Tribune*, June 2, 1917, 1.

7 Sugar v. United States, 252 F. 79 (6th Cir. 1918); "Plotters Here against Draft under Watch," *New York Times*, May 30, 1917.

8 See, for example, "College Girl, Youths Taken in Draft Raid," *San Francisco Chronicle*, June 1, 1917, and "Eligible Conscripts Halted at the Border," *Los Angeles Times*, June 1, 1917.

9 "Hit Draft Plots in Many States; Arrest 3 College Students Here; Police Display Quiets Antis' Meeting," *New York Times*, June 1, 1917.

10 "Anti-Draft Plotters Held," *New York Times*, May 23, 1917; Bryant v. United States, 257 F. 378 (5th Cir. 1919).

11 Bush, "The Green Corn Rebellion."

12 "Nip Draft Plots," *Kansas City* (MO) *Times*, June 1, 1917; "Overpower Plot Printer," *Kansas City* (MO) *Star*, June 1, 1917; "Anti-Draft Plot Here Results in Arrest of 5," *Topeka* (KS) *Daily*

The conventional view of World War II is that of Tom Brokaw, who wrote in *The Greatest Generation* that America's men "answered the call ... but they did not protest."[13] It is certainly true that men answered the call, but it is not true that there was no protest. It is well known that more than 6,000 draft resisters went to prison during World War II and that another 12,000 COs went to Civilian Public Service (CPS) camps. But the actual number of COs during World War II was 70,000, according to the Selective Service, and 95,000 to 105,000, according to other contemporaneous estimates.[14] Even less frequently acknowledged is that there were even more COs during World War II—33,000—than those numbers account for. More than 13,000 men were classified as COs at one time but were placed into higher classifications before they could be drafted; another 20,000 men claimed CO status upon registration but were immediately given a higher classification, so their CO claims were never considered or ruled on.[15] The conventional notion that the World War II generation "did not protest" is untrue in another way too. All the men who were classified as IV-E and sent to CPS camps had a relatively safe and secure status. Yet hundreds of these men were eventually so opposed to conscription that they felt morally obligated to resist anyway. These were the men—like Bent Andresen, Igal Roodenko, and Wally Nelson—who voluntarily gave up their IV-E status by walking out of camp or stopping work and going to prison.

Similarly, the Cold War era is widely viewed in popular culture as one of quiet conformity. This was the era of *The Man in the Gray Flannel Suit*, the TV shows *Ozzie and Harriet* and *Leave it to Beaver*, the era depicted in the popular contemporary show *Mad Men*. George Flynn wrote that "there was no [draft] protest during the Korean War" and so little opposition during the 1950s that "the major problem" during that time was "the overabundance of young men."[16] While there surely was much conformity in the 1950s, there was nevertheless considerable opposition to conscription during this era as well. During the Cold War, hundreds of draft resisters went to prison, and 14,727 men performed

Capital, June 1, 1917; "Make 3 Arrests in Topeka Today," *Lawrence* (KS) *Daily Journal*, May 31, 1917; "Courts Ducked Decision on Draft Law," *Daily* (*Lawrence, KS*) *Gazette*, June 4, 1917.

13 Tom Brokaw, *The Greatest Generation* (New York: Random House Trade Paperbacks, 1998), xxvii.

14 US Selective Service System, *Conscientious Objection:* Special Monograph No. 11 (Washington, DC: Government Printing Office, 1950), 1:313; Winslow Ames, "CO's in Uniform—Non-Combatant Duty," *Reporter* 3, no. 2 (September 16, 1944): 3; William Fifield, "Report to Harper's," *Pacifica Views* 2, no. 36 (February 9, 1945): 2.

15 US Selective Service System, *Conscientious Objection*, 1:314–15.

16 Sloan Wilson, *The Man in the Gray Flannel Suit* (New York: Simon and Schuster, 1955); George Q. Flynn, *The Draft, 1940–1973* (Lawrence: University Press of Kansas, 1993), 121, 168.

alternative service as COs.[17] Tens of thousands of other men believed that they were COs, but their claims were never addressed because their draft boards immediately placed them into higher classifications. In just the first few months after the 1948 draft law was enacted, more than a dozen people were criminally prosecuted for advocating or counseling draft refusal. Tens of thousands of cases of draft violations were referred to the Justice Department during this period, including more than ten thousand cases of induction refusal alone in 1952 and again in 1953.[18] What makes these last figures stand out is that these referrals came only after repeated remonstrances to the men had proved unsuccessful.

Even historians of the era understate the degree of opposition to conscription in this period. Staughton Lynd, Michael Ferber, and Andrew Hunt all said that four hundred or five hundred draft cards were burned or returned to the government in the February 1947 Break With Conscription actions. The actual figure is double that number.[19]

One of the most revealing facts about draft opposition during the Cold War period is that the National Council Against Conscription coordinated the efforts of eighty local anticonscription efforts in cities and towns and on college campuses across the country. These efforts were low key and quiet; they certainly did not grab newspaper headlines the way Emma Goldman's Madison Square Garden rally in June 1917 did (or that Break With Conscription did in February 1947). But to have so many grassroots chapters of NCAC active across the country betokens a movement and a viewpoint that was widely shared by ordinary Americans.

In her 1968 *Nation* article, Carper wrote, "For the first time in our history, the older men, themselves safe from the draft, have joined forces with the young." Julius Eichel, Evan Thomas, A. J. Muste, and the other overage public

17 J. Harold Sherk, "A Report from the NSBRO," January 16, 1963, Center on Conscience and War Records, RG 025, pt. II, ser. A, box 16, folder labeled "Reports by NSBRO," Swarthmore College Peace Collection (SCPC), Swarthmore College, Swarthmore, PA.

18 US Selective Service System, *Annual Report of the Director of Selective Service for the Fiscal Year 1952 to the Congress of the United States Pursuant to the Universal Military Training and Service Act as Amended* (Washington, DC: Government Printing Office, 1953) (hereafter, *Annual Report*, followed by the year), 37; *Annual Report*, 1953, 28; *Annual Report*, 1954, 30; *Annual Report*, 1955, 34; *Annual Report*, 1956, 32; *Annual Report*, 1957, 30; *Annual Report*, 1958, 25; *Annual Report*, 1959, 28; *Annual Report*, 1960, 35.

19 Michael Ferber and Staughton Lynd, *The Resistance* (Boston: Beacon Press, 1971), 3; Andrew E. Hunt, *David Dellinger: The Life and Times of a Nonviolent Revolutionary* (New York: New York University Press, 2006), 89–90; George Houser, "Brief Report," March 5, 1947, Central Committee for Conscientious Objectors (CCCO) Records, RG 073, box 10, folder labeled "Break With Conscription Committee," SCPC.

nonregistrants in the World War II draft would beg to differ. Carper also wrote that Spock's urging young men to resist the draft constituted an "unheard-of breach" of societal norms.[20] Carper's statement would come as a surprise to Ammon Hennacy, Emma Goldman, Kate Richards O'Hare, Larry Gara, Wirt Warren, and all the other women and men who went to prison during World Wars I and II and the Cold War era for urging men to resist the draft. In a similar vein, Michael Foley wrote of the Spock-Coffin indictment, "News of the indictments stunned the nation and especially the antiwar movement."[21] Foley was, of course, correct: people *were* stunned by the indictments. But the surprise and shock revealed a lack of knowledge of US history during the preceding fifty years and of the many men and women who for decades had been prosecuted for urging resistance to conscription.

1 Channeling

In 1967, the young antidraft activists of Students for a Democratic Society and the Resistance can surely be forgiven for believing that Hershey's "Channeling" memo, recently exposed, was somehow new and startling. But contemporary scholars are not so easily forgiven. In 1985, George Flynn wrote, incorrectly, that Hershey started touting the Selective Service's role in channeling in 1955. In 2019, citing only Flynn, Jeremy Kessler made the same mistake.[22] Amy J. Rutenberg, in her book *Rough Draft: Cold War Military Manpower and the Origins of Vietnam-Era Draft Resistance*, carried the inaccurate argument considerably further. A major theme that Rutenberg expounded is that the Selective Service's channeling, wholly new during the Cold War era, led to, and helped create, Vietnam era draft resistance. Her fourth chapter is subtitled "Manpower Channeling, 1955 to 1965." Rutenberg claimed that the Selective Service's interest in channeling became public in January 1958 testimony that Hershey gave to a House committee, and an article he published in the March 1958 issue of *Selective Service*. "Hershey's testimony and the follow-up letter marked one of

20 Carper, "Real Crime of Dr. Spock," 329, 328.

21 Michael S. Foley, *Confronting the War Machine: Draft Resistance during the Vietnam War* (Chapel Hill: University of North Carolina Press, 2003), 226.

22 George Q. Flynn, *Lewis B. Hershey: Mr. Selective Service* (Chapel Hill: University of North Carolina Press, 1985), 209, and *Draft, 1940–1973*, 140; Jeremy K. Kessler, "New Look Constitutionalism: The Cold War Critique of Military Manpower Administration," *University of Pennsylvania Law Review* 167, no. 7 (June 2019): 1777.

the first times that the director publicly acknowledged the Selective Service's policy of manpower channeling."[23]

This is all wrong. During World War I, the government was really clear that channeling was a central benefit of the Selective Service. When Congress was considering conscription in 1917, it heard from Captain Percy Reginald about the fiasco Britain had experienced when the war broke out and England at first permitted men to volunteer for the army: the "wrong men" were constantly volunteering.[24] This led to a serious problem: captains of industry were going to the front and dying as cannon fodder while uneducated proletarian workers were staying home. The United States was not about to make *that* mistake.

On May 18, when Woodrow Wilson signed the draft into law and issued his proclamation requiring men to register, he chose to make channeling the very heart of the announcement:

> The nation needs all men; but it needs each man, not in the field that will most [please] him, but in the endeavor that will best serve the common good. Thus, though a sharpshooter pleases to operate a trip-hammer for the forging of great guns and an expert machinist desires to march with the flag, the nation is being served only when the sharpshooter marches and the machinist remains at his levers. The whole nation must be a team, in which each man shall play the part for which he is best fitted. To this end, Congress has provided that the nation shall be organized for war by selection.[25]

In his first World War I report to the secretary of war, Crowder said, "The draft itself is an instrument of compelling force in controlling and distributing labor supply throughout the United States."[26] In his second report, Crowder explained the effect of Selective Service regulations: "Either they placed such registrants in the military service, or they forced them, through fear of such service, to engage in a useful industrial or agricultural undertaking."[27] That's

---

23 Amy J. Rutenberg, *Rough Draft: Cold War Military Manpower and the Origins of Vietnam-Era Draft Resistance* (Ithaca, NY: Cornell University Press, 2019), 96–97.

24 *Increase of Military Establishment of 1917: Hearings Before the Committee on Military Affairs, House of Representatives*, 65th Cong., 1st sess., 3–4 (1917).

25 "A Proclamation by the President of the United States," *New York Times* May 19, 1917.

26 US Office of the Provost Marshal General, *Report of the Provost Marshal General to the Secretary of War on the First Draft Under the Selective-Service Act, 1917, 1918* (Washington DC: Government Printing Office, 1918), 33.

27 US Office of the Provost Marshal General, *Second Report of the Provost Marshal General to the Secretary of War on the Operation of the Selective Service System to December 20, 1918* (Washington, DC: Government Printing Office, 1919), 10.

why, almost immediately after the World War I draft law started providing inductees for the army, the government abolished volunteering.

Exactly the same thing occurred in 1940, when Congress was considering the World War II draft. Both Hershey and Secretary of War Henry L. Stimson emphasized the central role of channeling in the draft. General William Shedd told Congress that conscription was necessary to prevent what happened in England during World War I: "England experienced difficulty …, when she was recruiting her army by voluntary enlistment and found she had to return from the army to industry and agriculture hundreds of thousands of men because her industry had been crippled when these men volunteered."[28] In World War II, just as in World War I, the United States banned volunteer enlistments for much of the war, thereby confirming the primacy of channeling.[29]

Although it is widespread, the notion that channeling was a new idea in the 1950s or during the Vietnam era is wrong.

2 Is the Draft Unconstitutional?

It is probable that when the Constitution was written in 1789, the section 1 power to "raise and support armies" meant doing so, as Hamilton wrote in the *Federalist* No. 23, "in the customary and ordinary modes practiced by other governments," which at the time most certainly did not include conscription. Thus, if one hews to an originalist interpretation of the Constitution, one would necessarily conclude that conscription is unconstitutional—because, according to *Webster's Collegiate* [*Online*] *Dictionary*, originalism is the "legal philosophy that the words in documents and especially the US Constitution should be interpreted as they were understood at the time they were written."[30]

Most people, however, are not originalists but instead recognize that constitutional understandings and interpretations can, should, and do change and evolve over time. That is why, for example, we do not believe that "racial segregation of public schools [is] constitutional" even though the authors of the Fourteenth Amendment probably fully accepted the doctrine of separate but equal.[31]

---

28 *Hearings Before the Committee on Military Affairs of the United States House of Representatives on H.R. 10132*, 76th Cong., 3d sess. (1940), 117, 121, 384, 536.

29 Exec. Order No. 9279, 7 Fed. Reg. 10177 (1942), sec. 4.

30 The substance of the argument in this section appeared in somewhat different form in Jerry Elmer, "Conscription after *Bostock v. Clayton County*," *Rhode Island Bar Journal* 69, no. 3 (November–December 2020): 7–9.

31 David A. Strauss, *The Living Constitution* (Oxford: Oxford University Press 2010), 12.

Nevertheless, regardless of what the framers may have thought or believed in 1787, an unbroken line of twentieth-century Supreme Court cases held that conscription is constitutional. Today, more than a century after it was decided, *Arver* (1918) remains good law.

During the twentieth century, many opponents challenged the constitutionality of the draft based on the Reconstruction era Thirteenth Amendment, which prohibits involuntary servitude. During World War I, Donald Stephens and others tried this approach. During World War II, Jack Bryant Ryals and others tried the approach. George William Crocker and many others tried it during the Vietnam War. The argument never worked. Courts uniformly explained that the Thirteenth Amendment was meant to, and did, ban chattel slavery in the immediate aftermath of the Civil War, which had been fought, in part, about the issue of slavery, but the amendment was simply not applicable in other contexts, such as conscription. Many courts that so held quoted a 1916 Supreme Court case, *Butler v. Perry*: "This amendment was adopted with reference to conditions existing since the foundation of our government, and the term involuntary servitude was intended to cover those forms of compulsory labor akin to African slavery.... [It] certainly was not intended to interdict enforcement of those duties which individuals owe to the state, such as services in the army, militia, on the jury, etc."[32] And, of course, these many courts were absolutely correct. The Thirteenth Amendment was meant to ban slavery. It was not intended to make conscription illegal.

Still, there may be a sound legal argument that the Thirteenth Amendment bars conscription. On June 15, 2020, the US Supreme Court handed down its decision in *Bostock v. Clayton County*. The question in the case was whether Title VII of the Civil Rights Act of 1964 protects gay and transgender people from workplace discrimination.[33] Both sides agreed that the 1964 statute had outlawed workplace discrimination based on race, color, religion, national origin, and sex.

In *Bostock*, both sides agreed that, by including sex, the drafters of the statute had meant to bar workplace discrimination against women and had not meant to bar discrimination against gay or transgender people. Both sides agreed that, since the enactment of the law, the statute had been widely understood not to protect gay or transgender people. In fact, both sides agreed

32 United States v. Stephens, 245 F. 956 (D. Del. 1917); United States v. Ryals, 56 F. Supp. 772 (N.D. Ga. 1944); United States v. Crocker, 294 F. Supp. 776 (D. Minn. 1969); Butler v. Perry, 240 U.S. 328, 332 (1916).

33 Bostock v. Clayton County, 140 S. Ct. 1731 (2020); Civil Rights Act of 1964, 42 U.S.C. § 2000e-2(a)(1) (1964).

that—precisely because the law was widely understood not to protect gay or transgender people—members of Congress had introduced sixteen separate bills over the years to add gay people to the enumerated classes protected by the statute. And both sides agreed that, for the same reason, members of Congress had introduced seven more bills over the years to add both gay and transgender people to the protected classes.

Both sides agreed that, had the original Civil Rights Act protected gay and transgender people, those twenty-three separate, proposed statutory amendments (all introduced by supporters of LGBT rights) would have been wholly unnecessary. As Justice Samuel Alito said in his dissent: "The Court tries to convince readers that it is merely enforcing the terms of the statute, but that is preposterous.... Our duty is to interpret statutory terms to mean what they conveyed to reasonable people at the time they were written. If every single living American had been surveyed in 1964, it would have been hard to find any who thought that discrimination because of sex meant because of sexual orientation—not to mention gender identity, a concept that was essentially unknown at that time."[34] Nevertheless, in *Bostock*, the Supreme Court ruled that Title VII of the Civil Rights Act of 1964 does protect gay and transgender people.

The Supreme Court's 6–3 majority, speaking through Justice Neil Gorsuch, grounded its ruling in the judicial doctrine of textualism, a method of statutory interpretation championed by the late justice Antonin Scalia. Scalia's view was that if the plain language of a statute is clear, that is the end of the inquiry; in that case, one need not inquire what a legislature (or 535 separate legislators) may or may not have been thinking at the time the statute was enacted. Similarly, one need not do a hypothetical poll of the entire US population years or decades earlier to find out how ordinary people may have interpreted various words at the time the law in question was enacted. Today, textualism may be the methodology most widely used by courts for interpreting constitutional (and statutory) texts. As Justice Elena Kagan famously said at a 2015 lecture at Harvard Law School, "We're all textualists now."[35]

Gorsuch acknowledged that the drafters of Title VII may not have had sexual orientation and gender expression in mind when they wrote it. He also acknowledged that the meanings of words evolve over time—yet in interpreting a statute, one still needs to look at plain meaning first and foremost:

34 Bostock, 140 S. Ct. at 1755 (Alito dissenting).

35 Judge Diarmuid F. O'Scannlain, "'We Are All Textualists Now': The Legacy of Justice Antonin Scalia," *St. John's Law Review* 91, no. 2 (Summer 2017): 304.

> Those who adopted the Civil Rights Act might not have anticipated their work would lead to this particular result. Likely, they weren't thinking about many of the Act's consequences that have become apparent over the years, including its prohibition against discrimination on the basis of motherhood or its ban on the sexual harassment of male employees. But the limits of the drafters' imagination supply no reason to ignore the law's demands. When the express terms of a statute give us one answer and extratextual considerations suggest another, it's no contest. Only the written word is the law.[36]

Gorsuch went on to describe at some length these (and other) multiple ways that the Title VII prohibition on workplace discrimination based on sex have changed and evolved in the decades since its passage in ways that the drafters of the Civil Rights Act of 1964 could scarcely have imagined: sexual harassment, hostile work environment, discrimination against men, and so forth. In each of these cases, the Supreme Court rooted its holding in the plain language of the text barring discrimination "on account of sex," while not being at all troubled that the latest gloss was almost certainly not what the drafters of the Civil Rights Act had envisioned decades earlier.

That is, societal norms and community understandings of words may change over time, but courts are still obligated to interpret the meaning of statutes in front of them. The court concluded its *Bostock* opinion this way: "Ours is a society of written laws. Judges are not free to overlook plain statutory commands on the strength of nothing more than suppositions about intentions or guesswork about expectations. In Title VII, Congress adopted broad language making it illegal for an employer to rely on an employee's sex when deciding to fire that employee. We do not hesitate to recognize today a necessary consequence of that legislative choice: An employer who fires an individual merely for being gay or transgender defies the law."[37]

Now let me return to the post–Civil War cases that challenged conscription based on the Thirteenth Amendment. The conscripts who brought those earlier cases would no doubt cheerfully agree that when the Thirteenth Amendment was adopted in 1865, Congress had meant for it to abolish chattel slavery. But the amendment does not say "chattel slavery." Instead, the amendment bars "involuntary servitude." One need not be especially radical to acknowledge that military conscription is involuntary servitude within any plain, ordinary meaning. The only reason that draftees were drafted is that they were being

36 Bostock, 140 S. Ct. at 1731.

37 Ibid., 1754.

conscripted *involuntarily*. And they were being involuntarily conscripted into the armed services. This was involuntary servitude.

The teaching of *Bostock* is that, even when the common understanding of a statute's words evolve over time, interpreting courts are still obligated to interpret the plain words of a statute whenever that is possible. In this context, if conscription were legislatively enacted today, and a Thirteenth Amendment challenge to it were mounted, such a challenge would have considerable merit.

This does not mean, of course, that such an argument would be likely to prevail in any federal court, including, but not limited to, the Supreme Court. From the embarrassingly shoddy reasoning of *Arver* during World War I to the widely criticized *O'Brien* of the Vietnam era, the federal courts have shown a remarkable consistency in upholding the constitutionality of conscription. Moreover, the courts have consistently rejected invitations, in contexts other than conscription, to expand the ambit of the Thirteenth Amendment beyond chattel slavery. For example, in 1992, when prochoice activists challenged abortion restrictions by arguing that forcing women to carry unwanted pregnancies to term involved a form of involuntary servitude, the court disagreed, saying that "the argument borders on the frivolous."[38] In short, merely because the argument that conscription violates the Thirteenth Amendment is respectable does not mean that the argument is likely to prevail in court. It is not.

3 Is there a Constitutional Right to Conscientious Objection?

Assuming that conscription is constitutional, is there a constitutional right to conscientious objection? The well-established case law is that there is no such constitutional right. The rule is that conscientious objection is a matter of legislative grace. If Congress enacts conscription, it can—but need not—make provisions for COs; if Congress does make such provisions, those provisions can be on virtually any terms it wishes.[39]

But several things have changed in the last century. First, the legal rule that there is no right to conscientious objection was imported by the courts into the *conscription* context from the antecedent rule in the naturalization context. But the early naturalization cases that established the rule—*Schwimmer* and

38 Jane L. v. Bangerter, 794 F. Supp. 1537 (D. Utah 1992), *aff'd*, 61 F.3d 1505, 1511–12 (10th Cir. 1995). For a contrary view, see Andrew Koppelman, "Forced Labor: A Thirteenth Amendment Defense of Abortion," *Northwest University Law Review* 84, no. 2 (Winter 1990): 480–535.

39 Hamilton v. Regents of the Univ. of Calif., 293 U.S. 245, 264 (1934); United States v. Newman, 44 F. Supp. 817 (D.C. Ill. 1942); Hammond v. Lenfest, 398 F.2d 705, 715 (2d Cir. 1968).

Macintosh—are no longer good law. The Supreme Court overruled those cases in 1946 in *Girouard v. United States*. Today, there *is* a constitutional right to be a conscientious objector in the naturalization context.[40] Thus, at the very least, the original doctrinal foundation for the rule that there is no constitutional right to be a CO to conscription has been severely weakened.

Also, during the Vietnam era, the Supreme Court dramatically widened the ambit of who qualifies to be a CO in the landmark *Seeger* and *Welsh* decisions. Where Congress had made a careful, conscious, deliberate decision to limit conscientious objection to those whose beliefs are religious *and* who believe in God, the court ruled, first, that COs need not believe in God (*Seeger*) and then that COs need not even be religious (*Welsh*). In both cases, the Supreme Court's rulings vitiated the plain and unambiguous language of the statute and the equally unambiguous intent of Congress.[41] Taken together, *Seeger* and *Welsh* may fairly be taken to mean that the Supreme Court has taken a capacious, expansive view of the right to conscientious objection.

There are at least two counterarguments to the assertion that *Seeger* and *Welsh* may open the door to finding a constitutional right to conscientious objection. Both decisions can be read as meaning that *if* Congress establishes a provision for COs, then Congress cannot invidiously discriminate against would-be COs based on the nature of their beliefs. This is different than saying that there is a constitutional right to conscientious objection or that Congress must establish a provision for COs in any draft law it enacts. Also noteworthy is that, although the lawyers for Seeger and Welsh plied the constitutional argument (the First Amendment's free exercise clause) in their briefs to the Supreme Court, in both cases the court skirted the constitutional issues and grounded its rulings solely on the statute.

Nevertheless, *Seeger* and *Welsh*, taken together, do create a powerful argument that the Supreme Court might find a right to conscientious objection at some time in the future—especially in the context of *Girourd*'s having overturned *Schwimmer* and *Macintosh*. It is impossible to read *Seeger* and *Welsh* today and not hear the echo of Walter Nelles's argument in *Arver*, that for Congress to establish conscription without also providing an exemption for those whose religion forbids them to take up arms violates the free exercise clause of the First Amendment. And while it is certainly true that the Supreme Court did not actually hold that in *Seeger* or in *Welsh*, the court came close to

---

40 Girouard v. United States, 328 U.S. 61 (1946).

41 Universal Military Training and Service Act, 50 U.S.C. App. § 451 et seq.; United States v. Seeger, 380 U.S. 163 (1965); Welsh v. United States, 398 U.S. 333 (1970).

doing so, and it is difficult to imagine these rulings were issued without at least a tacit understanding of the Nelles argument.

Again, however, noting the soundness of the argument is not the same as suggesting that the US Supreme Court would be likely to accept the argument.

If the Supreme Court ever did find a constitutional right to conscientious objection, that would immediately present a large, real-world issue: What is the definition of a CO? In the United States, that definition has long been contested, and US law has provided multiple, mutually inconsistent answers. In World War I, a CO was a member of a historic peace church. In World War II, a CO was any religious pacifist. During the Cold War, a CO was a religious pacifist who believed in God. After *Seeger* and *Welsh*, a CO could be a wholly secular pacifist.

And England, during World War II, had an entirely different definition. In England, a CO was a man who said he was one, and this included nonpacifists, that is, selective objectors who opposed a particular war. Quakers and others in the United States proposed this definition when Congress was considering the conscription bill in 1940, but the military strongly opposed it, and Congress did not adopt it. The military's argument at that time was that allowing every man to decide whether he was a CO would make it impossible to raise an army.

But that was obviously not true for England during World War II. England had exactly that provision, yet England proved every bit as successful at raising an army as the United States was. Equally obvious is that the United States would not have been able to raise an army for the Vietnam War if every man were permitted to decide whether he was a CO. Reconciling these two facts is not difficult. Despite the opposition of pacifists, World War II was a popular war; in the United States, after Pearl Harbor, men lined up outside army recruiting stations to enlist. In contrast, Vietnam was a deeply unpopular war; especially after the Tet Offensive in February 1968, men went to extraordinary lengths to avoid and evade the draft. This history suggests that adopting England's 1939 definition of who is a CO would make it more difficult to fight unpopular wars like Vietnam but would not materially impede national defense in circumstances such as World War II.

∵

In 1917, the argument that Walter Nelles made to the Supreme Court was farfetched for its time: that the free exercise clause of the First Amendment creates a constitutional right to conscientious objection if Congress enacts conscription. There was no chance that the Supreme Court would accept that argument in 1917, and it didn't. But three decades later, in *Girourd*, the Supreme

Court did find a right to conscientious objection, albeit in the naturalization context. And in *Seeger* and *Welsh*, the court came close to finding an analogous right in the conscription context.

Seeger and *Welsh* represented substantial, but not yet total, recognition of the right for which Walter Nelles argued in 1917 in his amicus curiæ brief to the Supreme Court on behalf of the ACLU in *Arver*. If the United States returns to conscription in the future, and if the Supreme Court were to find a constitutional right to conscientious objection in the conscription context, that would be a complete vindication of Nelles's argument.

Select Bibliography

Archives

Gerald R. Ford Presidential Library, University of Michigan, Ann Arbor
Charles E. Goodell Papers

Mennonite Central Committee Archives, Akron, PA
Mennonite Central Committee Reports, 1942–63
National Service Board for Religious Objectors, Minutes and Correspondence, 1940–69

Swarthmore College Peace Collection, Swarthmore College, Swarthmore, PA
Central Committee for Conscientious Objectors Records
Center on Conscience and War Records
Collected Document Group: Category A, misc. papers of organizations or individuals based in the United States, 1814–
Eichel Family Collection
National Council Against Conscription Records
Juanita Morrow Nelson and Wallace F. Nelson Papers
War Resisters League Records

Books, Journal Articles, and Pamphlets

Abbott, Leonard D. *The Trial and Conviction of Emma Goldman and Alexander Berkman.* https://www.lib.berkeley.edu/goldman/pdfs/PublishedEssaysandPamphlets_TheTrialandConvictionofEmmaGoldmanandAlexanderBerkman.pdf. Last accessed April 26, 2022.

American Civil Liberties Union. *Conscience and the War: A Report on the Treatment of Conscientious Objectors in World War II.* New York: American Civil Liberties Union, 1943.

American Civil Liberties Union. *Freedom in Wartime: A Report of the American Civil Liberties Union in the Second Year of War.* New York: American Civil Liberties Union, 1943.

An Account of the Sufferings of Friends of North Carolina Yearly Meeting, In Support of Their Testimony Against War, From 1861 to 1865. New Vienna, NC: Peace Association of Friends in America, 1872.

Anderson, Jervis. *Bayard Rustin: Trouble I've Seen.* Berkeley: University of California Press, 1998.

Anderson, Martin, ed. *The Military Draft.* Stanford, CA: Hoover Institution Press, 1982.

Astor, Gerald. *The Right to Fight: A History of African Americans in the Military.* Cambridge, MA: Da Capo Press, 1998.

Auman, William T. *Civil War in the North Carolina Quaker Belt: The Confederate Campaign against Peace Agitators, Deserters, and Draft Dodgers.* Jefferson, NC: McFarland, 2014.

Barbeau, Arthur E., and Florette Henri. *The Unknown Soldiers: Black American Troops in World War I.* Philadelphia: Temple University Press, 1974.

Barnes, David M. *The Draft Riots in New York: July 1863.* New York: Baker & Godwin, 1863.

Baskir, Lawrence M., and William A. Strauss. *Chance and Circumstance: The Draft, the War, and the Vietnam Generation.* New York: Alfred A. Knopf, 1978.

Bayard, Rustin. *Time on Two Crosses: The Collected Writings of Bayard Rustin.* Edited by Devon W. Carbado and Donald Wiese. San Francisco: Cleis Press, 2003.

Bennett, Scott H. "Conscience, Comrades, and the Cold War: The Korean War Draft Resistance Cases of Socialist Pacifists David McReynolds and Vern Davidson." *Peace & Change* 38, no. 1 (January 2013): 83–120.

Bennett, Scott H. "Conscientious Objectors: Conscience, Courage, and Resistance." In *A History of American Literature and Culture of the First World War*, edited by Mark Van Wienen and Tim Dayton, 220–32. Cambridge: Cambridge University Press, 2021.

Bennett, Scott H. *Radical Pacifism: The War Resisters League and Gandhian Nonviolence in America, 1915–1963.* Syracuse, NY: Syracuse University Press, 2003.

Bennett, Scott H., and Charles F. Howlett, eds. *Antiwar Dissent and Peace Activism in World War I America.* Lincoln: University of Nebraska Press, 2014.

Bernstein, Iver. *The New York City Draft Riots: Their Significance for American Society and Politics in the Age of the Civil War.* Lincoln: University of Nebraska Press, 1990.

Bernstein, J. L. "Conscription and the Constitution: The Amazing Case of *Kneedler v. Lane.*" *ABA Journal* 53, no. 8 (1967): 708–12.

Berrigan, Daniel, Barbara Deming, Jim Forest, William M. Kunstler, Staughton Lynd, Richard Shaull, and Malcolm Boyd. *Delivered into Resistance.* New Haven, CT: Advocate Press, 1969.

Brock, Peter, ed. *Liberty and Conscience: A Documentary History of the Experiences of Conscientious Objectors in America through the Civil War.* New York: Oxford University Press, 2002.

Capozzola, Christopher. *Uncle Sam Wants YOU: World War I and the Making of the Modern American Citizen.* New York: Oxford University Press, 2008.

Carlson, David. "The 'Loanly Runagee': Draft Evaders in Southern Georgia." *Georgia Historical Quarterly* 84, no. 4 (Winter 2000): 589–97.

Carper, Jean. *Bitter Greetings: The Scandal of the Military Draft.* New York: Grossman, 1967.

Catton, Bruce. *The Civil War*. Boston: Houghton Mifflin, 2005.
Chafee, Zechariah. *Freedom of Speech*. New York: Harcourt, Brace & Howe, 1920.
Chambers, John Whiteclay II. *To Raise an Army: The Draft Comes to Modern America*. New York: Free Press, 1987.
Chambers, John Whiteclay II, ed. *The Eagle and the Dove: The American Peace Movement and United States Foreign Policy, 1900–1922*. Syracuse, NY: Syracuse University Press, 1991.
Chatfield, Charles. *For Peace and Justice: Pacifism in America, 1914–1941*. Knoxville: University of Tennessee Press, 1971.
Clifford, J. Garry, and Samuel R. Spencer, Jr. *The First Peacetime Draft*. Lawrence: University Press of Kansas, 1986.
Cornell, Julien. *Conscience and the State*. New York: John Day, 1944.
Cornell, Julien. *The Conscientious Objector and the Law*. New York: John Day, 1943.
Cottrell, Robert C. *Smoke Jumpers of the Civilian Public Service in World War II: Conscientious Objectors as Firefighters for the National Forest Service*. Jefferson, NC: McFarland, 2006.
Crowder, Enoch. *The Spirit of Selective Service*. New York: Century, 1920.
Cushman, Barry. "Rethinking the New Deal Court." *Virginia Law Review* 80, no. 1 (February 1994): 201–62.
Dancis, Bruce. *Resister: A Story of Protest and Prison during the Vietnam War*. Ithaca, NY: Cornell University Press, 2014.
Davis, James W., Jr., and Kenneth M. Dolbeare. *Little Groups of Neighbors: The Selective Service System*. Chicago: Markham, 1968.
DeBenedetti, Charles. *Origins of the Modern American Peace Movement, 1915–1929*. Millwood, NY: KTO Press, 1978.
DeBenedetti, Charles. *The Peace Reform in American History*. Bloomington: Indiana University Press, 1980.
DeBenedetti, Charles, and Charles Chatfield. *An American Ordeal: The Antiwar Movement of the Vietnam Era*. Syracuse, NY: Syracuse University Press, 1990.
Dellinger, David. *From Yale to Jail: The Life Story of a Moral Dissenter*. New York: Pantheon Books, 1993.
Dickson, Paul. *The Rise of the G.I. Army, 1940–1941: The Forgotten Story of How America Forged a Powerful Army before Pearl Harbor*. New York: Atlantic Monthly Press, 2020.
Dietrich, Frank, and Albert Dietrich. *Army GI, Pacifist CO: The World War II Correspondence of Frank and Albert Dietrich*. Edited by Scott H. Bennett. New York: Fordham University Press, 2005.
Donahue, Charles, Jr. "The Supreme Court vs. Section 10(b)(3) of the Selective Service Act: A Study in Ducking Constitutional Issues." *UCLA Law Review* 17, no. 5 (June 1970): 908–74.
Dorf, Michael C. "Incidental Burdens on Fundamental Rights." *Harvard Law Review* 109, no. 6 (April 1996): 1175–251.

Dower, John W. *War without Mercy: Race and Power in the Pacific War.* New York: Pantheon Books, 1986.

Doyle, Michael. *Radical Chapters: Pacifist Bookseller Roy Kepler and the Paperback Revolution.* Syracuse, NY: Syracuse University Press, 2012.

Duffy, Nora. "Jim Crow behind Bars: Benjamin Davis and Early Prison Desegregation Efforts." *Columbia Undergraduate Journal of History* 2, no. 1 (Spring 2017): 3–16.

Early, Frances H. *A World without War: How U.S. Feminists and Pacifists Resisted World War I.* Syracuse, NY: Syracuse University Press, 1997.

Eisan, Leslie. *Pathways of Peace.* Elgin, IL: Brethren, 1948.

Ekirch, Arthur A., Jr. "A Political Prisoner in Wartime." *Peace and Change* 12, no. 1 (April 1987): 77–92.

Ekirch, Arthur A., Jr. *The Civilian and the Military: A History of the American Antimilitarist Tradition.* Colorado Springs, CO: Ralph Myles, 1972.

Elliot, Jonathan. *The Debates in the Several State Conventions, on the Adoption of the Federal Constitution, on the Adoption of the Federal Constitution, as Recommended by the General Convention at Philadelphia, in 1787.* Washington, DC: Published Under the Sanction of Congress, 1836.

Elmer, Jerry. "Conscription after *Bostock v. Clayton County.*" *Rhode Island Bar Journal* 69, no. 3 (November–December 2020): 7–9.

Elmer, Jerry. *Felon for Peace: The Memoir of a Vietnam-Era Draft Resister.* Nashville, TN: Vanderbilt University Press, 2005.

Farrand, Max, ed. *The Records of the Federal Convention of 1787.* New Haven, CT: Yale University Press, 1911.

The Federalist Papers. Edited by Clinton Rossiter. New York: Signet, 1961.

Ferber, Michael, and Staughton Lynd. *The Resistance.* Boston: Beacon Press, 1971.

Fite, Gilbert C., and Harriet C. Peterson. *Opponents of War: 1917–1918.* Seattle: University of Washington Press, 1957.

Fitzpatrick, Edward A. *Conscription and America: A Study of Conscription in a Democracy.* Milwaukee: Richard, 1940.

Flynn, George Q. *The Draft, 1940–1973.* Lawrence: University Press of Kansas, 1993.

Flynn, George Q. *Lewis B. Hershey, Mr. Selective Service.* Chapel Hill: University of North Carolina Press, 1985.

Foley, Michael S. *Confronting the War Machine: Draft Resistance during the Vietnam War.* Chapel Hill: University of North Carolina Press, 2003.

Forest, James H. *Delivered into Resistance.* New Haven, CT: Advocate Press, 1969.

Freeman, Harrop. "The Constitutionality of Peacetime Conscription." *Virginia Law Review* 31, no. 1 (1944): 40–82.

French, Paul Comly. *The Origins of Civilian Public Service.* Washington, DC: National Service Board for Religious Objectors, 1940.

Friedman, Leon. "Conscription and the Constitution: The Original Understanding." *Michigan Law Review* 67, no. 8 (June 1969): 1493–552.

Frost, William J. "Our Deeds Carry Our Message: The Early History of the American Friends Service Committee." *Quaker History* 81, no. 1 (Spring 1992): 3–51.

Gales, Joseph, ed. *Debates and Proceedings in the Congress of the United States.* Washington, DC: Gales and Seaton, 1834.

Gara, Larry, and Lenna Mae Gara, eds. *A Few Small Candles: War Resisters of World War II Tell Their Stories.* Kent, OH: Kent State University Press, 1999.

Geary, James W. *We Need Men: The Union Draft in the Civil War.* DeKalb: Northern Illinois University Press, 1991.

Gibson, Campbell, and Kay Jung. *Population Division: Historical Census Statistics on Population Totals by Race, 1790 to 1900."* Working Paper no. 56. Washington, DC: US Census Bureau, September 2002.

Giffin, Frederick C. *Six Who Protested: Radical Opposition to the First World War.* Port Washington, NY: Kennikat Press, 1977.

Gingerich, Melvin. *Service for Peace: A History of Mennonite Civilian Public Service.* Akron, PA: Mennonite Central Committee, 1949.

Glassgold, Peter, ed. *Anarchy! An Anthology of Emma Goldman's* Mother Earth. Washington, DC: Counterpoint Press, 2001.

Golden, Michael J. "The Dormant Second Amendment: Exploring the Rise, Fall, and Potential Resurrection of Independent State Militias." *William & Mary Bill of Rights Journal* 21, no. 4 (May 2013): 1071–79.

Gottlieb, Sherry Gershon. *Hell No We Won't Go! Resisting the Draft during the Vietnam War.* New York: Viking, 1991.

Gray, Francine du Plessix. *Divine Disobedience.* New York: Alfred A. Knopf, 1970.

Hagan, John. *Northern Passage: American Vietnam War Resisters in Canada.* Cambridge, MA: Harvard University Press, 2001.

Halpert, Stephen, and Tom Murrray, eds. *Witness of the Berrigans.* New York: Doubleday, 1972.

Harris, David. *Dreams Die Hard: Three Men's Journey through the Sixties.* San Francisco: Mercury House, 1982.

Harris, David. *Goliath.* New York: Avon Books, 1970.

Hassler, R. Alfred. *Conscripts of Conscience.* New York: Fellowship of Reconciliation, 1942.

Hennacy, Ammon. *The Book of Ammon.* Baltimore: Fortcamp, 1994.

Hershey, Lewis B. *Report of the Director of the Office of Selective Service Records: 1947–1948.* Washington, DC: Government Printing Office, 1950.

Hickey, Donald R. *The War of 1812: A Forgotten Conflict.* Urbana: University of Illinois Press, 1989.

Hirsch, Alan. "The Militia Clauses of the Constitution and the National Guard." *University of Cincinnati Law Review* 56, no. 3 (1988): 919–69.

Homan, Gerlof D. *American Mennonites and the Great War, 1914–1918.* Waterloo, ON: Herald Press, 1994.

Hough, Emerson. *The Web*. Chicago: Reilly & Lee, 1919.

Howlett, Charles F. "John Nevin Sayre and the Fellowship of Reconciliation." *Pennsylvania Magazine of History and Biography* 114, no. 3 (July 1990): 399–421.

Howlett, Charles F. "Julien Davies Cornell: Gentle Quaker, Determined Litigator." *Friends Journal* 53, no. 5 (May 1, 2007): 6–8. https://www.friendsjournal.org/2007042. Last accessed October 4, 2022.

Hunt, Andrew E. *David Dellinger: The Life and Times of a Nonviolent Revolutionary*. New York: New York University Press, 2006.

James, Rawn, Jr. *The Double V: How Wars, Protest, and Harry Truman Desegregated America's Military*. New York: Bloomsbury Press, 2013.

Jones, Rufus M. *A Service of Love in Time of War: American Friends Relief Work in Europe, 1917–1919*. New York: Macmillan, 1920.

Jones, Rufus M. Introduction to Cyrus Pringle, *The Record of a Quaker Conscience: Cyrus Pringle's Diary*, 5–19. New York: Macmillan, 1918.

Journal of the House of Representatives of the First Congress of the Confederate States of America, 1861–1865. 7 vols. Washington, DC: Government Printing Office, 1904–1905.

Kastenberg, Joshua. *To Raise and Discipline an Army: Major General Enoch Crowder, the Judge Advocate General's Office, and the Realignment of Civil and Military Relations in World War I*. DeKalb: Northern Illinois University Press, 2017.

Kazin, Michael. *War against War: The American Fight for Peace, 1914–1918*. New York: Simon & Schuster, 2017.

Keim, Albert N. *The CPS Story: An Illustrated History of Civilian Public Service*. Intercourse, PA: Good Books, 1990.

Keim, Albert N., and Grant M. Stoltzfus. *The Politics of Conscience: The Historic Peace Churches and America at War, 1917–1955*. Eugene, OR: Wipf & Stock, 2000.

Keith, Jeanette. *Rich Man's War, Poor Man's Fight: Race, Class, and Power in the Rural South during the First World War*. Chapel Hill: University of North Carolina Press, 2004.

Kellogg, Walter Guest. *The Conscientious Objector*. New York: Boni and Liveright, 1919.

Kennedy, David M. *Over Here: The First World War and American Society*. Oxford: Oxford University Press, 2004.

Kennedy, Kathleen. *Disloyal Mothers and Scurrilous Citizens: Women and Subversion during World War I*. Bloomington: Indiana University Press, 1999.

Kessler, Jeremy K. "The Administrative Origins of Modern Civil Liberties Law." *Columbia Law Review* 114, no. 5 (June 2014): 1083–166.

Kessler, Jeremy K. "New Look Constitutionalism: The Cold War Critique of Military Manpower Administration." *University of Pennsylvania Law Review* 167, no. 7 (June 2019): 1749–82.

King, Martin Luther, Jr. *The Radical King: Martin Luther King, Jr*. Edited by Cornel West. Boston: Beacon Press, 2015.

Kirchner, Earl L. *Enforcement of the Selective Service Law*. Selective Service Special Monograph No. 14. Washington, DC: Government Printing Office, 1950.

Kohn, Stephen M. *Jailed for Peace: The History of American Draft Law Violators, 1658–1985*. Westport, CT: Greenwood Press, 1986.

Krehbel, Nicholas A. *General Lewis B. Hershey and Conscientious Objection during World War II*. Columbia: University of Missouri Press, 2011.

Leach, Jack Franklin. *Conscription in the United States: Historical Background*. Rutland, VT: Charles E. Tuttle, 1952.

Levine, Peter. "Draft Evasion in the North during the Civil War, 1863–1865." *Journal of American History* 67, no. 4 (March 1981): 816–34.

Lincoln, Abraham. *The Collected Works of Abraham Lincoln*. Vols. 4 and 6. Edited by Roy Basler. New Brunswick, NJ: Rutgers University Press, 1953.

Lincoln, Abraham. *The Complete Works of Abraham Lincoln*. Edited by John George Nicolay and John Hay. 24 vols. New York: Francis D. Tandy, 1894.

Lofgren, Charles A. "Compulsory Military Service under the Constitution: The Original Understanding." *William and Mary Quarterly* 33, no. 1 (January 1976): 61–88.

Lonn, Ella. *Desertion during the Civil War*. Lincoln: University of Nebraska Press, 1998.

Lynd, Alice, ed. *We Won't Go: Personal Accounts of War Objectors*. Boston: Beacon Press, 1968.

Lynd, Staughton, ed. *Nonviolence in America: A Documentary History*. New York: Bobbs-Merrill, 1966.

Mackey, Mike, ed. *A Matter of Conscience: Essays on the World War II Heart Mountain Draft Resistance Movement*. Powell, WY: Western History Publications, 2002.

Macy, Jesse. *Jesse Macy: An Autobiography*. Edited by Katherine Macy Noyes. Springfield, IL: Charles C. Thomas, 1933.

Mason, Alpheus Thomas. "Harlan Fiske Stone: In Defense of Individual Freedom, 1918–1920." *Columbia Law Review* 51, no. 2 (February 1951): 147–69.

Matthews, James M., ed. *Public Laws of the Confederate States of America*. Richmond, VA: R. M. Smith, 1862–64.

McCague, James. *The Second Rebellion: The New York City Draft Riots of 1863*. New York: Dial Press, 1968.

McGoldrick, James M. "*United States v. O'Brien* Revisited: Of Burning Things, Waving Things, and G-Strings." *University of Memphis Law Review* 36, no. 4 (Summer 2006): 903–49.

McPherson, James M. *Battle Cry of Freedom*. New York: Oxford University Press, 1988.

Meconis, Charles A. *With Clumsy Grace: The American Catholic Left, 1961–1975*. New York: Seabury Press, 1979.

Mitchell, Memory F. *Legal Aspects of Conscription and Exemption in North Carolina, 1861–1865*. Chapel Hill: University of North Carolina Press, 1965.

Mollin, Marian. *Radical Pacifism in Modern America: Egalitarianism and Protest*. Philadelphia: University of Pennsylvania Press, 2006.

Moore, Albert Burton. *Conscription and Conflict in the Confederacy*. 1924. Columbia: University of South Carolina Press, 1996.

Moore, Howard W. *Plowing My Own Furrow*. Syracuse, NY: Syracuse University Press, 1993.

Muller, Eric L. *Free to Die for Their Country*. Chicago: University of Chicago Press, 2001.

Murdock, Eugene C. *One Million Men: The Civil War Draft in the North*. Madison: State Historical Society of Wisconsin, 1971.

Murphy, Paul L. *World War I and the Origin of Civil Liberties in the United States*. New York: W. W. Norton, 1979.

Naeve, Lowell. *Field of Broken Stones*. Glen Gardner, NJ: Libertarian Press, 1950.

National Advisory Commission on Selective Service. *In Pursuit of Equity: Who Serves When Not All Serve?* Washington, DC: Government Printing Office, 1967.

National Civil Liberties Bureau. *The Conviction of Mrs. Kate Richards O'Hare and North Dakota Politics*. New York: National Civil Liberties Bureau, 1918.

National Civil Liberties Bureau. *The Facts about Conscientious Objectors in the United States (under the Selective Service Act of May 18, 1917)*. New York: National Civil Liberties Bureau, 1918.

National Civil Liberties Bureau. *Political Prisoners in Federal Military Prisons*. New York: National Civil Liberties Union, November 21, 1918.

National Service Board for Religious Objectors. *Directory of Civilian Public Service, May, 1941 to March, 1947*. Washington, DC: Author, 1947.

Nelles, Walter. *Espionage Act Cases: New Law in the Making as to Criminal Utterances*. New York: National Civil Liberties Bureau, July 1918.

Nelles, Walter. *Seeing Red: Civil Liberty and Law in the Period following the War*. New York: American Civil Liberties Union, 1920.

O'Neil, Robert M. "Review of Selective Service Reclassifications." *George Washington Law Review* 37, no. 3 (March 1969): 536–63.

Peck, James. *Underdogs vs. Upperdogs*. New York: AMP&R, 1980.

Peck, James. *We Who Would Not Kill*. New York: Lyle Stuart, 1958.

Perri, Timothy J. "The Evolution of Military Conscription in the United States." *Independent Review* 17, no. 3 (Winter 2013): 429–39.

Peters, Shawn Francis. *The Catonsville Nine: A Story of Faith and Resistance in the Vietnam Era*. New York: Oxford University Press, 2012.

Polner, Murray, and Jim O'Grady. *Disarmed and Dangerous: The Radical Lives and Times of Daniel and Philip Berrigan*. New York: Basic Books, 1997.

President's Advisory Commission on Universal Training. *A Program for National Security: Report of the President's Advisory Commission on Universal Training*. Washington, DC: Government Printing Office, May 29, 1947.

Raley, Bill. "How Conscientious Objectors Killed the Draft: The Collapse of the Selective Service during the Vietnam War." *Cleveland State Law Review* 68, no. 2 (2020): 151–71.

Reiger, John F. "Deprivation, Disaffection and Desertion in Confederate Florida." *Florida Historical Quarterly* 48, no. 3 (January 1970): 279–98.

Rostker, Bernard. *I Want You! The Evolution of the All-Volunteer Force*. Santa Monica, CA: Rand, 2006.

Rubenfeld, Jed. "The First Amendment's Purpose." *Stanford Law Review* 53, no. 4 (April 2001): 767–832.

Rutenberg, Amy J. *Rough Draft: Cold War Military Manpower and the Origins of Vietnam-Era Draft Resistance*. Ithaca, NY: Cornell University Press, 2019.

Sacher, John M. "'Twenty-Negro,' or Overseer Law: A Reconsideration." *Journal of the Civil War Era* 7, no. 2 (June 2017): 269–92.

Salvatore, Nick. *Eugene V. Debs: Citizen and Socialist.* Urbana: University of Illinois Press, 1982.

Sandow, Robert M. *Deserter Country: Civil War Opposition in the Pennsylvania Appalachians*. New York: Fordham University Press, 2009.

Sauers, Richard A., and Peter Tomasak. *The Fishing Creek Confederacy: A Story of Civil War Draft Resistance.* Columbia: University of Missouri Press, 2012.

Schecter, Barnet. *The Devil's Own Work: The Civil War Draft Riots and the Fight to Reconstruct America.* New York: Walker, 2005.

Sellars, Nigel Anthony. "Treasonous Tenant Farmers and Seditious Sharecroppers: The 1917 Green Corn Rebellion." *Oklahoma City University Law Review* 27, no. 1 (Spring 2002): 1097–1141.

Sellars, Nigel. "With Folded Arms? Or with Squirrel Guns? The IWW and the Green Corn Rebellion." *Chronicles of Oklahoma* 77, no. 2 (Summer 1999): 150–69.

Shankman, Arnold M. *The Pennsylvania Antiwar Movement, 1861–1865*. Rutherford, NJ: Fairleigh Dickinson University Press, 1980.

Shaw, William L. "The Confederate Conscription and Exemption Acts." *American Journal of Legal History* 6, no. 4 (October 1962): 368–405.

Shaw, William M. "The Interrelationship of the United States Army and the National Guard." *Military Law Review* 31, no. 1 (January 1966): 39–84.

Sibley, Mulford Q., and Philip E. Jacob. *Conscription of Conscience: The American State and the Conscientious Objector*. Ithaca, NY: Cornell University Press, 1952.

Sibley, Mulford, and Ada Wardlaw. *Conscientious Objectors in Prison*. Philadelphia: Pacifist Research Bureau, 1945.

Stoltzfus, Duane. "Nurse without a Country: When a Mennonite Who Refused to Bear Arms Prevailed in Her Bid for Citizenship." *Mennonite Quarterly Review* 93, no. 3 (July 2019): 335–62.

Stoltzfus, Duane C. *Pacifists in Chains: The Persecution of Hutterites during the Great War.* Baltimore: Johns Hopkins University Press, 2013.

Stone, Harlan Fiske. "The Conscientious Objector." *Columbia University Quarterly* 21, no. 4 (October 1919): 253–72.

Strauss, David A. *The Living Constitution.* Oxford: Oxford University Press, 2010.

"The Supreme Court, 1969 Term: Delinquency Status and Punitive Reclassification." *Harvard Law Review* 84, no. 1 (November 1970): 222–29.

Tatum, Georgia Lee. *Disloyalty in the Confederacy.* Chapel Hill: University of North Carolina Press, 1934.

Thomas, Norman. *The Conscientious Objector in America.* New York: B. W. Huebsch, 1923.

Tracy, James. *Direct Action: Radical Pacifism from the Union Eight to the Chicago Seven.* Chicago: University of Chicago Press: 1996.

US Census Bureau. *Sixteenth Census of the United States. Volume 1: Population, General Report and Analysis.* Washington, DC: Government Printing Office, 1913.

US Census Bureau. *Population of the United States in 1860.* Washington, DC: Government Printing Office, 1864.

US Census Bureau. *Population of the United States in 1860: Recapitulation of the Tables of Population Nativity and Occupation.* Washington, DC: Government Printing Office, 2021.

US Census Bureau and the Bureau of Labor Statistics Staff. *Recent Trends in Social and Economic Conditions of Negroes, July 1968.* Washington, DC: Government Printing Office, 1968.

US Courts. *Annual Report of the Director of the Administrative Office of the U.S. Courts.* Washington, DC: Government Printing Office, 1968, 1971, 1974, 1975.

US Office of the Provost Marshal General. *Final Report Made to the Secretary of War, by the Provost Marshal General, of the operations of the Bureau of the Provost Marshal General of the U.S.:* […] *March 17, 1863 to March 17, 1866,* […]. Washington, DC: Government Printing Office, 1866.

US Office of the Provost Marshal General. *Final Report of the Provost Marshal General to the Secretary of War on the Operations of the Selective Service System to July 15, 1919.* Washington, DC: Government Printing Office, 1919.

US Office of the Provost Marshal General. *Report of the Provost Marshal General to the Secretary of War on the First Draft under the Selective-Service Act, 1917, 1918.* Washington DC: Government Printing Office, 1918.

US Office of the Provost Marshal General. *Second Report of the Provost Marshal General to the Secretary of War on the Operation of the Selective Service System to December 20, 1918.* Washington, DC: Government Printing Office, 1919.

US Secretary of War. *Statement Concerning the Treatment of Conscientious Objectors in the Army.* Washington, DC: Government Printing Office, 1919.

US Selective Service System. *Annual Report of the Director of Selective Service for the Fiscal Year 19__ to the Congress of the United States Pursuant to the Universal Military*

Training and Service Act as Amended. Washington, DC: Government Printing Office, 1952–67.

US Selective Service System. *Conscientious Objection:* Special Monograph No. 11. 2 vols. Washington, DC: Government Printing Office, 1950.

US Selective Service System. *The Operation of Selective Service.* Special Monograph No. 17, vol. 1. Washington, DC: Government Printing Office, 1955.

US Selective Service System. *Quotas, Calls and Inductions: Special Monograph No. 12.* Washington, DC: Government Printing Office, 1948.

US Selective Service System. *Selective Service and Victory: The Fourth Report of the Director of Selective Service, 1944–1945, With a Supplement for 1946–1947.* Washington, DC: Government Printing Office, 1948.

US Selective Service System. *Selective Service as the Tide of War Turns: The Third Report of the Director of Selective Service.* Washington, DC: Government Printing Office, 1945.

US Selective Service System. *Selective Service in Peacetime: First Report of the Director of Selective Service, 1940–41.* Washington, DC: Government Printing Office, 1942.

US Selective Service System. *Selective Service in Wartime: Second Report of the Director of Selective Service, 1941–41.* Washington, DC: Government Printing Office, 1943.

US Selective Service System. *Selective Service under the 1948 Act.* Washington, DC: Government Printing Office, 1951.

US Selective Service System. *Selective Service under the 1948 Act Extended, July 9, 1950–June 19, 1951.* Washington, DC: Government Printing Office, 1951–67.

US Selective Service System. *Semi-Annual Report of the Director of Selective Service ... to the Congress of the United States pursuant to the Military Selective Service Act of 1967.* Washington, DC: Government Printing Office, 1968–1973 (but many are not dated).

US War Department. *1912 Report of the Chief of Staff U.S. Army to the Secretary of War. Annual Reports, War Department Fiscal Year Ended June 30, 1912.* Vol. 1. Washington, DC: Government Printing Office, 1912.

US War Department. *The War of the Rebellion: A Compilation of the Official Records of the Union and Confederate Armies*, 128 vols. Washington, DC: Government Printing Office, 1880–1901.

Van Zandt, Jennifer. "Confederate Conscription and the North Carolina Supreme Court." *North Carolina Historical Review* 72, no. 1 (January 1995): 327–63.

Weinrib, Laura. "Freedom of Conscience in War Time: World War I and the Limits of Civil Liberties." *Emory Law Journal* 65, no. 4 (2016): 1051–137.

Wells, Tom. *The War Within: America's Battle over Vietnam.* New York: Henry Holt, 1994.

Westheiter, James E. *Fighting on Two Fronts: African Americans and the Vietnam War.* New York: New York University Press, 1997.

Williams, David. *A People's History of the Civil War: Struggles for the Meaning of Freedom.* New York: New Press, 2005.

Wilson, E. Raymond. "Evolution of the c.o. Provisions in the 1940 Conscription Bill." *Quaker History* 64, no. 1 (Spring 1975): 3–15.
Wittner, Lawrence S. *Rebels against War: The American Peace Movement, 1933–1983*. Rev. ed. Philadelphia: Temple University Press 1984.
Wright, Edward Needles. *Conscientious Objectors in the Civil War*. New York: A. S. Barnes, 1931.
Wright, Robert K., Jr. *The Continental Army*. Washington, DC: Center of Military History, US Army, 2006.
Zaroulis, Nancy, and Gerald Sullivan. *Who Spoke Up? American Protest against the War in Vietnam, 1963–1975*. New York: Holt, Rinehart & Winston, 1984.

Dissertations and Theses

Bateman-House, Alison. "Compelled to Volunteer: American Conscientious Objectors to World War II as Subjects of Medical Research." PhD diss., Columbia University, 2014.
Bush, Charles C. "The Green Corn Rebellion." Master's thesis, University of Oklahoma, 1932.
Schoonover, Lynn Ira. "The History of the Civil War Draft in Wisconsin." Master's thesis, University of Wisconsin, 1915.
Sterling, Robert E. "Civil War Draft Resistance in the Middle West." PhD diss., Northern Illinois University, 1974.
Wood, John W. III. "Wally and Juanita Nelson and the Struggle for Peace, Equality, and Social Justice: 1935–1975." PhD diss., Morgan State University, 2008.

Legal Cases

American Beauty Homes Corp. v. Louisville & Jefferson City, 379 S.W.2d 450 (Ky. 1964).
American Socialist Society et al. v. United States, 266 F. 212 (2d Cir. 1920).
Anderson v. United States, 264 F. 75 (8th Cir. 1920).
Antrim's Case, 1 F. Cas. 1062, 5 Phila. 278 (E.D. Pa. 1863).
Atherton v. United States, 176 F.2d 835 (9th Cir. 1949).
Barbour v. Camden, 51 Me. 608 (1865).
Berman v. United States, 156 F.2d 377 (9th Cir. 1946).
Bertalini v. Mitchell, 248 F. 997 (E.D.N.Y. 1918).
Billings v. Truesdale, 135 F.2d 505 (10th Cir. 1943).
Billings v. Truesdale, 321 U.S. 542 (1944).
Booth v. Woodbury, 32 Conn. 118 (1864).

Bostock v. Clayton County, 140 S. Ct. 1731 (2020).
Breen v. Selective Service, 396 U.S. 460 (1970).
Bronemann v. United States, 138 F.2d 333 (8th Cir. 1943).
Bryant v. United States, 257 F. 378 (5th Cir. 1919).
Bucher v. Selective Service, 421 F.2d 24 (3rd Cir. 1970).
Burroughs v. Peyton, 57 Va. (16 Gratt.) 470 (1864).
Butler v. Perry, 240 U.S. 328 (1916).
Coldwell v. United States, 256 F. 805 (1st Cir. 1919).
Commers v. United States, 66 F. Supp. 943 (D. Mont. 1946).
Connell v. Leonard, 33 Ga. Supp. 58 (1864).
Cox v. Wood, 247 U.S. 3 (1918).
Daly v. Harris, 33 Ga. Supp. 50 (1864).
Debs v. United States, 249 U.S. 211 (1919).
District of Columbia v. Heller, 554 U.S. 570 (2008).
Doty v. United States, 218 F.2d 93 (8th Cir. 1955).
Druecker v. Saloman, 21 Wisc. 621 (1867).
Estep v. United States, 327 U.S. 114 (1946).
Ex parte Barton, 39 Ala. 452 (1864).
Ex parte Beck, 245 F. 967 (D. Mont. 1917).
Ex parte Billings, 46 F. Supp. 663 (D. Kan. 1942).
Ex parte Burke 4 F. Cas. 731 (E.D. Pa. 1863).
Ex parte Cain, 39 Ala. 440 (1864).
Ex parte Cohen, 254 F. 711 (E.D. Va. 1918).
Ex parte Coupland, 26 Tex. 386 (1862).
Ex parte Furston, 253 F. 90 (E.D. Tenn. 1918).
Ex parte Hill, 38 Ala. 429 (1863).
Ex parte Lockhart, 39 Ala. 450 (1864).
Ex parte Stringer, 38 Ala. 457 (1863).
Ex parte Watkins, 28 U.S. 193 (1830).
Falbo v. United States, 320 U.S. 549 (1944).
Fein v. Selective Service System, 405 U.S. 365 (1972).
Firth v. United States, 253 F. 36 (4th Cir. 1918).
Fitzgerald v. Harris, 33 Ga. Supp. 38 (1864).
Fletcher v. United States, 129 F.2d 262 (5th Cir. 1942).
Franke v. Murray, 248 F. 865 (1918).
Frohwerk v. United States, 249 U.S. 204 (1919).
Gaitlin v. Walton, 60 N.C. 333 (1864).
Gara v. United States, 178 F.2d 38 (6th Cir. 1949).
Gates v. McManus, 33 Ga. Supp. 67 (1864).
Gatlin v. Walton, 60 N.C. 333 (1864).

Giese v. United States, 143 F.2d 633 (D.C. Cir. 1944).
Giese v. United States, 323 U.S. 682 (1945).
Gilchrist v. United States, 430 F.2d 631 (10th Cir. 1970).
Gillette v. United States, 401 U.S. 437 (1970).
Girouard v. United States, 328 U.S. 1 (1946).
Goff v. United States, 135 F.2d 610 (4th Cir. 1943).
Goldman and Berkman v. United States, 245 U.S. 474 (1918).
Gonzales v. United States, 348 U.S. 407 (1955).
Gutknecht v. United States, 396 U.S. 295 (1970).
Hamilton v. Regents of the University of California, 293 U.S. 245 (1934).
Hammond v. Lenfest, 398 F.2d 705 (2d Cir. 1968).
Harris v. Nelson, 394 U.S. 286 (1969).
Henderson v. United States, 339 U.S. 816 (1950).
Hoagland v. Delaware, 17 N.J. Eq. 106 (1864).
Holmes v. United States, 391 U.S. 936 (1968).
In re Barrett, 25 How. Pr. 380, 42 Barb. 479 (N.Y. 1863).
In re Blum, 3 F. Cas. 752 (D. Mass. July 1863).
In re Bradshaw, 60 N.C. 454 (1864).
In re Ellsberg, 446 F.2d 954 (1st Cir. 1971).
In re Hopson, 40 Barb. 34 (N.Y. 1863).
In re Kemp, 16 Wisc. 359 (1863).
In re Roseman, 60 N.C. (Win.) 443 (1864).
In re Spangler, 11 Mich. 298 (1863).
In re State *ex rel.* Graham, 39 Ala. 437 (1864).
In re Summers, 325 U.S. 561 (1945).
In re Wyrick, 60 N.C. 450 (1864).
Jane L. v. Bangerter, 794 F. Supp. 1537 (D. Utah 1992).
Jeffers v. Fair, 33 Ga. 347 (1862).
Katzenbach v. McClung, 379 U.S. 294 (1964).
Kneedler v. Lane, 45 Pa. 238 (1863).
Korematsu v. United States, 232 U.S. 214 (1944).
Kramer v. United States, 147 F.2d 756 (6th Cir. 1945).
Laparouse v. Rice, 13 La. Ann. 567 (1858).
Masaaki Kuwabara, 56 F. Supp. 719 (N.D. Cal. 1944).
McClung v. Katzenbach, 233 F. Supp. 815 (1964).
McCulloch v. Maryland, 17 U.S. 316 (1819).
Miller v. United States, 386 U.S. 972 (1967).
Mulloy v. United States, 398 U.S. 410 (1970).
National Federation of Independent Business v. Sebelius, 567 U.S. 519 (2010).
National Student Association v. Hershey, 412 F.2d 1103 (D.C. Cir. 1969).

New Negro Alliance v. Sanitary Grocery, 303 U.S. 552 (1938).
O'Hare v. United States, 253 F. 538 (9th Cir. 1918).
Oestereich v. Selective Service, 393 U.S. 233 (1968).
Opinion of the Justices, 8 Mass. (8 Tyng) 548 (1812).
Patterson v. Cole, 205 U.S. 454 (1907).
Pearson *ex parte* Walton, 60 N.C. (Win.) 425 (1864).
Peter v. United States, 324 F.2d 173 (9th Cir. 1963).
Petition of Soberman, 37 F. Supp. 522 (E.D.N.Y. 1941).
Plessy v. Ferguson, 163 U.S. 537 (1896).
Roodenko v. United States, 147 F.2d 752 (10th Cir. 1944).
Schenck v. United States, 249 U.S. 47 (1919).
Scott v. Lazenby, 33 Ga. Supp. 134 (1864).
Selective Draft Law Cases (Arver), 245 U.S. 366 (1918).
Shimola v. Local Board No. 42 for Cuyahoga County, 40 F. Supp. 808 (N.D. Ohio 1941).
Simmons v. United States, 348 U.S. 397 (1955).
State *ex rel.* Beckett v. Collector of Bordentown, 32 N.J.L. 192 (1867).
State v. Hannibal, 51 N.C. (6 Jones) 57 (1858).
State v. Jackson, 33 N.J.L. 450 (1867).
Stilson v. United States, 250 U.S. 583 (1919).
Stone v. Christensen, 36 F. Supp. 739 (D. Ore. 1940).
Sugar v. United States, 252 F. 79 (6th Cir. 1918).
Toussie v. United States, 397 U.S. 112 (1970).
United States *ex rel.* Bergdoll v. Drum, 107 F.2d 897 (2d Cir. 1939), *cert. denied*, 310 U.S. 648 (1940).
United States *ex rel.* Brandon v. Downer, 139 F.2d 761 (2d Cir. 1944).
United States *ex rel.* Koopowitz v. Finley, 245 F. 871 (E.D.N.Y. 1917).
United States *ex rel.* Morton v. McBee, 310 F. Supp. 328 (N.D. Ill 1972).
United States *ex rel.* Pascher v. Kinkead, 250 F. 692 (3d Cir. 1918).
United States *ex rel.* Phillips v. Downer, 135 F.2d 521 (2d Cir. 1943).
United States *ex rel.* Reel v. Badt, 141 F.2d 845 (2d Cir. 1944).
United States *ex rel.* Zucker v. Osborne, 54 F. Supp. 984 (W.D.N.Y. 1944).
United States v. Bouziden, 108 F. Supp. 395 (W.D. Okla. 1952).
United States v. Bowles, 131 F.2d 818, 819 (3d Cir. 1942).
United States v. Butler, 297 U.S. 1 (1936).
United States v. Cornell, 36 F. Supp. 81 (D. Idaho 1940).
United States v. Crocker, 294 F. Supp. 776 (D. Minn. 1969).
United States v. Emerson, 270 F.3d 203 (5th Cir. 2001).
United States v. Emery, 168 F.2d 454 (2d Cir. 1948).
United States v. Garst, 39 F. Supp. 367 (E.D. Pa. 1941).
United States v. Grieme, 128 F.2d 811 (3d Cir. 1942).

United States v. Henderson, 180 F.2d 711 (7th Cir. 1950).
United States v. Herling, 120 F.2d 236 (2d Cir. 1941).
United States v. Jakobson, 325 F.2d 409 (2d Cir. 1963).
United States v. Jones, 468 F.2d 454 (3d Cir. 1972).
United States v. Kauten, 133 F.2d 703 (2d Cir. 1943).
United States v. Lambert, 123 F.2d 395 (3d Cir. 1941).
United States v. Macintosh, 283 U.S. 605 (1931).
United States v. Masaaki Kuwabara, 56 F. Supp. 716 (N.D. Cal. 1944).
United States v. Morrison, 529 U.S. 598 (2000).
United States v. Mroz, 136 F.2d 221 (7th Cir. 1943).
United States v. Newman, 44 F. Supp. 817 (E.D. Ill. 1942).
United States v. Nugent, 346 U.S. 1 (1953).
United States v. O'Brien, 391 U.S. 367 (1968).
United States v. Palmer, 223 F.2d 893 (3d Cir. 1955).
United States v. Rappeport, 36 F. Supp. 915 (S.D.N.Y. 1941).
United States v. Ryals, 56 F. Supp. 772 (N.D. Ga. 1944).
United States v. Sauler, 139 F.2d 173 (7th Cir. 1943).
United States v. Schenck, 253 F. 212 (E.D. Pa. 1918).
United States v. Schwimmer, 279 U.S. 644 (1929).
United States v. Seeger, 326 F.2d 864 (2d Cir. 1964).
United States v. Seeger, 380 U.S. 163 (1965).
United States v. Spock, 416 F.2d 165 (1st Cir. 1969).
United States v. Stephens, 245 F. 956 (D. Del. 1917).
United States v. Stilson, 254 F. 120 (E.D. Pa. 1918).
United States v. Vincelli, 215 F.2d 210 (2d Cir. 1954).
Wells v. United States, 257 F. 605 (9th Cir. 1919).
Welsh v. United States, 398 U.S. 333 (1970).
Wolff v. Selective Service, 372 F.2d 817 (2d Cir. 1970).
Youngstown Sheet & Tube Co. v. Sawyer (The Steel Seizure Case), 343 U.S. 579 (1952).

Newspapers, Magazines, and Newsletters

Chicago Tribune
Cincinnati Enquirer
Columbia (University) Spectator
Congressional Record
Conscientious Objector (New York)
Daily (Lawrence, KS) Gazette
Des Moines Register

Direct Action for a Nonviolent World (Voluntown, CT)
Hartshorne (OK) *Sun*
Indianapolis News
Indianapolis Star
Jackson (TN) *Sun*
Kansas City (MO) *Star*
Kansas City (MO) *Times*
Latimer County News-Democrat (Wilburton, OK)
Lawrence (KS) *Daily Journal*
Life
Los Angeles Times
Muskogee (OK) *Times-Democrat*
Nation
New Left Notes (Chicago)
New York Review of Books
New York Times
New York Tribune
NSBRO *Reporter* (National Service Board for Religious Objectors, Washington, DC)
Pacifica Views (Glendora, CA)
Peacemaker (Cincinnati)
Philadelphia Daily News
Philadelphia Inquirer
Politics (New York)
Providence (RI) *Journal*
Ramparts (Menlo Park, CA)
Resist Newsletter (Cambridge, MA)
San Francisco Examiner
Selective Service, later *Selective Service News* (Washington, DC)
St. Louis Post-Dispatch
St. Petersburg (FL) *Times*
Topeka (KS) *Daily Capital*
Tucson (AZ) *Citizen*
Tulsa (OK) *Democrat*
WIN *Magazine* (New York)
WRL *News* (New York)

Index

Note: Page numbers in italics indicate figures; page numbers followed by "*t*" indicate tables.

www.ingramcontent.com/pod-product-compliance
Lightning Source LLC
Chambersburg PA
CBHW071353220926
51613CB00009B/174
9789004515284